The Quest for Forbidden Lands

Eurasian Studies Library

History, Societies & Cultures in Eurasia

VOLUME 10

The titles published in this series are listed at *brill.com/esl*

The Quest for Forbidden Lands

Nikolai Przhevalskii and his Followers on Inner Asian Tracks

By

Alexandre Andreyev
Mikhail Baskhanov
Tatiana Yusupova

BRILL

LEIDEN | BOSTON

Cover illustration: A caravan in Tsaidam, photo from Petr K. Kozlov's expedition in 1907.

The Library of Congress Cataloging-in-Publication Data is available online at http://catalog.loc.gov
LC record available at http://lccn.loc.gov/2018026232

Typeface for the Latin, Greek, and Cyrillic scripts: "Brill". See and download: brill.com/brill-typeface.

ISSN 1877-9484
ISBN 978-90-04-30398-0 (hardback)
ISBN 978-90-04-37626-7 (e-book)

This book is printed on acid-free paper and produced in a sustainable manner.

Contents

Preface

This book gives a historical overview of the epoch-making Russian exploration of Inner Asia during the late nineteenth – early twentieth centuries with the view of introducing a wide circle of Western readers to the most outstanding explorers through a series of biographical sketches. This format allowed us not only to present the life stories of the travelers, with all their enlightening details, but also to fit these narratives into the broad context of Russia's public life and scholarship of the epoch, by focusing on the circumstances which shaped the personalities and scholarly interests of these men.

The essays include the biographies of Nikolai M. Przhevalskii, Mikhail V. Pevtsov, Vsevolod I. Roborovskii, Petr (Pyotr) K. Kozlov, Bronislaw L. Grombchevskii and Grigorii E. Grumm-Grzhimailo, who led a number of major expeditions into the region, mounted by the Imperial Russian Geographical Society (IRGO) with the assistance of the Main Staff (*Glavnyi Shtab*) of the War Ministry. All of these travelers were army officers, with the exception of Grumm-Grzhimailo, although the latter is known to have been accompanied on his two expeditions to the Pamir and the Nanshan Mountains by his brother Mikhail who was a military man. It is owing to their titanic efforts and enormous enthusiasm that the many blank spots were erased from geographical maps of the day, an accomplishment which should never be forgotten.

The Imperial Russian Geographical Society started sending its exploratory missions to Inner Asia in 1856, when the young and ambitious geographer Petr Petrovich Semenov, a follower of Alexander von Humboldt and Carl Ritter, led an expedition to the little-known Tian Shan Mountains. Yet the most active stage of exploration of the region began a decade later and was inextricably linked with the name of Nikolai Mikhailovich Przhevalskii, a pioneer explorer who set an example for many other travelers. He outlined an extensive program of explorations and swept intrepidly into the depth of unknown lands where no other European had trod before. Przhevalskii worked out methods of field work, largely under the influence of Semenov's system of 'preliminary reconnaissance', and obtained a huge amount of accurate data, which substantially increased our knowledge of Inner Asia. It is no surprise then that his travels, in the 1870–80s, were equated to a 'geographical feat' by his contemporaries, and these two decades are often spoken of nowadays as 'the epic era' of explorations by Przhevalskii. This geographical conquest of Inner Asia became possible owing to the vigorous support of the same Semenov who acted as the Society's Vice-Chairman between 1873 and 1914.

Przhevalskii's main priority was Tibet, then a terra incognita, though he was unable to reach its capital, Holy Lhasa. Roborovskii and Kozlov were Przhevalskii's disciples and companions on his last two expeditions; after his sudden death (in 1888) they traveled together under Pevtsov, as his assistants, on his Tibet journey (1889–1890), and then led their own expedition into the region (1893–1895), following their mentor's example. Their joint exploration of the Nanshan Mountains was another substantial contribution to geography. Subsequently, Kozlov made three more very successful journeys to Mongolia and Eastern Tibet, in 1899–1901, 1907–1909, and 1923–1926, which earned him world fame. On the other hand, the brothers Grum-Grzhimailo made some important discoveries in the Tian Shan, in 1889–1890, while Grombchevskii is best known for his exploration of the Pamir in the same years.

These six travelers were men of different social and educational background, temperament, and worldviews, yet they had one thing in common – their keen interest in Inner Asia, its nature and peoples, which eventually made them into scholar-travelers. This was a hard lot, but they were ambitious enough, self-motivated and daring to take upon themselves the noble task of pioneer explorers – for the sake of science and their own country. The proof of the universal appreciation of the activities of these Russian pioneers were the high awards they were given by the leading European geographical societies. Thus Przhevalskii and Kozlov were awarded the gold medals by the Italian Geographical Society and the Royal Geographical Society of London, while Grumm-Grzhimailo received a prize of the French Academy of Sciences.

Regrettably, out of this host of Russian travelers only Nikolai Przhevalskii and his follower Petr Kozlov have been merited properly by contemporary Western scholarship; the rest are little known outside Russia, despite their outstanding contributions to many different branches of learning which were once hailed enthusiastically throughout the world. Thus, the British explorer and naturalist Douglas Carruthers noted in his *Unknown Mongolia* (1913) that "The romance of exploration has no more enthralling chapter than the Russian penetration of Central Asian and Far Western China. It was a great undertaking, well suited to the Muscovite mind. The Russian does everything on a large scale. His journeys are colossal undertakings".[1]

The journeys of Przhevalskii and his followers were semi-military ventures, in terms of their organization, mode of travelling, and performance of the field work. They were designed ultimately as *preliminary scientific reconnaissance* of the uncharted territories of Imperial China's Western periphery, the so-called 'forbidden lands', inaccessible or of difficult access to Western travelers. The

1 Carruthers 1913, 15.

essays presented in this book provide an insight into the lives and travels of the Russian pioneer explorers of Inner Asia. Their authors, Alexandre Andreyev, Mikhail Baskhanov and Tatiana Yusupova, used miscellaneous sources, Russian and Western, including the geographical and military archives (those of the Russian Geographical Society and of the Russian State Military Historical Archive) to sketch the portraits of the heroic cohort of Przhevalskii and his followers and assess the major results of their travels.

The introduction to these essays provides the historical, ethno-geographical, and geopolitical contexts within which these expeditions are located, with a focus on the role of the Imperial Russian Geographical Society and the Main Staff as their organizers and sponsors, and the Anglo-Russian rivalry in the region, the Great Game, as a major deterrent for exploration of the 'forbidden lands'.

The last two and a half decades, following the collapse of the Soviet Union, have witnessed a surge of interest, both public and scholarly, in Russian history as well as in Central Asia as a distinct geopolitical region, where new independent states emerged in the post-Soviet period, which were formerly the Central Asian republics of the USSR. As a result, a wide range of literature has been produced in the new Russia in an attempt to rethink the country's turbulent past and its imperial and Soviet legacies, some of which has dealt specifically with Central / Inner Asian history, the East-West cultural dialogue over the centuries, and, of course, the intriguing tale of Russian penetration and expansion into the region concurrent with its scientific exploration – an expansion which remains, up to the present moment, a popular topic for intense academic discourse.

Of the recent publications devoted to the subject two collections of articles are worthy of mention. The first one discussed the Russian archeological explorations in Eastern Turkestan; it was published by the Institute of Oriental Manuscripts in St. Petersburg under the title *Rossiiskie ekspeditsii v Tsentral'nuiu Aziiu v kontse XIX – nachale XX veka* (Russian Expeditions to Central Asia at the Turn of the 20th Century, ed. by I.F. Popova, St. Petersburg, 2008). The second collection of essays was entitled *Rossiiskie ekspeditsii v Tsentral'nuiu Aziiu: Organizatsiia, polevye issledovaniia, kollektsii, 1870–1920-e gg'* (Russian expeditions to Central Asia: Their organization, field observations and collections, 1870–1920s, ed. by A.I. Andreyev, St. Petersburg, 2013) and focused specifically on the geographical exploration of Inner Asia. The essays were written by science historians, curators of the P. Kozlov Memorial Museum (an entity affiliated with the Institute for the History of Science and Technology, St. Petersburg branch) A. Andreyev, T. Gnatiuk, and M. Kozhevnikova. One more important work, of relevance to this study, was the biobibliographical dictionary of Russian

Military Orientologists compiled by M.K. Baskhanov, expert on the history of the Imperial Russian Army.[2] The dictionary was intended to create a data base of Russian military orientology, as full as possible, within the chronological period of 1750–1917, by tracing the life stories of the persons included in the dictionary up to 1950s.

A most valuable contribution to scholarship was also the publication of varied documentary sources – travel diaries, memoirs, scientific reports and notes, letters, etc., belonging to the travelers. These included the full texts of Kozlov's Mongolia-Sichuan and Mongolia-Tibet expedition diaries of 1907–1909 and 1923–1926;[3] the diary of Kozlov's trip to Urga in 1905;[4] and finally the travel journals and photographic collections of B.L. Grombchevskii's Central Asian expeditions.[5]

Furthermore, several new biographies of Russian travelers – Przhevalskii, Grombchevskii, and Kozlov were penned by V.M. Gavrilenkov, E.P. Gavrilenkova, A.A. Kolesnikov, T.I. Yusupova, and N.N. Kravklis.[6]

As for non-Russian scholars, their main attraction remains Przhevalskii, largely due to his international renown of long standing, charismatic figure, and the fact that some of his works were translated into European languages (English, German and French). The first substantial Western biography of the traveler by D. Rayfield, published in 1976,[7] has been supplemented of late by several more insightful studies, such as a chapter on Przhevalskii in D. Schimmelpenninck van der Oye's *Toward the Rising Sun: Russian Ideologies of Empire and the Path to War with Japan*,[8] and an essay by D. Brower, *Imperial Russia and Its Orient*, to name a few.[9] The basic study of the IRGO and the Main Staff cooperation in field research in Inner Asia was produced by Alex Marshall in his comprehensive and valuable work, *The Russian General Staff and Asia, 1800–1917*.[10]

In general, there is clearly a dual approach to the subject among international scholarship nowadays – while natural historians (zoologists and botanists), ethnographers, archaeologists and Orientalists focus on the purely

2 Baskhanov 2005.

3 Kozlov 2003; Kozlov 2015.

4 Kozlov 2004, 97–127.

5 Baskhanov, Kolesnikov, Matveeva 2015; Baskhanov, Kolesnikov, Matveeva, Glukhov 2017.

6 Gavrilenkov V.M. 1989; Gavrilenkova E.P. 1999; Kolesnikov 1997, 2004, 2006; Yusupova 2017; Kravklis 2006; 2007; 2013.

7 Rayfield 1976.

8 Schimmelpenninck van der Oye 2001, 24–41.

9 See Brower, 1994.

10 Marshall 2006.

scientific aspects of the Russian expeditions, namely the natural collections and cultural artefacts gathered by travelers, scholars involved in the social and political studies are primarily concerned with the socio-political, geopolitical and imperial contexts, which certainly makes difficult to produce an unbiased and well-balanced picture of these missions. For this reason, we chose not to go deep into the discussion of the geopolitical and imperial aspects linked with the Russian geographical exploration of Inner Asia in the late nineteenth century, but rather to concentrate on the biographies of the geographical heroes of the day, their travels and achievements.

The travelers portrayed in this book were a special kind of people; a handful of devotees to geographical science, men of great enthusiasm, perseverance, and courage. They were endowed with a peculiar 'nomadic spirit' (in P. Kozlov's words), a wanderlust that made them turn away from the comforts of sedentary life for the sake of journeying in distant and inhospitable lands, at the risk of their lives, to be rewarded ultimately by many remarkable discoveries and finds. They mapped the vast expanses of Inner Asia, with all its major orographic and hydrographic systems, and described the unique wildlife and multi-ethnic population of the region. It is to these travelers that the Imperial Russian Geographical Society owes much of its world fame, so it is no surprise that all of them were given the Society's highest award, the Konstantin Gold Medal: N. Przhevalskii (1874), G. Potanin (1886), M. Pevtsov (1891), V. Roborovskii (1897), V. Obruchev (1900), P. Kozlov (1902), and G. Grumm-Grzhimailo (1907), with the exception of B. Grombchevskii who was awarded the Small Gold Medal (1889).

The activities of these pioneer travelers, their travelogues as well as their vast naturalistic, ethnographic and archeological collections enjoyed great popularity in the fin-de-siècle Russia, having thus acquainted the Russian public with the amazing nature and little-known peoples of the 'forbidden lands' of Inner Asia.

The authors of the book, Alexandre Andreyev and Tatiana Yusupova, are research fellows at the Institute for the History of Science and Technology, St. Petersburg Branch, Russian Academy of Sciences; and Mikhail Baskhanov is a researcher based in Glasgow, United Kingdom.

Alexandre Andreyev, Mikhail Baskhanov and Tatiana Yusupova

Acknowledgements

First, we would like to thank Prof. Charles W.J. Withers from the University of Edinburg for his taking interest in the personality of Petr (Pyotr) Kozlov and encouraging us (Tatiana Yusupova and Alexandre Andreyev) to write a biographical essay about him for the GEOGRAPHERS Biobibliographical Studies. The essay was published in 2015 in volume 34 of the GBS (edited by Hayden Lorimer and Charles W.J. Withers), and we later used it, in the slightly revised form, for this book. Incidentally, while working on this essay we came up with an idea of writing a series of popular essays about the most outstanding Russian explorers of the Asian heartland intended for Western (English speaking) readers. One more researcher, an expert on Russian military Orientalists Mikhail Baskhanov, joined us then, having volunteered to write essays about M. Pevtsov and B. Gromchevskii. And this was how the project of the *Quest for Forbidden Lands* emerged.

We are also thankful to the Program for Fundamental Studies of the Russian Academy of Sciences (2015 – 2017) and personally to one of the Program managers, a corresponding member of the Academy Yury Mikhailovich Baturin, for giving some financial support to our project.

Otherwise, kind help was provided by the Russian Geographical Society, particularly the longtime head of the Society's Archive Maria Fedorovna Matveeva, who supplied us with a large number of scanned photographs and drawings to serve as illustrations for this volume.

Finally, our thanks are to the former professor of the University of North Dakota Ralph Koprince and the British researcher and writer Catriona Bass for their being the first English readers of the essays and for their making some valuable suggestions to improve the narratives stylistically.

Alexandre Andreyev and Tatiana Yusupova
St. Petersburg, January 2018

List of Illustrations

Figures

Maps

List of Abbreviations

AN SSSR	*Akademiia Nauk SSSR*, Academy of Sciences of the USSR
GUGSh	*Glavnoe upravlenie General'nogo shtaba*, Main Directorate of the General Staff, St. Petersburg
IRGO	*Imperatorskoe Russkoe Geograficheskoe Obshchestvo*, Imperial Russian Geographical Society, St. Petersburg
JRAS	Journal of the Royal Asiatic Society, London
MID	*Ministerstvo innostrannykh del*, Russian Ministry of Foreign Affairs
TNA	The National Archives, Kew
NA RGO	*Nauchnyi arkhiv Russkogo geograficheskogo obshchestva*, Scientific Archive of the Russian Geographical Society, St. Petersburg
OGIKM	*Omskii gosudarstvennyi istoriko-kraevedcheskii muzei*, Omsk State Museum of Local Lore, Omsk
OIOC	*Oriental and India Office Collections*, London
PRGS	*Proceedings of the Royal Geographical Society, London*
RGADA	*Rossiiskii gosudarsvennyi arkhiv drevnikh aktov*, Russian State Archive of Old Documents, Moscow
RGO	*Russkoe Geograficheskoe Obshchestvo*, Russian Geographical Society, St. Petersburg
RGS	Royal Geographical Society, London
RGVIA	*Rossiiskii gosudarstvennyi voenno-istoricheskii arkhiv*, Russian State Military History Archive, Moscow
SPbF ARAN	*Sankt-Peterburgskii filial arkhiva Rossiiskoi Akademii Nauk*, St. Petersburg Branch of the Archive of the Russian Academy of Sciences, St. Petersburg
SMA	*Sbornik geograficheskikh, topograficheskikh i statisticheskikh materialov po Azii*, Selections from Intelligence on Geography, Topography & Statistics of Asia, St. Petersburg
TsGARU	*Tsentral'nyi gosudarstvennyi arkhiv Respubliki Uzbekistan*, Central State Archive of Republic of Uzbekistan, Tashkent
VSO	*Voenno-statisticheskii otdel*, Military Statistical Section of the Main Staff
VTO	*Voenno-topograficheskii otdel*, Military-Topographical Department
VUA	*Voenno-uchenyi arkhiv*, Military and Scientific Archive, St. Petersburg
VUK	*Voenno-uchenyi komitet*, Military-Scientific Committee of the Main Staff, St. Petersburg
ZSO	*Zapadno-Sibirskii otdel*, West Siberia Branch, Omsk

Archival Citations

f., r. (F., R.)	*fond*, *razriad* collection
op. (Op.)	*opis'*, subgroup within the collection
d. (D.)	*delo*, file
l. (ll.)	*list* (y), page (s)
ob.	*oborotnaia storona*, reverse side of the page

A Note on Transliteration, Terminology and Dates

The transliteration of place-names (geographical locations) in Mongolia, Western China and Northern Tibet presents much difficulty for someone writing about Central (Inner) Asia due to great variations of their spelling. Moreover, practically all major cities mentioned in this book changed their names in the twentieth century. Thus, for example, Sinkiang (name of province) is spelled nowadays as Xinjiang; Sinin – as Xining, Peking – as Beijing, Ulan-Bator – as Ulaanbaatar, Hwang Ho (Yellow River) – as Huang He, etc. To solve the problem we rendered all place-names (Russian, Mongolian, Tibetan, Chinese and other) as they occur in source materials (travel diaries, books, articles, and correspondence of the explorers) according to the modified U.S. Library of Congress system, giving the modern equivalents of the most commonly used place-names in brackets. As for Huang He, the name of the river was spelled like that by all Russian travelers in the nineteenth century and the form is still in common use in today's Russian cartography, so we chose to use this particular form and not the old one, Hwang Ho.

Another problem we came across is that a number of place-names were spelled by Russian and Western travelers alike in the corrupt forms, the way they heard them from native speakers. Some of these places were nomad encampments, wells, or small Buddhist monasteries where the travelers camped, and quite naturally these cannot be found in modern maps, therefore we used the old spellings of the travelers themselves. Also, it is well known that N. Przhevalskii and P. Kozlov gave some Russian names to their newly discovered geographical objects (mountain ranges, lakes) on the Tibetan plateau. However these were subsequently discarded by geographers and are no more in use.

Russian Christian names were not rendered in Anglicized forms and retain their original Russian spelling, such as Nikolai, Petr, Mikhail (not Nicholas, Peter, and Michael), etc. The only exception is the names of Russian Emperors Nicholas and Alexander (not Nikolai and Aleksandr) well-established in the English-language historical literature. Transliteration of Turkic, Arabic, Persian, and Chinese terms, geographical and personal names, phrases, and bibliographical entries is generally left to the discretion of contributors.

The Russian historical terms are given in English translation and, when used for the first time, are duplicated in transliteration in brackets; explanations of the most essential terms are given in the footnotes.

Dates in this book, unless otherwise indicated, are given according to the Julian calendar, used in the Russian Empire, which was twelve days behind the Gregorian calendar in the nineteenth century and thirteen days in the twenties century.

INTRODUCTION

Into Inner Asia, in Search for the Unknown: The Russian Explorations and Explorers of the Region

Alexandre I. Andreyev

We have accomplished our task to the very end –
we have passed through and explored those places in Inner Asia
where the foot of a European has never stepped.

N.M. PRZHEVALSKII

•••

The Russian expeditions to Inner Asia
opened up an entirely new world to bio-geographers.

V.L. KOMAROV

•••

Settled life for a traveler is like a cage for a bird.
The mysterious voice of distant lands wakes up his soul
and is mightily calling him …

P.K. KOZLOV

1 A Glimpse of History: Early Travelers in the 'Forbidden Lands' (From the 13th to the Mid-19th Century)[1]

Before telling the story of Russia's exploration of Inner Asia, we would like to provide the reader with the historical background to answer the questions –

1 On the early Russian travelers in Inner Asia see Murzaev 1948, Lebedev D.M. 1949; idem 1950, Shchukina 1955, Magidovich I.P, Magidovich V.I. 1983–1986.

 | DOI 10.1163/9789004376267_002

what was Inner Asia to Russia, and why did the state and the Russian emperors in particular support the initiatives of the Imperial Russian Geographical Society and the War Ministry of those days, the heyday of European imperialism.

To begin with, we would like to make some remarks on geographical terminology. The term 'Inner Asia' as used in this book is well-established in English-language geographical literature, as distinct from a much broader term 'Central Asia'.[2] The latter was introduced in 1843 by the eminent German geographer Alexander von Humboldt to describe the intracontinental territory stretching from the Caspian Sea in the west to the Greater Khingan (Hinggan) Range in northeast China. Inner Asia, or High Asia, is a part of this greater Central Asia and refers normally to the vast region of the Asian heartland, consisting of Mongolia, Western China, formerly known as Chinese or Eastern Turkestan (today's Xinjiang-Uyghur Autonomous Region, with adjacent Qinghai and Gansu Provinces), and Tibet (Tibetan AR), the outlying westernmost parts of the Chinese Manchu Empire. Strictly speaking, it was mainly this territory – Chinese Central Asia – that was thoroughly explored by Russian travelers.

In Russia, Humboldt's broad term 'Central Asia' (*Asie centrale*) was in wide usage until the end of the nineteenth century. Since then, up to the present time, the term has been applied only to Chinese Central Asia (i.e., Inner Asia), while the western part of the region known as Western or Russian Turkestan (*Turkestanskii Krai*, since 1886) became Middle Asia (*Sredniia Aziia*). This included after the formation of the USSR the autonomous republics of Kazakhstan, Uzbekistan, Turkmenia, Tajikistan and Kirghizia, often spoken of as Russia's "soft under-belly".

When speaking of the nineteenth century Imperial Russia, one should remember that it was a unique multi-ethnic Eurasian giant-state, split between two continents, still in the process of expanding its eastern and southern limits.[3] This explains to a large extent the country's permanent concern for its long and troublesome Asian frontier, her neighbours there, the Caucasians, Turks, Mongolians, Manchus and Chinese, and, of course, her unremitting imperial ambition. The Russo-Chinese frontier alone extended some 6000 *versts* (6420 km) and stretched from Semirechye ('Seven Rivers Land', today's south-eastern Kazakhstan) to the Pacific Ocean.

---

2 See, for example, L.I. Miroshnikov, "A note on the meaning of the term 'Central Asia' as used in the book", in History of civilizations of Central Asia (A.H. Dani, V.M. Masson, eds), Vol. I, Paris: UNESCO Publishing, 1992, Appendix, 467–470.

3 On the intellectual discourse of the Russian European – Asian – Eurasian identity in the 18th – 20th centuries, see Bassin 1991, 1–17.

Today the territory of Inner Asia covers about 6 million sq. km. Its surface is made of numerous sand and rubble-covered plains (desert steppes) surrounded or crossed by high-altitude mountain ridges. The relief of the region can be roughly divided into three distinct west-to-east-oriented geographical zones, namely 1) the northern mountain zone, including the Tian Shan Mountains, the Mongolian Altai and Khangai; 2) the middle zone of arid sand plains – the Gobi (Shamo) desert and the Kashgar depression occupied by the Takla Makan desert; and 3) the Tibetan plateau, flanked by the Himalayas in the south, Karakoram in the west, Kunlun in the north, and the Sino-Tibetan Chain in the east. It is in Inner Asia that the great rivers of the continent – Huang He (Yellow River), Yangtze, Mekong, Salween, Brahmaputra, Indus, Amur, and others – have their sources, and here also are situated a great number of lakes the largest of which is Lake Qinghai (Kokonor, the Blue Lake) with a surface area of 4,317 sq. km.

Early knowledge of Inner Asia reached Europe in the thirteenth century, when two Franciscan monks, Plano (Pian) Carpini and Willem (William) Rubruck, envoys to the Mongol Khans, brought back from their journeys the first descriptions of the land routs leading to the remote Oriental countries they had visited. And thanks to the Venetian merchant Marco Polo who journeyed to China in 1271–1275 the Europeans learned for the first time of the interior part of Gobi, Takla Makan, Tibet and the ancient oases-towns (Lop, Cherchen, Hami, Shazhou). This incipient and succinct information found its reflection in the medieval maps which in most cases were inaccurate and rudimentary, showing the fantastic contours of geographical objects, as evidenced, for example, by the world map of the Italian monk Fra Mauro (c. 1450), one of the most renowned mapmakers of the Middle Ages. As a result Inner Asia began to be regarded by European scholars, from the sixteenth century on, as "a high plateau of great dimensions, with a mountain juncture in the center from which emerged the rivers Indus, Ganges, Huang He, Amu Darya, and Syr Darya".[4]

More information about the region was provided in the course of subsequent centuries by Western travelers in China and Tibet (diplomats, tradesmen and Catholic missionaries), but more importantly by their Russian counterparts, of whom we will speak below.

Muscovy's conquest of the Siberian khanate and subsequent colonization of Siberia, in the late sixteenth – seventeenth centuries, put Russians in touch with the many peoples inhabiting the region, commonly referred to as *inorod-*

4 Shchukina 1955, 10–11.

60°
70°
80°
90°
100°
110°
Salekhard
Arctic Circle
Igarka
Turukhansk
Sredne Sibirskoye Ploskogor'ye
Niz. Tunguska
Tura
Zapadno Sibirskaya Nizmennost
Khanty Mansiysk
Pod. Tunguska
Tobol'sk
Tym
Ob
Yenisey
Irtysh
Kolpashevo
Chulym
Yeniseysk
Angara
UNION OF
SOCIALIST
Tara
Tomsk
Achinsk
Bratsk
Ust Kut
Petropavlovsk
Omsk
Tatarsk
Kemerovo
Krasnoyarsk
Oz. Chany
Novosibirsk
Kokchetav
Karasuk
Prokopyevsk
Novokuznetsk
Tulun
Slavgorod
Barnaul
Sayan
Minusinsk
Aksu
Abakan
Cheremkhovo
Irkutsk
Tselinograd
Biysk
Zapadnyy
Gorno Altaysk
Kyzyl
Munku Sardyk
Selenduma
Karaganda
Semipalatinsk
Tannu Ola
Hövsgöl Nuur
Ust' Kamenogorsk
Karkaralinsk
Balukha 4620
Uvs Nuur
Mörön
Selenge Mörön
Ulaan
KAZAKH S.S.R.
Altai
Khangai Nuru
Mointy
Ayaguz
Oz. Zaysan
Har Us Nuur
Uliastay
Balkhash
Tarbagatay
Irtysh
Hovd
Oz. Balkhash
T'a-ch'eng
Urungu
MONG
Chu
R. Ili
Taldy Kurgan
Kapchagay
I-ning (Kuldja)
Ma-na-ssu
Urumchi (Wu-lu-mu-ch'i)
Alma Ata
Frunze
KIRGIZ S.S.R.
Issyk Kul
Shan
Turfan Depression -154
Ha-mi
Naryn
Khan Tengri 6995
Tien
Baghrash Köl
Alayskiy Khr.
P. Kommunizma 7495
T'a-li-mu Ho
Ka-shih
Chinese
SINKIANG
An-hsi
Pamir
Yarkand
Lo-pu Po (Lop Nor)
Yü-men
Chiu-ch'üan
AFG.
So-ch'e
Turkestan
Shan
Nan-Shan
chin
Khotan
A-erh
Tsaidam
Ch'ing Hai (Koko Nor)
Lan-ch
8611 Mt K2
PAKISTAN
Karakoram Ra.
Huang Ho
KASHMIR
Srinagar
Kun-lun Shan
Islamabad
Ch'iang-tang (Plateau of Tibet)
Indus
Lahore
Himalaya
Yangtze
TIBET
Tang-ku-la Shan-mo
Amritsar
Ka-erh ya-sha
Ch'i lin Hu
Simla
Tatu Ho
Chandigarh
Ch'ang-tu
Mekong
Saharanpur
Nam Hu
Pa-tang
Meerut
Dhaulagiri
Jih-k'a-tse
Lhasa (La-sa)
Delhi
Bareilly
NEPAL
Mt Everest 8848
Farrukhabad
Mountains
Jaipur
Agra
Kathmandu
Sadiya
Chambal
Kanpur
Lucknow
Faizabad
Kanchenjunga 8598
SIKKIM
BHUTAN
Gwalior
Ghaghra
Darjeeling
Brahmaputra
Patkai Hills
Yamuna
Ganga
Patna
Khasi Hills
Shillong
Myitkyina
Allahabad
Varanasi
INDIA
Sagar
Murshidabad
BANGLA DESH
Bhamo
Vindhya Ra.
Jabalpur
Dacca
Chindwin
Irrawaddy
Calcutta
Jamshedpur
Howrah
Lashio

MAP 0.1
Map of Inner Asia, a fragment of the Map of the Orient, from *The Times Atlas of China* (Times Books, 1974).

sty (members of ethnic minority groups). One of these peoples was the Buryats, of Mongolian origin, who since time immemorial inhabited an area around Lake Baikal in the southern part of East Siberia. The Buryats, originally shamanists, embraced Tibetan Buddhism, which was propagated among them by travelling Mongolian and Tibetan lamas. From that time on, they began to make regular pilgrimages to Lhasa by a special route, also used by their Mongol brethren.[5] However, their lively liaison with Tibet, or the 'Land of Joo',[6] was cut short towards the very end of the eighteenth century when the Tibetan rulers closed their country's doors to all western visitors, which made Tibet the 'Forbidden Land' for more than a century.

The subjugation of Siberia to the power of Muscovy met with strong resistance by some of the local tribes as well as by the Mongolian chieftains in Outer (Khalkha) Mongolia, who laid claim to the same land. By the 1660s the entire territory of Transbaikalia was conquered by the Russian state that would make the annexation legitimate through the Nerchinsk Treaty concluded with the Peking court in 1689. But it was only in 1727 that the border between the Russian and the Manchu Chinese Empires was finally delimited (though not demarcated) by the Kiakhta Treaty. The Nerchinsk Treaty was the first diplomatic agreement between the Russian tsars and the Qing Manchu dynasty and it was also the first one ever to be concluded by China with a foreign power, which permitted the subjects of both counties to engage in trade.

While Russia was steadily expanding eastwards, towards the Pacific Ocean, the Celestial Empire, under its Manchu rulers, was advancing in the opposite (western) direction. Khalkha Mongolia was incorporated into the Manchu's Qing China in the late seventeenth century. By that time the Mongols had already adopted Tibetan Buddhism as their religion and moreover had the head of their own Buddhist church established, with the due approval of the Manchus (who thus sought to pacify the seemingly unruly and warlike Mongols). This was the incarnate lama (Grand Lama) named Bogdo-Gegen, or Jebtsundamba Khutuktu, who was ranked third in the Buddhist ecclesiastical hierarchy, after the Dalai Lama and the Panchen Lama.

5 The Buryat pilgrims' route to Tibet (Urga – the encampment of Alasha van – Lake Kukunor – Lhasa) was minutely described by the Russian consul in Mongolia Y.P. Shishmarev, see [Y.P. Shishmarev] 1873, 185–191; also Potanin 1893, 38–43. On Shishmarev see Andreyev 2003, 119–123. On Buryats' pilgrimages to Tibet and their colony in Lhasa see Andreyev 2001, 349–362.

6 Tib. *Jo-bo*, the Buddha.

Another western borderland that later came to be known as Chinese Turkestan or Xinjiang, situated southwest of Mongolia, was also subjugated to the rule of the Manchu dynasty, in the mid-eighteenth century. The area consisted of two distinct geographic, historic and ethnic territories – Dzungaria and the Tarim Basin (Kashgaria) divided by the Tian Shan Mountains. To the north of Tian Shan dwelt a steppe nomadic Oirat Mongol people who were Buddhists, and to the south, a sedentary Muslim Uigur people. Both Dzungaria and Kashgaria were later unified under the common name of Xinjiang.

As for Tibet, the country had entered, in the latter half of the seventeenth century, into a very special type of relationship with the Manchu imperial court, known as the *Chö-Yön* (Patron - Priest) alliance, which can be translated into Western legal terms as a loose protectorate. According to this arrangement, the Manchu dynasty (the Patron) was to provide the necessary military protection to theocratic Tibet (the Priest), if needed, and had its two representatives, or *ambans*, stationed in Lhasa with a small escort. Formally this did not turn Tibet into a vassal territory, as was the case with Khalkha Mongolia or Xinjiang, yet the Manchu authorities as well as the later Chinese republican government tended to interpret the arrangement precisely this way.[7]

•••

The early contacts with China were made in 1618 when the Tsar Mikhail Fedorovich, the first of the Romanov dynasty, sent two Cossacks across Mongolia and the Great Wall to Kalgan with a view to making contact with the Ming Court. Several more attempts were made in the subsequent years, especially after the Manchu dynasty had replaced the Mings in 1644, yet it was only in 1675 that the first Russian ambassador Nikolai Spathary arrived in Peking and was accepted there, the first plenipotentiary to be received by China from a western power. Spathary (Nicolae Milescu-Spataru, Nikolai Gavrilovich Spafarii, Prince of Moldavia 1636–1708), of Greek origin, was an able diplomat and scholar who collected and brought back to Russia some important geographical data from his journey; thus he was the first to give a description of Lake Baikal and he also compiled an orographic scheme of East Siberia and determined a number of geographical latitudes in the region. Spathary's travel records, written in the Russian language, made three separate works: "The

7 On the ambiguous status of Tibet vis-à-vis China see Praag 1987.

journey through Siberia up to the borders of China", "The travel records" and "The description of China".[8]

The names of early Russian travelers (Cossack freebooters, merchants, diplomats, missionaries, and adventurers) in Mongolia and China in the seventeenth and eighteenth centuries are many and some of them left rather interesting accounts of their travels, later to be translated into Western languages. It goes without saying that these travels had little to do with any scientific or geographical exploration of the unknown lands. Still, an important thing to note in this connection, the Russian authorities, since the seventeenth century, had obliged the leaders of the missions, or 'embassies' they dispatched to the Peking court, to make hand-drawn sketches of their itineraries. For this purpose one or several specially trained persons capable of land-surveying and equipped with various measuring instruments were usually attached to these missions, including a draughtsman, as was the case with Spathary's mission.[9] Another mission leader, Evert Isbrandt Ides (a Danish merchant at the Russian service), is known to have been the first to determine a number of geographical coordinates en route to China, in 1692–1695. However these early Russian road surveys can hardly be regarded as maps in the proper sense as they presented rather crude and schematic drawings.

It is worthy of mention here that land surveying, as well as geographic, geodetic, orographic, hydrographic and meteorological observations became possible only after the appearance of special terrestrial measuring instruments and devices, such as barometers, thermometers, astrolabes, telescopes, verriers, levels, etc., all invented in the sixteenth and the seventeenth centuries. The scientific progress in Russia owed much to the personal efforts of the Tsar Peter the Great (1672–1725) who was not only a great reformer of Russia, but also an enthusiastic sponsor of various geographic and cartographic projects and incidentally the founder of the Russian Academy of Sciences in his new Russia's capital, Saint Petersburg (1724). Under his reign and much owing to the young tsar's ambitious aspirations, Russia's knowledge of Central Asia, its landscape, nature and peoples, had considerably increased. Personally, the emperor showed much interest in the Caspian Sea and the Transcaspian area, being agitated by rumours of large gold deposits situated there, but he was also intent on finding direct land and sea routes to China and especially to India. Towards that end he dispatched to India, then part of the Mughal Empire, in 1694, his envoy, a merchant Semen Malen'kii.

8 Spafarii 1882.

9 Shchukina 1955, 11.

The enigmatic Orient had a strong and long-standing appeal for the tsars, hence they used every opportunity to increase their knowledge of Asiatic lands and peoples, primarily China, their close neighbour and the main trade partner on the continent. They were largely assisted in that task by the Russian Orthodox Church's Ecclesiastical Mission (*Russkaia dukhovnaia missia*) established in Peking in 1715 and active for over two centuries, until 1954.[10] Some of the mission leaders and staff are known to have become outstanding scholars, who left their mark mainly in Oriental studies, especially sinology. Thus the head of the 9th mission (1805–1822) Iakinf Bichurin penned several works about Mongolia which he had crossed frequently, the major one being *Zapiski o Mongolii, sochinennye monakhom Iakinfom s prilozheniem karty Mongolii i raznykh kostiumov* (Records of Mongolia compiled by monk Iakinf with a map of Mongolia and pictures of various costumes attached), in four parts. In this book Iakinf described his journey made in 1821, providing some interesting statistic and historical facts about the country. The geographical information was included in his other work, *Statisticheskoe opisanie Kitaiskoi imperii* (A Statistic account of the Chinese Empire).[11]

E.F. Timkovskii, the police-officer (*pristav*) attached to the 10th mission under P. Kamenskii, passed through Mongolia in 1820, 1821 and 1824 and published a three-volume travelogue titled *Puteshestviia v Kitai cherez Mongoliiu* (Travels to China across Mongolia). This work attracted the attention of Western scholars of the day; it was translated into French and English and quoted abundantly by the German geographer Karl Ritter in his masterpiece *Die Erdkunde* (Geography).[12]

The 11th Ecclesiastical mission under archimandrite Veniamin (1830–1840) was accompanied by astronomer Georg Albert Fuss and botanist Alexandre A. Bunge, both of German extraction. The former took upon himself the task of conducting astronomic, geodetic and geophysical explorations in Siberia and China. A.A. Bunge, on the other hand, made botanic and entomological collections and described the Mongolian steppes along the Urga[13] – Kalgan high road (*Kalganskii trakt*). Both Fuss and Bunge also measured the heights of

10 On the history and activities of the Orthodox Ecclesiastical Mission, see Tikhvinskii, Miasnikov et al. (eds) 1997.

11 On Bichurin and his activities, see Denisov 1997.

12 The full title of this 19-volume work by K. Ritter is Die Erdkunde im Verhältniss zur Natur und zur Geschichte des Menchen (Geography in relation to Nature and the history of mankind), 1816–1859. The German-born Karl Ritter is known as one of the founders of modern geography, along with Alexandre von Humboldt, and the person who established the Berlin Geographical Society (*Gesellschaft für Erdkunde zu Berlin*) in 1828.

13 Urga or Ikh Huree was the capital of Khalkha Mongolia since the 17th century.

some of the places they traversed, such as Urga and Bogdo (Khan)-Ula Mountains in the vicinity of Urga. As for Bunge's botanic collection of the Mongol flora, this remained for a long time the only one in the world herbarium, featuring some new specimens and even genera.[14]

Archimandrite Palladii (Petr Ivanovich Kafarov, 1817–1878), head of the 13th mission (1850–1858), was another celebrated scholar, a Sinologist and researcher of Inner Asia. Palladii, who first came to China as a member of the 12th mission (1840–1849), made some important geographic observations during his travels, such as a description of the ancient Uliasutai road. Still, his best known works are the 2-volume Chinese-Russian dictionary, compiled jointly with P.S. Popov (1888); the first Russian translation of the Mongol epic 'The Secret History of the Mongols', published in the *Trudy russkoi dukhovnoi missii v Pekine* (vol. 4, St. Petersburg, 1866); and a commentary on Marco Polo's travels (*Kommentarii na puteshestvie Marko Polo po severnomu Kitaiu*, 1876, 1902).

Summing up these early explorations of Inner Asia in the seventeenth – mid-nineteenth centuries one of the leading Soviet geographers, E.M. Murzaev, noted their obvious methodological setbacks due to the undeveloped state of natural sciences. The characteristics given by travelers to various geographical objects were "too general, descriptive and superficial". It was only in the latter half of the nineteenth century that travels in the East and Central Asia could yield some substantial and fruitful scientific results. The single routes of individual travelers, of no great length, across the territory of Mongolia, failed to provide any reliable material for the cognizance of the vast expanses of Central Asia. Murzaev also pointed out that the majority of these early travelers, both Russian and foreign, were not geographers and naturalists and therefore their major contributions were made in such disciplines as history, ethnography and linguistics. Besides, the exploration of Mongolia was conducted only in passing, as the country was generally regarded as an outlying 'borderland' of China, having no political importance of its own. However, these early travels were not in vain as they paved the way for a new round of explorations in the latter half of the nineteenth which proved extremely successful.[15]

The best Russian maps of Asia at that time were maps produced by the Military-Topographic Depot in St. Petersburg. These showed Central Asia as a whole or its separate parts and incorporated the results of the latest land surveys (Russian and Western) in the region. Still, the mountain ranges were usually shown on these maps as interconnected, with many complicated branches, and one could also see there non-existent lakes and rivers, as evidenced, for

14 Murzaev, op. cit, 34–35; Shchukina, op. cit, 37.

15 Murzaev 1948, 48, 49.

example, by the *Map of Asiatic Russia* (1860). The interior of Central Asia was depicted largely according to Chinese sources which in most cases were unreliable. Thus, the *New Atlas of all parts of the world* (*Novyi atlas vsekh chastei sveta*), printed in 1870 by the MT Depot, showed more or less accurately only the parts of Central Asia which had been previously explored by Russian travelers, while the interior of the continent still appeared as having "fantastic outlines of all oro- and hydrographic objects" and numerous blank spots of unexplored areas.[16]

2 The Imperial Russian Geographical Society, P.P. Semenov, and the Onset of Exploration of Inner Asia

The nineteenth and early twentieth centuries were quite a turbulent period for Russia. The country's steady southward expansion resulted in some important territorial acquisitions during the reign of the emperors Alexandre II and Alexandre III – the Caucasus (1780s-1864); the Middle Asian khanates, in the 1860–1880s; the left bank of the Amur in the Far East, whose earlier occupation by Russian Cossacks was legalized by the Treaty of Aigun. This finally settled Russia's Asian borders.[17] The Aigun and Peking Treaties, concluded with the Manchu Government in 1858 and 1860, were of particular importance as they provided the basis for the demarcation of the lengthy Russo-Chinese frontier and reanimated the slackening trade between the two neighboring empires; much more, they allowed Russia to have her diplomatic representatives posted in Peking and Urga (today's Ulaanbaatar) in Outer Mongolia and start postal communication between Kiahta, on the Mongolian border, and Peking. At the same time Russia's vigorous advance in Central (Middle) Asia brought her more gains, namely the occupation of Tashkent and founding of some new settlements, such as Verny (Alma-Ata) and Semipalatinsk on the approaches to Xinjiang. And in 1851 a local trade agreement opened Kuldja (Ili) and Tarbagatai to Russian trade. All this created a very favourable situation for Russia's further penetration into Inner Asia.

This advance, in the opinion of A.V. Postnikov, was seen by the rulers of the Empire as "involuntary" and "practically inevitable". It was motivated primarily by economic reasons – a desire to expand the country's commercial market. "The trade with Eastern peoples significantly broadened the geographical horizons of the Russian rulers in the Asian borderlands, offered them possibilities

16 Shchukina 1955, 48, 49.

17 On the formation of the Russian frontiers in Central and Middle Asia, see Postnikov 2007.

for penetrating deeper into the hitherto unknown parts of Asia, and provided information about the allegedly inexhaustible riches of those lands…". The first attempts to search for those riches were made under Peter the Great (who ruled between 1689 and 1725), yet despite the failure of his Oriental missions, "an illusion had firmly gripped the minds of Russian politicians, for nearly two centuries, namely that Central Asia is the easy way to conquer India, the 'Jewel in the British Crown'".[18] It is worthy of mention here that the Tsar Peter was not only a great reformer of Russia, but also an enthusiastic sponsor of various geographic and cartographic projects and incidentally the founder of the Russian Academy of Sciences in his new Russia's capital, Saint Petersburg (1724). Under his reign and much owing to the young tsar's ambitious aspirations, Russia's knowledge of Central Asia, its landscape, nature and peoples, had considerably increased. Personally, the emperor showed much interest in the Caspian Sea and the Transcaspian area, being agitated by rumours of large gold deposits situated there, but he was also intent on finding direct land and sea routes to China and especially to India.

The Soviet geographer E.M. Murzaev pointed to "a temporal paradox" in the history of geographical cognizance of our planet – the travelers first explored the world's oceans, then islands and coastal territories, and finally they turned their gaze to the interiors of continents, difficult of access. "The systematic scientific exploration of intracontinental regions of Asia, Africa, America and Australia was started only in the nineteenth century".[19]

And indeed the exploration of Asia's heartland by means of specially mounted geographical expeditions began rather belatedly, in the last three decades of the nineteenth century, on the initiative of the Imperial Russian Geographical Society (*Imperatorskoe Russkoe Geograficheskoe Obshchestvo,* IRGO or RGO). The IRGO was established in 1845, under imperial patronage, with the Grand Duke Konstantin Nikolaevich Romanov acting as its honorary President or Chairman. The founders of the Society were a group of leading geographers and naturalists, such as Admiral Feodor P. Litke, Ivan F. Kruzenstern, Karl M. Baer (Karl Ernst von Baer), and Konstantin I. Arseniev, to name a few.[20] By the end of the first year the Society numbered 144 active members, the majority of whom were military and naval officers. It was located at first in the private

18 Postnikov 2007, 349. On the preparation of the Russian military expedition to British India in 1878, see Zagorodnikova 2014. See also Lebedev V.T. 1898; Schimmelpenninck 2014, 1–10.

19 Murzaev 1995, 15.

20 On the history and activities of the IRGO see Semenov 1896; Berg 1946; Kolesnik 1970; Isachenko 1995.

FIGURE 0.1 The initial seat (1862–1907) of the Imperial Russian Geographical Society, the building of the 6th Gymnasium at the Chernyshov Square. ALBUM OF PHOTOGRAPHS BY S.M. PROKUDIN-GORSKII. NA RGO, R. 117, OP. 2, D. 2–1.

apartments at the house of Pushchin on the Moika Embankment and in the 6th Gymnasium on the Chernychev Square (since 1862). It was only in 1908 that the IRGO moved into its own building designed in the art nouveau style by the well-known architect G. Baranovskii, in Demidov (today's Grivtsov) Lane 7, right in the center of the city, where it remains to this day.

The Society was actually run by its 'brain trust' – the Vice-Chairman,[21] the supervisors of the IRGO's four main sections (those of general geography, geography of Russia, statistics and ethnography), and the secretary. During its early (formative) period, the IRGO focused mainly on the exploration of the territories and peoples within the Russian Empire, as prescribed by its charter, yet the Society had also contemplated, right from the start, geographical explorations far beyond Russia's borders. As Admiral F.P. Litke put it unambiguously in his speech made at the Society's first meeting in October 1845:

---

21 The post of the IRGO's Vice-Chairman was held successively by Count M.N. Muravioev (1845–1856), Count F.P. Litke (Assistant Chairman, 1845–1850, Vice-Chairman, 1857–1873) and P.P. Semenov-Tian-Shanskii (1878–1914).

FIGURE 0.2
The interior of the Society's main office. ALBUM OF PHOTOGRAPHS BY S.M. PROKUDIN-GORSKII. NA RGO, R. 117, OP. 2, D. 2–2.

FIGURE 0.3
The new building of the Imperial Russian Geographical Society at the Demidov Lane. ALBUM OF PHOTOGRAPHS BY S.M. PROKUDIN-GORSKII. NA RGO, R. 117, OP. 2, D. 2–3.

FIGURE 0.4
The new building of the Russian Geographical Society at the Grivtsov (Demidov) Lane, a present-day view.
PHOTO BY A. ANDREYEV (2018).

> Our fatherland itself is a separate continent. Let us say more, a continent which has been poorly explored thus far. The geography of Russia is the main objective of the Society's activities. But it is also of much importance for us to acquaint ourselves with the geography of the adjacent territories. Turkey, Persia and Khiva, China and Japan, the United States of America and the possessions of the Hudson Company (Canada) also make an important agenda....[22]

Yet it took the IRGO two and half decades before it was able to organize, in 1870, its first expedition to Inner Asia – to Mongolia and China – under a young and ambitious military officer Nikolai Przhevalskii. By that time the Society had turned into a well-run machine, with several local branches. One of these, named the East-Siberian Department (*Vostochno-Sibirskii Otdel*) of the IRGO based in Irkutsk (1851), would henceforth send its expeditions to the north of Russia, as well as to Mongolia and China. (By the end of the century, in 1894, one more IRGO branch office would emerge in Troitskosavsk, near the trade

22 Quoted from Aldan-Semenov 1965, 54.

FIGURE 0.5 The Russian consulate in Urga, the starting point of many expeditions into Inner Asia. COPIED FROM *MOSKOVSKAIA TORGOVAIA EKSPEDITSIIA* (MOSCOW, 1912).

settlement of Kiahta on the Russo-Mongolian border, on the initiative of G.N. Potanin and D.A. Klements, the Troitskosavsk-Kiahta Department.)

The establishment, in 1861, of the Russian consulate in Urga (Ikh-Huree), the capital of Mongolia, gave an additional impetus to the expeditionary activities of the IRGO in the region as the entity would provide a permanent base and the starting point for the expeditions arriving from Transbaikalia by the Kiahta-Urga route. The Russian consul-general, Yakov Parfenievich Shishmarev, who had held this post for almost half a century (1864–1911), assisted the travelers, as best as he could, and was their actual caretaker. He hosted them at his home and helped them to hire Mongolian and Chinese speaking interpreters and guides. (Shishmarev set up, right after he had taken over his office, a special interpreters' school at the consulate.) Having joined the ranks of the IRGO, he himself was an enthusiastic explorer of Mongolia. He collected a great amount of information about the country – its population, administrative system, trade, etc.[23] and thoroughly investigated the routes connecting this borderland with Russia. Shishmarev was, incidentally, the first European traveler to visit

23 On Shishmarev see Kuzmin 1997, 1998; Edinarkhova 2000; idem 2001; Andreyev A.I. 2003, 118–124.

FIGURE 0.6
Petr Petrovich Semenov Tian-Shanskii, 1827–1914, the Vice-President of the IRGO.
NA RGO.

the little known Uliasutai area in Western Mongolia (1868) and to report on the Khangai's highest mountain peak, Othon-Tengri.

The precursor of the Russian geographical exploration of Inner Asia was Petr Petrovich Semenov (1827–1914), the pioneer explorer of the Tian Shan Mountains and one of the IRGO long-time key administrators, known as Semenov-Tian-Shanskii since 1906.[24] His own travels, particularly his complex method of collecting and interpreting raw material about little-known geographic areas which he would call 'reconnoitering', set an example for many a traveler in the region. Semenov ushered in a new era of the 'scientific reconnaissance' of Inner Asia. His career as a geographer was closely linked with the IRGO, where he had supervised the section of physical geography for over a decade, in 1860–1873, and where subsequently, until his death, he had acted as the Society's Vice-Chairman.

24 In 1906, when the fiftieth anniversary of his journey to Tian Shan was celebrated at the IRGO, the elderly Semenov was awarded by the Emperor's decree a geographic 'title' – Tian-Shanskii (lit. 'of / from Tian Shan'), he would proudly add to his last name. Henceforth he would be addressed as Petr Petrovich Semenov-Tian-Shanskii.

FIGURE 0.7
P.P. Semenov-Tian-Shanskii, portrayed by G.I. Pridantsev. LOCATION UNKNOWN.

Semenov's biography has been well researched by Russian and Western scholars,[25] so we will not recount it here. An important thing to mention is that Semenov emerged as a geographer under a strong influence of the two outstanding German geographers of the day, Karl Ritter and Alexander von Humboldt, whose theories he wanted to test in the field.

Ritter's groundbreaking *Erdkunde* captivated him so much that he volunteered to translate this brilliant multivolume work into Russian, at the request of the IRGO. The first volume of the translation, with Semenov's own extensive supplement, introducing new geographical data on Central Asia, was published in 1856. And in the preface to the second volume he expressed his inmost philosophical ideas on the role of science, as a tool of self-knowledge (*samopoznanie*) of the environment and the forces of nature, in the hands of scientists. Having befriended Ritter, Semenov discussed with him various geographical topics, including his pioneer travel project for 'reconnoitering' the Celestial Mountains of Tian Shan, to solve their many enigmas.

Petr Petrovich also made the acquaintance of another geographical genius of the day, Humboldt. In 1843 Humboldt published in French his magnum opus '*Asie centrale*' (1843),[26] which contained abundant orographic material

25 See Aldan-Semenov 1965; Dostoevskii A.A. 1928; Lavrov 1962, 460–468; Kozlov I.V., Kozlova A.V. 1991; Lincoln 1982.

26 A. Humboldt's work "Asie centrale: Recherches sur les chains de montagnes et la climatologie compare" (Paris, 1843. P. I-III) was published in the Russian translation in 1915, see

based on a great variety of sources. In this work he refuted the popular concept of High Tartary – a vast plateau allegedly filling up the space of the entire Inner Asia, up to the Gobi, the view then prevailing in European geography. However, Humboldt's own scheme of mountain ranges, although being "a great step forward in the cognition of the surface structure of Central Asia", according to V.A. Obruchev,[27] was nonetheless largely a fiction, an abstract construct, far from reality. Humboldt also put forward the volcanic theory which explained the nature of the mountain formation in Asia and sketched a rather intricate system of mutually perpendicular ranges, forming the mountain junctures.

Humboldt's theory, as well as his orographic scheme, would be later discarded by geographers, mainly thanks to Nikolai Przhevalskii's groundbreaking discoveries during his travels. Still, his ideas were a great contribution to the geography of his day. According to I.V. Mushketov, "having published his *Asie centrale*, Humboldt provided the basis, method and direction for explorers of Central Asia".[28] The 87-year old Humboldt was delighted to meet the enthusiastic Russian student of geography, the more so because he himself had long dreamt of travelling to the Tian Shan, so he gave him his go-ahead.

Thus encouraged by both Ritter and Humboldt, Semenov, submitted his project to the IRGO's Vice-Chairman, General M.N. Muraviov, and having obtained his approval, headed for High Asia, in the spring of 1856. His two-year exploration of the Tian Shan (1856–1857) proved a most successful scientific enterprise. Semenov drew up a scheme of the Tian Shan ranges, reconnoitered Lake Issyk-Kul, located the upper reaches of the Syr Darya, known as Yaksartes to the old Greeks, observed the rugged Tengri Tag mountain cluster, with the majestic pyramid of Khan Tengri ('Lord of the Spirits') therein, the second-highest peak in the Tian Shan, and ascended to the Alpine glaciers originating in the Tengri Tag.

At the same time, Semenov was unable to find any volcanoes or volcanic activity in the Tian Shan, as postulated by Humboldt, which was one of the most significant results of his journey. He also ascertained that the eternal snow line lay only at a great altitude and discovered the vertical natural waists in the mountains. On top of that, the Russian traveler made some collections – one of these, botanical, numbered several dozen species of plants, totally

Gumboldt 1915. On Humboldt's contacts with Russia see Shcherbakov 1962, 119–121; Oberman 1981, 87–91; Suckow 2001, 247–264.

27 Quoted from Murzaev 1948, 44.

28 Mushketov 1915, 149.

FIGURE 0.8
Admiral Fedor Petrovich Litke, 1797–1882, one of the founders of the Imperial Russian Geographical Society. Copied from *Istoriia poluvekovoi deiatel'nosti Imperatorskogo russkogo geograficheskogo obshchestva. 1845–1895*. P.I. St. Petersburg, 1896.

unknown to European botanists.[29] Overall, Semenov's journey was a real breakthrough in geographical knowledge of the Tian Shan region.

The Celestial Mountains were a typical terra incognita in those days. As its first European explorer, Semenov surveyed the territory and collected as much useful information (geographical, geological, botanical and other) as possible. His system of scientific reconnaissance that he employed during his two journeys in the Tian Shan was rather innovative. One of its most remarkable elements was the versatile character of field observations as noted above. Also, Semenov chose his routes rather carefully, trying to make them pass *across* mountains, rather than *through* mountain valleys, which allowed the traveler to obtain some valuable data on the mountain configuration. In one of his letters to the IRGO he explained his system by saying: "Of course, this journey of mine, being made with much haste caused by dangers and privations that surrounded me, can only be a *scientific reconnaissance*, and not a real scientific

29 See N.G. Fradkin, "P.P. Semenov-Tian-Shanskii i ego puteshestvie, 1856–1857", in Semenov-Tian-Shanskii 1958, 3–38.

exploration; but even so, it will yield some good results for the geography of Asia".[30]

Back in St. Petersburg, Semenov was welcomed enthusiastically by his colleagues at the IRGO and by the Russian academic elite. The full account of his travels Semenov penned, however, only half a century later, already in declining years, yet he did it most thoroughly and in a masterly literary fashion. The book was intended as the second volume of his memoirs, yet it came out only after his death, as a posthumous edition, in 1946.[31] Semenov planned to continue his explorations of the Tian Shan in 1860, along a totally new route – from Lake Issyk-Kul towards Urumchi (Urumqi) and further to Kuldja in the Ili region, that is across Chinese Turkestan. His project was supported by Litke, yet the IRGO failed to organize his new expedition due to the lack of necessary funds and also because of the strong opposition of the Ministry of Foreign Affairs (MID). The latter, in Semenov's words, "guarded jealously the Asian countries beyond the Russian borders from the intrusion of Russian geographical science in the person of Russian travelers".[32] The actual reason was the Anglo-Russian geopolitical rivalry in Asia which made tsarist diplomats extremely cautious, especially with regard to the RGO expedition activities in the areas where the interests of the two competing empires clashed.

In 1873 Semenov succeeded Litke as head of the IRGO, an important landmark in the Society's history. Over the next four decades he would be instrumental in organizing a great number of geographical expeditions and individual trips around the world, including those to Inner and Central Asia. Semenov's outstanding role as organizer of exploration of Inner Asia was a most remarkable phenomenon in the history of Russian geographical science, one that brought about the flowering of the IRGO towards the close of the nineteenth century.

3 The IRGO – War Ministry Cooperation. The 'Epic Era' of Przhevalskii's Expeditions

Nikolai Przhevalskii (1839–1888), like Semenov, was a true product of his time, the age of positivism, astounding geographical discoveries and colonialism. He also read the books of Ritter and Humboldt and shared with Semenov the

30 The letter was published in the *Vestnik IRGO*, XVIII, St. Petersburg, 1856; quoted in Fradkin, op. cit., 11.

31 Semenov-Tian-Shanskii 1958, prefaced by G.N. Fradkin.

32 See Fradkin, op. cit., 41.

same passion for geography and travelling. Being a military officer, he was looking for a place, some virgin lands in Africa or Asia, where he could show his worth as an explorer. On Semenov's advice Przhevalskii would travel first, in 1867–1869, to the Ussuri basin (*Ussuriiskii krai*), recently annexed by the Russian Empire, to survey the region geographically and statistically, and then would lead, with the backing of the IRGO and partly the War Ministry, four more expeditions into the wilds of Inner Asia, between 1870 and 1888, when death suddenly interrupted his brilliant career of a scholar-traveler. These two decades marked the beginning of Russia's exploration of Inner Asia, having laid the foundation for further, more substantial, explorations following his tracks. It will not be an exaggeration to say that geography, or rather regional geography (*stranovedenie*, in Russian), and cartography were two indispensable scientific instruments in the service of colonial or military expansion in the nineteenth century. This explains to a large extent the close intellectual cooperation between the IRGO and the War Ministry, namely the Main Staff[33] – *Glavnyi Shtab*, and its divisions, such as the military-topographic and military-statistical departments, abbreviated in Russian as VTO and VSO, and the military-scientific committee, or VUK,[34] which joined their efforts in mounting a series of exploratory missions to Inner Asia.

There was yet another military entity closely involved in the IRGO projects – the General Staff (*General'nyi Shtab*). In the Imperial Russian Army, the blueprint for the General Staff was drawn up by Peter the Great who established the institute of Quartermasters. The proper General Staff was established after the emperor's death, in 1763, and for more than half a century had been changing its name and structure – the Department of the General Staff (1774), The Quartermaster Section of the Retinue of His Imperial Majesty (1797), and the Department of Quartermaster-General (1815). The General Staff was re-organized in 1832, within the War Ministry, under the name of the Department of the General Staff, and received authority over war planning and military administration in Imperial Russia. The officers who graduated from the Nicholas Academy of the General Staff were incorporated into the Officer Corps of the General Staff under the direction of the Chief of the Main Staff. In 1863 the

33 The Main Staff of the War Ministry was the organ of central military administration in charge of military planning, mobilization, dislocation, communication, intelligence service, armament and army's basic equipment, quartering of troops, etc.

34 The VUK was originally established with a view to coordinating the military-scientific activity of the officers cops of the General Staff and the Corps of the military topographers, as well as for collecting military-statistic information on the armed forces of Russia and foreign states (military intelligence), the study of Oriental languages by the staff officers, compilation of programs for military colleges and publication of various manuals.

FIGURE 0.9 The Main Staff of the War Ministry with a view on the Arch of Triumph. St. Petersburg, 1913. COPIED FROM *VOENNAIA STOLITSA ROSSIISKOI IMPERII V FOTOGRAFIIAKH KONTSA XIX – NACHALA XX VEKOV* (ST. PETERSBURG, 2004, ILL. NO. 34).

Department of the General Staff was transformed into the Main Directorate of the General Staff and two years later the entity merged with the Inspectorate Directorate of the War Ministry to form the Main Staff. The newly-established Main Staff possessed broad military-administrative powers, including jurisdiction over war planning, intelligence, personnel, structure, deployment, and service of troops. The General Staff would be re-established some four decades later, in 1905, as a supervising body of imperial military planning and mobilization, while the Main Staff retained authority in matters related to military personal assignments, pensions, Cossacks service and civil-military government

FIGURE 0.10
General of Infantry Count Fedor Logginovich Geiden, Chief of the Main Staff, 1866–1881.
COPIED FROM *VOENNAIA ENTSIKLOPEDIIA* (T. VII. ST. PETERSBURG, 1912).

for selected provinces.[35] The Russian expeditions to various corners of Asia with the participation of the military prior to 1905 were organized by the IRGO with the assistance of the Main Staff superseded by the General Staff after that date.

The cooperation of the Imperial Russian Geographical Society and the Main Staff calls for a brief comment.[36] As already noted, a considerable part, if not the majority, of the IRGO members, from the very start, were military and naval officers, men with a keen interest in geography and statistics, some of whom had previously left their mark in the field of geographical studies and exploration. Nine out of seventeen of IRGO's founding members belonged to the military elite, including three admirals (F.P. Litke, P.I. Rickord, and F.P. Wrangel). The military also held the administrative reins of the Society in the capacity of its Vice-Chairman (M.N. Muraviov, F.P. Litke, 1850–1873), Chairman's assistants, heads of the scientific sections and the Council members. The Council, at different periods, enlisted the services of some prominent military figures, such as Count D.A. Miliutin (war minister), Count F.L. Geiden (head of the Main Staff), Baron N.V. Kaulbars (head of the military-scientific committee), and

35 This brief note on the history of the General Staff was kindly provided to me by M. Baskhanov (*A. Andreyev*).

36 On the connections between the IRGO and the Russian military elite see Marshall 2006.

FIGURE 0.11
General-Adjutant Nikolai Nikolaevich Obruchev, Chief of the Main Staff, 1881–1897. COURTESY OF M. BASKHANOV.

O.E. von Shtubendorf (head of the military-topographic department).[37] This galaxy of famed 'geographical generals' (to use M.K. Baskhanov's term) would shape to a large degree, in the decades to come, the Society's exploratory priorities and strategy.

The formal alliance of Russia's first-rate geographers and top military under the aegis of the IRGO was by no means incidental. Military science in the process of its evolution, long before Miliutin's military reforms of 1860s–1870s, began to look to geography as a discipline vitally important for military-strategic purposes and as one of great practical use for reconnoitering little-known territories, both within Russia and beyond. Geography was taught in military colleges and it was in the curricula of the Imperial Military Academy (to become, in 1855, the Nicholas Academy of the General Staff, NAGSh for short). A participant of the Caucasian War, Miliutin moved from the war theatre to St. Petersburg, where he lectured, in 1845–1856, on military geography and military statistics at the Academy. The former, as he understood it, focuses on the study

37 Count Dmitrii Alekseevich Miliutin (1816–1912) was a member of the IRGO Council in 1849–1852, 1855–56; Feodor Logginovich Geiden (1821–1900) in 1867–1870; Nikolai Vasilievich Kaulbars (1842–1905) was a member of the Council in 1889–1894 and the secretary of the physical geography sector in 1873; Otto Eduardovich Shtubendorf (1837–1918), headed the mathematical geography sector in 1871–1875; he was a member of the Council in 1876–1880 and 1881–1882, and the assistant of IRGO's Vice-Chairman in 1882–1886.

of "lands or parts of the land surface from a military perspective".[38] In other words, military geography is, in fact, regional geography, a sub-discipline of general geography. While studying military geography at the NAGSh, according to N.V. Andreev, "the officers were obliged to compile some descriptions and sketches of different states. The Academy, of course, did not prepare specialists in geography, but it provided enough practice to allow officers to compile descriptions [of this kind]".[39] Thus, as a student of the Military Academy, Przhevalskii compiled, by using all available sources, 'A military-statistical sketch of the Priamurskii district'. The paper received a high appraisal from geographers; hence he was elected an IRGO member in 1864, an event which marked the beginning of his career as a military geographer.

It is important to note that geography and geographical explorations meant more for Przhevalskii than his military career, as he confided in a letter to Baron F.R. Osten-Saken in 1870, prior to his first journey to Inner Asia. "If the conditions of my service become an obstacle for my intended trip, I am prepared to resign any time and devote myself entirely to the service of science".[40] Yet, as D. Schimmelpenninck van der Oye justly noted, "While a geographer, he (Przhevalskii) was first an officer, and this military affiliation strongly colored his view about the regions he explored. As he saw it, Chinese Inner Asia was not just an unknown region waiting to be opened to science, but virgin territory to be conquered for the glory of Russia".[41]

The contribution of the military to the IRGO expeditions, however, has been overly exaggerated by some contemporary authors, who tend to see these expeditions as basically military reconnaissance missions. Yet it was not so. The expeditions were organized by the IRGO with some assistance from the Main Staff, and, judging by their programs and ultimate results, their main focus was on geographical (landscape) observations and collecting samples of flora and fauna, and partly ethnographic artefacts. The assistance of the Main Staff was primarily the provision of equipment (arms and measuring instruments) and paying the travelers – those who were servicemen – their annual service allowances, plus their travel fares from St. Petersburg (Moscow) to the expedition point of departure (at Kiakhta, Zaisan or elsewhere) and back. Like the IRGO, the Main Staff also allocated certain sums to the expedition leaders, but these were rather modest and not sufficient for the entire period of travelling, as shown by the financial difficulties Przhevalskii faced during his first Inner

38 Miliutin 1846, 4.

39 Andreev N.V 1962, 16.

40 Quoted in Dubrovin 1890, 91 (letter dated 29 March 1870).

41 Schimmelpenninck van der Oye 2001, 34.

FIGURE 0.12 The Imperial Nicholas Military Academy (former the Nicholas Academy of the General Staff), St. Petersburg, 1909. COPIED FROM *VOENNAIA STOLITSA ROSSIISKOI IMPERII V FOTOGRAFIIAKH KONTSA XIX – NACHALA XX VEKOV* (ST. PETERSBURG, 2004, ILL. NO. 99).

Asian journey. Apart from the IRGO and the War Ministry, the geographical expeditions received quite substantial funding from private sponsors, "the patrons of geography", including the representatives of the Romanov dynasty – the grand dukes and the emperor himself, while academic institutions (the Zoological museum and the Botanical Gardens) also gave some financial support, though in a limited way, to the IRGO expedition projects.[42]

In general, the expedition members, as servicemen, were formally detached from their military units (regiments) and put under direct command of the Main Staff for the entire period of their travels, regarded as 'official journeys' and paid for from the military coffers.

The IRGO – War Ministry peculiar cooperation in mounting and managing their 'joint' expeditions to Inner Asia was discussed by M.K. Baskhanov in his

42 For more detail on this see Andreyev A.I (ed.) 2013, 56–82.

article examining the phenomenon of Russian "geographical generals".[43] In Baskhanov's opinion, the routes of the "joint IRGO – Main Staff expeditions" were determined by the Geographical Society (!), therefore the military (i.e. Main Staff) could only obtain information from a particular "fixed" route. No "improvised" changes of routes in the interests of the MS were possible. The military aspect of these "joint expeditions" consisted mainly in transmitting the originals of all route surveys, plans (those of castles, towns, road junctions etc.), as well as astronomical and barometric measurements (needed for latitude and altitude determination) directly to the Military-topographic department. Sometimes the expedition leaders also submitted to the MS their reports on "military-political questions", as Przhevalskii often did. In this way the MS employed the method of "passive" reconnaissance in these "joint expeditions". This was well justified since the expedition areas were too far from the Russian territory and the zones of responsibility of the frontier Asiatic military districts. Besides, there were no significant military objects in these areas, such as large economic and military centers or any considerable concentration of military forces, or strategic communications, etc. For this reason, the War Ministry focused mainly on cartographic issues. But that is where the interests of the geographers and the military largely overlapped – the geographers too needed accurate and fully-reliable – scientific – geographical maps of Inner Asia and other continental regions.

The most obscure aspect of the IRGO expeditions relates to the special commissions given by the Main Staff and other governmental departments to the expedition leaders and connected with gathering various kinds of information (mainly political, economic, and military-statistical). These commissions, however, were only auxiliary, side-line tasks which did not interfere with their predominantly scientific priorities. On the other hand, the War Ministry, as is well-known, sent its special agents to different Asiatic countries (China and Japan in particular) for intelligence-gathering purposes.

In general one may discern two types of expeditions mounted by the War Ministry: 1) military reconnaissance expeditions, and 2) military-diplomatic missions or 'embassies', such as the missions of generals A.V. Kaulbars and A.N. Kuropatkin to the ruler of Kashgaria Yakub Beg (in 1871 and 1876 respectively) or N.G. Stoletov's mission to the Afghan Emir Shir Ali-Khan (1878). These missions had nothing to do with the IRGO expeditions in question, although they too conducted en route some topographic surveys and geographical observations. Also, not infrequently staff officers (mainly topographers) were attached

43 M.K. Baskhanov, *"Ne kovrom byla postlana nam doroga vglub' Azii"*, in: Chistiakov, Yusupova, Gnatiuk (eds.). 2014, 297–318.

to trade caravans, as was the case with Pevtsov's first Dzungaria expedition (1876), or Potanin's first Mongolian (Tarbagatai) journey (1876–1878).

Therefore, it would be perhaps too far-fetched to perceive the IRGO expeditions as 'military-geographical' missions. The predominant interest of their organizers and leaders, no matter if they were army officers or not, lay in the field of geography and natural sciences, as evidenced by the tasks they set themselves and successfully accomplished in the course of their journeys. As P.P. Semenov once put it, "the Geographical Society, despite its enterprising spirit, always organized its expeditions on the initiative of the travelers themselves, by regulating their activity with a view to yielding the greatest possible benefit for science".[44]

Przhevalskii borrowed from Semenov his scientific methods of route reconnaissance and versatile field work, including instrumental observations (astronomical, barometric, meteorological) and collecting samples of local flora, fauna, minerals and ethnographic artefacts. However he supplemented these with something of his own – the semi-military system of organization of the expeditionary team. The latter normally consisted of a leader, two or three assistants, including a professional preparator (to expertly preserve samples of flora and fauna), an armed grenadier-and-Cossack escort numbering about a dozen or more men, and, on top of that, several Mongolian- and Chinese-speaking interpreters and guides. This was a well-tried system based on two principal elements – strict discipline and the undivided authority of the expedition leader.

The system worked fine under Przhevalskii[45] and it was generally adopted by his immediate followers (Pevtsov, Roborovskii and Kozlov), though none of them had Przhevalskii's great charisma, his energy and ambitions, not to mention his notorious Sinophobia which strongly discredited him in the eyes of Oriental scholars of his day. Pevtsov, who took over, after Przhevalskii's death, as the leader of his fifth expedition, was Przhevalskii's antipode in many ways. Unlike Przhevalskii, who was stern, ardent and strong-willed, Pevtsov was softhearted and unambitious. Upon return from his Tibet journey, he turned into a typical armchair scientist, totally absorbed in his favourite astronomical and geodetic studies. The overly modest and unassuming Roborovskii was more like Pevtsov, whereas Kozlov, with his innate "nomadic" instincts and ambi-

44 Semenov 1896–1898, x.

45 While contemplating his expedition to Tibet in 1876, Przhevalskii thought that an 'ideal team' for traveling in Asia should be small and mobile, consisting of no more than six men – a leader, his assistant and four Cosacks as escort. See Przhevalskii 1947, 53.

tions, stands closest to Przhevalskii, whom he nearly worshipped throughout his lifetime and tried to imitate.

What primarily attracted Przhevalskii to Inner Asia were its many unexplored and uncharted lands, the largest of which was the mysterious Tibet, the forbidden land to European travelers. In the opinion of British geographer Thomas Holdich,

> Through all the ages Tibet has held a paramount position amongst those regions of the world which have been popularly invested with a veil of mystery because they are inaccessible and unknown.
>
> Tibet is so isolated, so lofty, so irresponsive to outside influences, has held herself so far apart from the meddling interference of the busy, commercial world, as to provoke the enterprise of generations of speculative geographers, who, accepting "omne ignotum pro magnifico", have startled the world with small installments of truth surrounded by wide embroideries of decorative fancy.[46]

The chief obstacles to Tibetan exploration, as Holdich justly noted, "have ever been the mountain barriers which surround the plateau, rather than the plateau itself". These formed a series of gigantic walls, the ranges and ridges, on all sides of Tibet, impeding access to the plateau from any direction.

Quite naturally, Tibet, from the first, became Przhevalskii's mightiest attraction, his obsession, and greatest challenge. His cherished dream was to ascend the highland and become its pioneer explorer and even more than that – to be the first Russian to come to Lhasa and get in touch with Tibet's theocratic ruler, the Dalai Lama. Przhevalskii succeeded in ascending the plateau for the first time in 1872; he was looking for passes leading to Tibet from Kashgaria during his second one (1876–1877); and his next two missions (1879–1880, 1883–1885) were also Tibet-bound. This unusual persistence of his can be easily understood – Tibet was not only the largest 'terra incognita' but at the same time presented a huge natural buffer isolating British India from the north, a geopolitical factor which eventually turned the country into one of the 'fields' of the Great Game, the Anglo-Russian race for supremacy in Central Asia. So Przhevalskii, being a proponent of Russia's forward policy in the region, could not but become a Great Gamer as well.

Apart from Tibet, there were many other attractions in Inner Asia for Przhevalskii, which were mainly places unexplored, uncharted, and of difficult

46 Holdich 1994 (repr.), 1. The expression *omne ignotum pro magnifico* (Everything unknown is taken as grand) belongs to Tacitus (Agricola, XXX, 3).

access to travelers, such as Lake Kokonor; the upper Yellow River (Huang He), the east Tian Shan, the Tsaidam country (on the northern margin of the Tibetan plateau), Lop nor and the lower Tarim Basin, the Gobi Desert, and Kashgaria in Chinese Turkestan, discovered earlier for Europeans by Chokan Valikhanov.[47] This was a vast area, the exploration of which was more than a single scholar-traveler could normally hope to achieve during his lifetime.

After Przhevalskii's death in Karakol in 1888, the IRGO, with the assistance of the Main Staff, organized several more expeditions into Inner Asia to continue their systematic and deliberate exploration of the region, the most prominent being: Pevtsov's Tibetan Expedition (1889–1891), Grombchevskii's two journeys to the Pamir and northwestern Tibet (1888 and 1889–1890), Roboroskii and Kozlov's joint expedition, known as the 'Expedition of Przhevalskii's Companions' (1893–1895), and three of Kozlov's own travels (in 1899–1901, 1907–1909, and 1923–1926). And there were also a number of other brilliant IRGO-sponsored expeditions led by civilians, such as Grigorii N. Potanin (who made five journeys altogether between 1876 and 1899), Vladimir A. Obruchev (1892–1894) and Grigorii E. Grumm-Grzhimailo (1889–1890 and 1903).

Potanin and Obruchev were both civilians, and their expeditions, albeit organized by the same IRGO, followed quite a different pattern of exploration. Potanin, for example, staffed his team with specialists in topography, geology, zoology, and Oriental studies, which allowed him, together with his wife, Alexandra, his permanent companion, to concentrate on their favourite botanical and especially ethnographic research. They closely associated with the indigenous population, paying much attention to their traditional way of life, religion, and folklore. As for Obruchev, he was a professional geologist who, although being formally on the staff of Potanin's fourth Inner Asian expedition (1892–1894), worked entirely on his own.

The routes of all these travelers traversed the territory of Asia's heartland in many directions in the last three decades of the nineteenth century. And there were many others whose names are well worth mentioning here – D.A. Klements, N.A. Severtsov, Yu.A. Sosnovskii, Z.L. Matusovskii, P.A. Rafailov, D.V. Putiata, G.A. Fritsche, N.M. Yadrintsev. Some of them traveled on their own, others were sponsored by the IRGO or its local departments, the War Of-

47 Chokan Valikhanov (1835–1865), a Kazakh by origin, was a historian, ethnographer, folklorist, a member of the IRGO and an officer of the General Staff. The latter often used him for intelligence gathering purposes. Thus, in 1858–1959 he traveled to Kashgaria, disguised as a merchant, to find out about the political situation there. On him, see Strelkova I.I., *Valikhanov*, Moscow: Molodaia Gvardiia, 1983.

fice, and the Russian Academy, yet all of them brought back a large amount of valuable geographical and topographical material.

Concurrently, the territory of Inner Asia, primarily Chinese Turkestan, Tibet and the Pamir region, was explored by a host of Western travelers – Thomas Douglas Forsyth, Robert B. Shaw, George and Teresa Littledale, Cecil Rawling, Béla Széchenyi, William Rockhill, Dutreuil de Rhins, Fernand Grenard, Gabriel Bonvalot, Sven Hedin, to name a few. All of them also made important contributions to the geography of the region. However, their individual efforts were not part of one and the same comprehensive project coordinated by either the leaders of the geographical societies of their countries (Britain, France, USA, Hungary, and Sweden) or the military, and the length of their routes as well as the amount of information they collected yielded in general to those of their Russian counterparts. Still the work of the Indian scouts (so-called pundits)[48] and British military topographers in the Himalayan trans-frontier area and western and southern Tibet sponsored by the government of British India was very productive. The pundits alone succeeded in reaching Lhasa, something which no Russian traveler could do, and gathered a great amount of accurate knowledge of Tibet and its capital.

The ultimate result of all these extensive explorations, Russian and Western, was new physical maps of Inner Asia, with all its complicated orographic and hydrographic systems, produced by Russian and European cartographers. The best Russian maps, as in the early half of the century, were those published by the Military-Topographic Department (Depot, until 1862) of the Main Staff[49] and scaled to 40 and 100 versts per inch. As for Western maps, they were largely based on the topographic material obtained by Russian travelers.

48 On pundits and their secret exploration of Tibet see: Peter Hopkirk, Trespassers on the Roof of the World: The race for Lhasa, London, John Murray Ltd., 1983; Derek Waller, The Pundits: British Exploration of Tibet and Central Asia, The University Press of Kentucky, 1990.

49 See Glushkov 2007. All Central Asian maps were edited at the MTD, in 1877–1897, by one of the department's best topographers A.A. Bol'shev. On his service career, see Glushkov 2007, 421; on Przhevalskii's topographic surveys from his four expeditions pp. 355–358. A member of the IRGO, Bol'shev compiled a Map of Asia for the Geographical Society based on the results of Russian explorations (1892) and showing the routes of the IRGO major expeditions.

4 Exploration and the Great Game

The time when Przhevalskii started his 'scientific reconnaissance' of Inner Asia was extremely unpropitious for Russian travelers, their two major obstacles being the Dungan (Muslim) uprising and the Great Game.[50] The anti-Qing revolt of the Chinese Muslims, or Dungans (Tungans), which broke out in mid-1860s, seized Eastern Turkestan (to become Xinjiang in 1884) from where it quickly spread to the adjacent Gansu and other provinces. By some strange coincidence, it was precisely at this time that the Great Trigonometrical Survey at Dehra-Dun began to dispatch its well-trained surveyor-scouts, or pundits, disguised as Buddhist pilgrims or traders, to the Trans-Himalayan area, Kashgaria and Tibet, which eventually touched off a new round of Anglo-Russian rivalry in that part of Asia. In the opinion of contemporary researcher Alexander Morrison, "For Russia herself the rivalry with Britain was important primarily in ideological terms: Britain was the leading 'Great Power' against which she measured her own imperial and diplomatic achievements, and whose global reach and clout she sought to emulate".[51]

The penetration of Przhevalskii into Mongolia and China with his first expedition in 1870 occurred against the backdrop of concurrent vigorous expansion of Russia into Central Asia (Middle Asia or Western Turkestan), with the result that the three Muslim khanates – Bukhara, Khiva and Kokand – had fallen into the Russian hands, one by one, by mid-1875. The military campaign started in 1864 by Colonel M.G. Cherniaev was strongly opposed by the influential minister of foreign affairs, Count A.M. Gorchakov, who apprehended that any military advance into the region would seriously aggravate Russia's relations with Great Britain. And so it did. Moreover, when the Russian troops crossed into Afghanistan and seized, in 1885, the Afghan town of Penjdeh, military confrontation between the competing empires seemed imminent as Penjdeh opened to Russia the road to Herat, the 'gateway to India'.

On the other hand, the Dungan uprising became Russia's major concern and a challenge to their military, as it had interrupted the existing Russo-Chinese trade via East Turkestan and was threatening its newly established Asiatic frontier. (Not to mention that it cut short the traditional Urga – Lhasa link, making it impossible for Mongolian pilgrims to travel safely through Xining in

50 Re the Great Game, see Morgan 1981, Hopkirk 1990; in Russian: Sergeev 2012; Zagorodnikova 2014 (a collection of archival documents); on the pundits see Waller 2004.

51 Alexandr Morrison, Introduction: Killing the 'Cotton Canard' and getting rid of the 'Great Game'. Rewriting the Russian conquest of Central Asia, 1814–1895, see:www.nur.nu.edy.kz/.../Intoduction_Killing_the_Cotton_Canard ...3.

Gansu). Therefore, while helping IRGO to mount Przhevalskii's first Inner Asian expedition, the Main Staff tasked its leader to collect information on the Dungans and their activities in Western China. Yet in June 1871, when Przhevalskii was already trekking across China, the Russian troops under General G.A. Kolpakovskii occupied the Ili or Kuldja district (sultanate) in Eastern Turkestan to restore order in the mutinous area. As the Russian emperor Alexander II explained in his dispatch to the Peking Manchu court, Russia's invasion "pursues only the goal of rendering China assistance towards the restoration of her influence in the [Dungan]-seized Western provinces of the Empire".[52] At the same time, one of the Russian military commanders, A.N. Kuropatkin, assured the Ili local population that their country "would never be returned to China".[53] However, a decade later (1881), the Russians did return the Ili district to China's Manchu rulers.

Tsarist Russia's part in the Great Game in Inner Asia proper, i.e. in the Western borderlands of the declining Qing Manchu Empire, as the country's policymakers of the late nineteenth century saw it, consisted of two basic tasks – the extension of Russian influence in the region without any territorial acquisitions, and counteraction to British expansion. This dual policy was formulated neatly by the Russian minister of finance, S.Iu. Witte, in the following words:

> For our future plans it would be (no less) important to place China in some sort of dependence upon us and prevent England from spreading its influence over this country. England dominates in the south of Asia, and we are not going to bother her there; however Central Asia must belong to us, not in the sense of a material conquest, but to make it serve our ends and our interests.[54]

It is noteworthy that concurrently with Przhevalskii's first journey the General Staff, well-aware of the British exploratory activities in Asia, commissioned the military geographer Colonel Mikhail Veniukov (incidentally a good friend of Przhevalskii) to provide a military-statistical overview of Russia's extended southern frontier and her Asian borderlands, using all available data.[55] The result of this laborious work was Veniukov's "Military Survey of the Russian Bor-

52 Prokhorov 1975, 154.

53 Kuropatkin 1913, 47.

54 Quoted from Andreyev, A.I. 2006. 74 (Witte's letter to the Russian ambassador in Paris A.P. Morengeim, 11 May 1895).

55 Esakov 2002, 60–72.

ders in Asia" (*Opyt voennogo obozreniia voennykh granitz v Aziit*) published in two parts in 1873 and 1876, with two maps enclosed – that of the main roads leading from Russia into China and the ethnographic map of Asiatic Russia. The author described thoroughly all the thirteen major borderlands from the Pacific Ocean to the Caspian Sea which was a most valuable contribution to the regional geography of the day.

Meanwhile, Przhevalskii was getting deeply involved in the Great Game. In subsequent years he would be gathering more intelligence on the Dungans (for whom he even showed some sympathy as they had risen against their oppressors, the Manchus), and concurrently, in the course of his second Lopnor expedition (1876–1877), on the leader of the newly founded Muslim khanate of Dzetyshaar (Yettishaar or 'Seven Cities' kingdom), in Kashgaria. The latter emerged in 1867 under Yakub Beg, Badaulet Khan, a Tadjik adventurer and confident of the ruler of the Kokand khanate, who succeeded in uniting the separate territories of the disintegrated Eastern Turkestan under his own rule. Yakub Beg was consequently recognized as a legitimate ruler by both the British and the Russians who signed commercial treaties with him in 1872 and 1873 respectively, trying hard to win him over, each to its side. And in 1876 Przhevalskii was received by Yakub Beg in a private audience at Kashgar. This and other valuable information that the Russian officer was able to collect on the "tsar of Kashgaria" – his career, psychological portrait and current activities he reported to both the RGO and the Main Staff.[56] A year later the Kashgarian ruler, whom Przhrvalskii perceived as "no more than a political rogue", suddenly died, however his death did not stop the Great Game in Kashgaria.

About the same time, Tibet began to figure prominently as the Gamers' most attractive field in Inner Asia. Beginning in the mid-1860s, the Main Staff and the IRGO turned their gazes to the mysterious "Land of the lamas" in connection with the secret explorations of the pundits. Their detailed travel accounts and sketch maps were published much later (in the 1880s), under their code names, by the Great Trigonometrical Survey of India and the Royal Geographical Society in London, and it was through these publications that world geographers learned of the remarkable work of the pundits. Still, Przhevalskii spoke rather critically of some of the surveys of pundit A – k (Kishen Singh) in

56 AN RGO, F. 13, Op. 1, D. 26, ll. 1–6 (O sovremennom sostoianii Vostochnogo Turkestana, 6 June 1877); see also: RGVIA, F. 400, Op. 1, D. 438, ll. 132–38; published in Dubrovin 1890, 570–577 (Appendix 12).

his last travelogue as they were clearly at variance with his own first-hand reconnaissance.[57]

Concurrently, one of the IRGO members, Lieutenant-Colonel P.A. Gelmersen, profoundly impressed by the crafty methods of route survey employed by the pundits, came up, as early as 1869, with a daring project – that of sending to Lhasa a Buryat native, "someone of European education", with a Mongolian religious 'embassy'. In his memo, submitted to the Vice-Chairman of the IRGO F.P. Litke, Gelmersen wrote:

> Eastern embassies usually consist of many people and are often joined by trade caravans. It will not be difficult for a person of Mongolian appearance to join the latter, particularly if the traveler is disguised as a trading lama-pilgrim. Few risks are involved even if the traveler's ethnic identity is discovered, as the Buryat lamas from Transbaikalia frequently travel to Tibet. <...>.
>
> The possibility of carrying out accurate geographical work in Tibet was brilliantly proved by the journey of the Indians in 1866, which serves as a perfect example for future travelers, esp. in terms of instruments, which were skillfully adapted to the Buddhist prayer devices.[58]

We do not know if the IRGO was able to dispatch their agent to Lhasa, following the example of the pundits, yet the scheme proposed by Gelmersen proved to be of little use, the more so that the strategic Urga – Lhasa route was still controlled by the Dungans. As a result, the IRGO and the Main Staff chose what seemed to be a far better solution for geographers, as soon as Przhevalskii appeared on the scene, that of sending their exploratory missions into Inner Asia.

Still, some three decades later two Russian 'pundits', one Kalmuck (Ovshe Norzunov), and one Buryat (Gombozhab Tsybikov), succeeded in penetrating to Lhasa, disguised as Mongolian pilgrims (the former in 1899 and 1901; and the latter in 1900). Both were supplied with French-made portable cameras and glass plates by the IRGO which enabled them to take many pictures of

57 See Przhevalskii 1888, 143–145. In this travelogue Przhevalskii referred mainly to two publications: "Report on the explorations in Great Tibet and Mongolia, made by A – k, in 1879–82", prepared by J.B.N. Hennessey (Dehra Dun, 1884) and "Four Years' Journeyings through Great Tibet, by one of the Trans-Himalayan Explorers of the Survey of India", published in the Proceedings of the Royal Geographical Society and Monthly Record of Geography (1885, no. 2, 65–92).

58 See Andreyev, A. 2001, 163–180.

FIGURE 0.13 A view of the Marbori Hill with the Potala Palace of the Dalai Lama in Lhasa. PHOTO BY OVSHE NORZUNOV, 1901. PICTURE LIBRARY OF THE ROYAL GEOGRAPHICAL SOCIETY, PR/029660.

Lhasa, its environs, and Tibet's major monasteries, sanctuaries and historical sites, the earliest ones ever taken by travelers in the no-man's land.[59]

As for Przhevalskii, he surveyed the northeastern borderland of Tibet during his two Tibet expeditions (in 1879–1880 and 1883–1885). He gathered a large amount of general information about the country, its administrative, political and social systems, population and its fabulous capital (and he incidentally brought back to St. Petersburg the first hand-drawn plan of Lhasa[60]), all these details incorporated in his published travelogues. The former journey had a clearly political dimension – as Przhevalskii put it in his memo to the General Staff in 1878, "scientific research ... will camouflage the political goals and ward off all suspicions of our adversaries". The idea was fully endorsed by Kuropatkin, then head of the Asiatic Section of the Main Staff, who underscored in his report to the Emperor that the expedition intended "to gather intelligence about the political system of Tibet, its relations to its neighbor, and, finally, to engender a possibility of entering into and strengthening our relations with

59 For the story of these travelers and their photographs see Andreyev, A. 2013.

60 On this in more detail see Andreyev, A. 2014, 41–55.

the Dalai Lama. Such a policy, with a little bit of luck, may reward us with influence over all of Inner Asia, right through to the Himalayas".[61]

Another Russian Great Gamer of distinction was Bronislav Grombchevskii (Bronislaw Grabczewski, 1855–1926),[62] who, like Przhevalskii, conveniently combined geographical and topographical surveys with intelligence gathering. He became notable for his explorations of the Pamir, the Cis-Pamir area and the southwest corner of Tibet in 1888–1890. Grombchevskii's two Pamir expeditions synchronized well with Przhevalskii – Pevtsov's Tibet expedition and G.E. Grumm-Grzhimailo's first Inner Asian journey of 1889–1890. The latter covered a vast area, from East Tian Shan across the Turfan depression and the Gobi to northern and eastern Nanshan.

In 1889, in the Pamir, this Roof of the World, Grombchevskii happened upon his rival, a British army officer Francis Younghusband, who was out on a special mission of his own in the area. (Of this the Russian explorer was then unaware, whereas Younghusband knew quite well what Grombchevskii was up to thanks to a notice that revealed his plans, published incautiously by his colleague, M.I. Veniukov, in the Proceedings of the Royal Geographical Society.)

In the course of his friendly chat with the Briton, Grombchevskii confirmed some of the worst presentiments of the British concerning Russia's plans for an alleged assault on India. As Younghusband's biographer, Patrick French, recounted this episode, Grombchevskii declared boastfully over their repast in his camp at Khaian Aksai that "nearly half a million Russian troops were ready to march south at the drop of a hat. He called some of the Cossacks over to the door of the tent and received "a rousing cheer" when he asked them if they would like to invade India".[63] A year later, when Grombchevskii had learnt about the genuine purpose of Younghusband's mission connected with the partition of the Pamir, from the friendly *amban* of Yarkand, he began to openly incite the Chinese to occupy the Pamir – not to let the Afghans, and hence the British, place this territory, adjacent to the Middle Asian possessions of the Russian Empire, under their control.[64]

Younghusband, however, behaved himself rather ungentlemanly when his host asked him for ways leading to the British-controlled Ladakh,[65] the coun-

61 Quoted in Schimmelpenninck van der Oye 1997, 214–215.

62 On him see Postnikov 2001, 163, 164, 168–176 and ff., 346–351 (Grombchevskii's short biographical sketch), also Baskhanov, Kolesnikov, Matveeva 2015.

63 French 1994, 75.

64 Postnikov 2001, 174, 175.

65 On Russian travelers in Ladakh in the 19th – 20th centuries, see Andreyev, A. 2009, 25–41.

try he wanted to visit, by giving him purposefully wrong directions to make certain that "he would not complete the journey".[66]

The final and most dramatic third phase of the Great Game took place in the early twentieth century. It was then that the Viceroy of India and arch-Russophobe, Lord Curzon, sent the British military-diplomatic mission to Lhasa, under Brigadier-General J. Macdonald and Fr. Younghusband, allegedly to thwart the "Russian intrigue" in Tibet "while there is yet time".[67] The occasion was provided by alarming reports in the Russian press on the two visits to St. Petersburg, in 1900 and 1901, of the Dalai Lama's "Russian emissary", a Buryat lama Agvan Dorzhiev. Dorzhiev's activities – his formal audiences with the tsar and negotiations with the tsarist top ministers, including Kuropatkin, indeed triggered a Russo-Tibetan rapprochement, albeit a short-lived one. The Russians established, in 1901, their consulate at Tachienlu (Darchendo), outside Tibet proper; however, this was no more than a listening post to monitor British activities in that part of the continent. (The entity had functioned for about a year and was abolished by the Ministry of Foreign Affairs in October 1904.[68]) The Russians did not dare (and were not willing in fact) to go any further than that and, of course, sending their troops or a supply of arms to Tibet was out of question. Despite the rapprochement and the secret arrangements made by Dorzhiev to facilitate Kozlov's coming to Lhasa at the head of his Mongolia and Kham expedition (1899–1901), the traveler was unable to enter the holy city, being halted by vigilant Tibetans, very much like Przhevalskii, on the close approaches to Lhasa.

As for the Younghusband mission of 1903–1904, it had most detrimental consequences for Tibet – the flight of the Dalai Lama from Lhasa and his nearly five-year absence from his country (an absence which saw him moving from place to place between Mongolia and China), and a humiliating bilateral treaty the British forced the Tibetans to sign (1904) which effectively placed Tibet under British protection, though not for a long time.

The Russian response to this British challenge presented by the Younghusband mission also came from military quarters – in January 1904 the war minister Kuropatkin dispatched a secret reconnaissance mission to Lhasa under a Kalmuck subaltern, Naran Ulanov.[69] Shortly prior to the event, in November 1903, Petr Kozlov proposed sending an "expeditionary force" to Tibet "to coun-

66 French, op. cit., 77.

67 On this mission see Mehra 2005.

68 On the Russo-Tibetan rapprochement and the Russian consulate at Tachienlu see Andreyev, A. 2003, 35.

69 On the Ulanov mission see Andreyev, A.I. 1997, 9–60.

ter the steps" of the British-Indian government. The Russian mission, consisting of Buryat Cossacks, was to persuade the Tibetan government, following the British example, to open their country to Russian pilgrims and goods by resorting to either diplomatic negotiations, or, if need be, to "appropriate" military action.[70]

Kozlov was, by far, the most prominent Russian Great Gamer of the day. Having met the Dalai Lama twice (in Urga, in 1905 and in the Kumbum monastery, in 1909), he easily won sympathy and confidence of Tibet's ruler and was personally invited by him to come to Lhasa for scientific research. Yet, despite all his efforts, Kozlov was unable to realize his and his mentor Przhevalskii's long cherished dream. His project for a "Russian escort" (that of Transbaikal Cossacks) to accompany the Dalai Lama on his return journey from Urga to Lhasa in 1905 was turned down by the Russian diplomatic office, despite the support given to it by the military. And then, when the Great Game was seemingly over in 1907, after Russia and Britain had carved out their spheres of influence in Central Asia (Persia, Afghanistan and Tibet), Kozlov made some more venturesome plans, in 1911, for a joint Anglo-Russian "rescue operation". This was intended to bring the ill-fated Dalai Lama back to Lhasa and to guarantee him that he would be protected from China (whose troops had earlier invaded Tibet, making him flee from Lhasa again) by the finally reconciled great powers.[71]

One more name deserves to be mentioned here – that of Carl Gustaf Mannerheim (1867–1951), "one of the last secret agents of the Tsar in the Great Game", in the words of Eric Enno Tamm.[72] A baron and cavalry officer of Finnish descent, in the service of the Russian army (as the Grand Duchy of Finland made up part of the Russian Empire at that time), Mannerheim was later to become a leading military and political figure in his native country, the Commander-in-Chief and President of Finland. Mannerheim was dispatched in 1906 to China by the General Staff on a special reconnaissance mission under the cover of the scientific expedition led by a renowned French archeologist and Orientalist Paul Pelliot. His task was to collect information on the current situation in the Celestial Empire after the Empress Dowager Cixi had started a number of vigorous reforms of the administration, army, educational system

70 Ibid, 42–43.

71 On Kozlov and his part in the Great Game, see Andreyev, A.I. "Nesbyvshiesia mechty o Lhase: Tibet v issledovaniiakh P.K. Kozlova (nauchnie i politicheskie aspekty) (The unrealized dreams of Lhasa: Scientific and political aspects of the exploration of Tibet by P.K. Kozlov), in Chistiakov, Yusupova, Gnatiuk (eds) 2014, 278–296.

72 Eric Enno Tamm, "A Secret Agent on the Silk Road", <http://steppemagazine.com/issue/winter-2008/>.

etc., his special focus being the adjacent Western China (Chinese Turkestan). There the Chinese military forces were being reorganized on the Japanese model with a view "to reinforcing the Chinese colonization of the region and to connect the latter closer with China proper", in the opinion of the General Staff.[73] While preparing for his secret mission, Mannerheim studied thoroughly the reports of Przhevalskii and Pevtsov and also met with Kozlov. He traveled together with Pelliot from Tashkent through Russian and Chinese Turkestan, up to Kashgar, and thence continued on his own, mainly on horseback, across the entire territory of Central and Northern China, including Inner Mongolia, towards Peking.

Mannerheim's "epic journey of archeology and espionage" (as Eric Tamm saw it) lasted for more than two years and yielded a large amount of valuable information of military-strategic, archeological and geographic character, as he traversed some little-known countries on the way. He mapped over 3,000 km of road, sketched twenty garrison towns, took 1,370 photographs, collected and purchased ancient manuscripts and artifacts (for the Finno-Ugrian Society) and also took notes of the places he visited and the people he encountered. Incidentally, like Kozlov in 1905, Mannerheim had a chance to meet and converse with the Dalai-Lama at the Wutai-shan monastery, in 1908, and he used the opportunity to sustain the Tibetan ruler's pro-Russian feelings. Moreover, from Peking he made a two-week trip to Japan to fulfill another secret commission of the Russian military. Upon his return to St. Petersburg, Mannerheim reported on his travels to the General Staff[74] and personally to the Emperor. However, his travel diary was not published until 1940,[75] and its publication was hardly noticed then by the general public.

To conclude, the Great Game certainly had a strong impact on the scientific pursuits of the IRGO in the region. Therefore, while mounting its expeditions, the Society had to take into account the current international and regional political climate and the potential reaction of Britain to their initiatives, especially when the expeditions were bound for the highly sensitive 'strategic areas', such as Eastern Turkestan, Tibet, and the Pamir. Not infrequently the crucial decisions regarding new expedition proposals were made at the highest governmental level of the War and Finance Ministries and the Ministry of For-

73 Report of the GUGSh (Chief Administration of the General Staff) to the Emperor, dated 13 June 1906, see RGVIA, F. 2000, Op. 1, D. 1567, ll. 1–4 ob.

74 Mannerheim's "preliminary report" was published in *Sbornik geograficheskikh, topograficheskikh i statisticheskikh materialov po Azii*, LXXXI, St. Petersburg, 1909.

75 C. Mannerheim, From Across Asia: From West to East in 1906 – 1908, in 2 volms, Oosterhout, the Netherlands, 1940 (repr. 1969).

eign Affairs (MID). Thus, in 1894 the head of the Main Staff and member of the State Council General N.N. Obruchev turned down the project for a joint Russo-French expedition to Tibet (Lhasa) by arguing that "all our scientific expeditions undertaken [thus far] in the close vicinity of India only encouraged England to make new conquests and extend its political influence, and to justify these steps by making false accusations that Russia has bellicose designs... So it is better to leave Lhasa alone, not to facilitate the access of England to Tibet".[76] And five years later the IRGO had to exert much effort to obtain the approval of the war minister (Kuropatkin) for Kozlov's new Tibet expedition project, with Lhasa as its destination.

5 Ethnographic Research and the 'Imperial Visions'

Ethnographic research constituted an important aspect of field exploration, as the travelers were trying to produce an integral picture of Central Asian nature and man. Moreover their travel accounts, apart from the great cognitive value, also shaped popular images of unknown lands and peoples for a vast audience – the readers of travel literature and visitors to their public lectures and exhibitions of collected materials (zoological, botanic, and ethnographic). This geographical boom inevitably triggered the "imperial visions", to use M. Bassin's term, in the minds of Russian "empire builders" of the late colonial era, the dreamers of Tsarist Russia's further expansion in the Asian direction.

Ethnographic observations and collecting of ethnic artefacts were part of the scientific programs of travelers from Przhevalskii's time on, though not all of them were particularly keen on ethnography the way Grigorii Potanin and Petr Kozlov were. There is one peculiar feature of Russian ethnography – the discipline originated and for a long time evolved within the framework of geographical studies, in close connection with natural history. As noted by A.A. Sirina, "Russian ethnography focused on the study of man, his life and activities, based on the achievements of natural history. Ethnography, having emerged as a discipline of its own, continued to be closely related to geography and eventually gave rise to an attempt to create a new scientific discipline as a combination of the two, ethno-geography".[77]

The term 'ethnography' appeared first in Russia in the early nineteenth century and it became widely used due to the journal *Severnyi Arkhiv* (Northern

---

76 RGVIA, F. 401, D. 52, l. 1 (1894).

77 A.A. Sirina, "Zabytye stranitsy sibirskoi etnografii: B.E. Petri", in Tumarkin (comp.) 2002, 59–60.

Archive) that carried, starting in 1826, a column under the same heading. Ethnography finally established itself as a separate branch of science in 1836 when the *Kunstkamera museum*,[78] a part of the Russian Academy of Sciences, split into several museums – Mineralogical, Botanic, Zoological, Anatomic, Asiatic, Ethnographic, Numismatic as well as the Peter I Study (*Kabinet Petra Pervogo*). In 1878, at the initiative of academicians K.M. Baer and A.A. Schifner, heads of the Anatomic and Ethnographic museums respectfully, these two institutions merged into one Museum of Anthropology and Ethnography, the first public museum in Russia.

The IRGO contributed significantly to making ethnography a scientific discipline. As already noted, the Society, from its very start, included an ethnographic department (along with the departments of mathematical geography, physical geography and statistics). Moreover, the IRGO was the first scientific society in Europe to have set itself the task of studying various ethnic groups inhabiting the multi-ethnic Russian Empire. The main guidelines for ethnographic studies were formulated by K.M. Baer and N.I. Nadezhdin, the first two heads of the IRGO's ethnographic department, and these made the basis of the special instructions compiled for the early IRGO expeditions within the borders of Russia.[79] Thereafter, all ethnographic material was collected in strict accordance with these instructions. The Society, however, did not provide the travelers with any funds for the purchase of ethnic artefacts. At the same time the expedition budget invariably included a special column for "presents, receptions, entertainment of guests, and unforeseen expenses". While meeting on their way any important officials – the Chinese local administrators (*taoties*, *tyns*, etc.), the Mongolian princes and influential religious figures, whose services and assistance they needed, the travelers would give presents to them, on behalf of the IRGO,[80] and not infrequently would receive some gifts in return. In Buddhist Mongolia and Tibet these were usually the *tangkas* (painted icons),

78 *Kunstkamera museum* (from Germ. *Kunstkammer*, today's Museum of Anthropology and Ethnography named after Peter the Great or MAE) was established in St. Petersburg by Peter the Great between 1718–1727. It was the first museum in Russia, originally conceived as one of 'natural and human curiosities and rarities'. For the history of the Kunstkamera see Its 1989.

79 See Baer 1849, 64–81; also see T.Yu. Gnatiuk, M.N. Kozhevnikova, A.I. Andreyev, "Etnograficheskie issledovaniia i kollektsii", in Andreyev, A.I. (ed.) 2013, 292.

80 In 1905, while in Urga, Kozlov presented a number of Western gadgets to the Dalai Lama and his numerous entourage, such as golden and silver watches, binoculars, microscopes, scissors, and, of course, the samples of Russian weapons. See NA RGO, f. 18, op. 1, d. 201 (Kozlov's Notebook, 1905).

burkhans (figurines of Buddhist deities), and snuff-boxes which thus replenished their ethnographic collections, small though they were.

This was a well-tested method of making friends in the East, yet also the way to boost the attractive image of Imperial Russia among the Asiatic natives by spreading word of the country's might and its powerful 'white tsar', the protector of Russia's 'small peoples', Muslim and Buddhist alike. It should be no surprise then that the stories the travelers told about Russia were taken at their face value by the simple-hearted indigenes and eventually turned into popular myths which circulated around the places they visited, and the travelers themselves soon enough began to be looked upon as super humans, Russian heroes (*baatars*) and 'miracle-workers', as was the case with Przhevalskii. In this way exploration went hand in hand with what can be regarded as a pro-Russian agitation at the grass roots level. These contacts fitted well in the context of gradually expanding Russian – Inner Asian ethno-cultural and civilizational interface by which Imperial Russia was unobtrusively, yet steadily, extending its influence over new lands and peoples of China's western periphery.

Przhevalskii fully realized the importance of ethnographic research, still, the research was only accessory to his high-priority topographical and geographical surveys. As he explained in his essay "How to travel in Central Asia" (1888), the collecting of ethnographic artefacts "in passing" was a rather difficult job for travelers who did not speak any Asian languages, but also because of the general "suspiciousness" of the locals towards Western foreigners. Therefore the research could not be conducted with the "desired accuracy".[81] Still, Przhevalskii's books abound in valuable ethnographic notes of various kinds which evidence that he took genuine interest in the way of life of the indigenous population of the countries he explored. These notes, marked by no small literary talent on the part of the author, were written rather lively according to a well thought-out scheme, including a brief anthropological sketch and a description of the traditional clothing, lodging, food, home utensils, occupation, language, religious beliefs, and even most common illnesses of the natives. In this way Przhevalskii sketched in his travelogues many ethnic groups – the Khalkha, Sunni and Chakhar Mongols, the Dungans, the Machins (inhabitants of the Keria Mountains), the dwellers of the Tarim Basin and Lopnor, of Northern Tibet (the Tanguts and Tibetans), and other minor peoples. Many of these were expressively hand-drawn by his assistant Vsevolod Roborovsky and later served as illustrations for his third travelogue *Iz Zaisana cherez Khami v Tibet* (1883).

81 Przhevalskii 1888, 56–57.

FIGURE 0.14
Dwellers of Inner Asia, drawn by V.I. Roborovskii: The Tibetan ethnic types.
FROM PRZHEVALSKII'S *IZ ZAISANA CHEREZ KHAMI V TIBET* (ST. PETERSBURG, 1883).

FIGURE 0.15
A Tsaidam chieftain Dzun-dzasak, in the center, with his favourites.

FIGURE 0.16
The Dalai Lama's envoys.

FIGURE 0.17
Khara-Tangutans.

FIGURE 0.18
A Tibetan woman.

No less remarkable are the photos by G.E. Grumm-Grzhimailo, Przhevalskii's another follower, taken during his own journey across Central Asia and vividly recreating the peculiar lifestyle of the then little-known peoples of the Oriental 'forbidden lands'.

Oddly enough, some of Przhevalskii's literary sketches are imbued with undisguised Eurocentrism and chauvinism, being clearly a sign of the epoch of colonial wars, social Darwinism and popular racial theories. His first book which was translated into English in 1876 under the title "Mongolia, the Tangut Country, and the solitudes of northern Tibet" provides ample evidence of this type of writing.

One of the most striking examples is a description of the Khalkha-Mongol women found in the volume:

> The appearance of the women is not attractive. The typical features of the race, the flat face and high cheekbones, spoil their looks; and the rough life in the yurta, exposure to the weather and dirt, deprive them of any feminine grace and delicacy, and attractiveness to European eyes.[82]

82 Ibid., 71.

FIGURE 0.19 A Turkic family from Hami, photo from G.E. Grumm-Grzhimailo's major expedition, 1889–1890. NA RGO, F. 32, OP. 1, D. 57A.

FIGURE 0.20 A group of the Dungan musicians from the Xining area, 1889–1890. NA RGO, F. 32, OP. 1, D. 57A.

Przhevalskii's admirers from the Orientalist circles might have been embarrassed by such utterances and attributed these, as did S.F. Oldenburg, partly to his misanthropy, and partly to the hostility of the local population. On the other hand, Semenov asserted in his *History of the 50-year activity of the IRGO* that Przhevalskii "had always sustained friendly and humane relations with the indigenes" in whom he saw the "ingenious children of mother-nature".

This was true to some extent, the more so that Przhevalskii himself had a penchant for nomadism and even called himself a "civilized nomad"; yet concurrently he was also a staunch Sinophobe who had a strong aversion for China, the Chinese people and everything Chinese in general (food, drinks, clothing, etc.). In his aforementioned travelogue he spoke rather disdainfully of the Celestial Empire and its inhabitants – the Chinese entertain "a general hatred for all Europeans", to whom they apply "the nickname of Yang-kzuei'-tsz' (foreign devils)"; they "impose upon the simple-minded Mongols in the most scandalous way", and "the dirt and smells" are "inseparable adjuncts of every town in the Celestial Empire", etc.[83]

Przhevalskii's long-time travels in Inner Asia convinced him beyond all doubt that the Qing Manchu Empire was on the decline, as evidenced the popular anti-Qing activities he personally witnessed. He believed that the regime would soon collapse as a result of an imminent Russian-Chinese military confrontation, an idea he tried to bring home to his superiors at the War Ministry. And, indeed, Russian-Chinese relations markedly deteriorated after the Peking authorities refused to ratify the Livadia Treaty of 1879, by which Russia formally agreed to restore the Ili district to Chinese control, retaining, however, a large part of the area, including the city of Kuldja, in its own hands. The Manchu ruling elite strongly opposed the treaty and even wanted to win the district back by force of arms in 1880.

Back in Russia from his third Central Asian journey, Przhevalskii, now posing as a colonel of the General Staff, would submit, in 1880 and 1881, two memoranda to the Military-Scientific Committee at the Main Staff which contained his well elaborated scheme for a "war with China", with a detailed assessment of the war theatres, the major one being that of Northern Mongolia.[84] The Przhevalskii's proposals must have struck a sympathetic chord in the hearts of the 'hawks' at the Main Staff, yet his memos became irrelevant in the light of the 1881 St. Petersburg Treaty that finally settled the Ili dispute and therefore were shelved by his less hawkish superiors.

83 Ibid, 37, 40–42.

84 See Przhevalskii 1883, 293–321.

Imperial Russia was seen by Przhevalskii as a 'pacifier' of the troublesome regimes in the conquered Western Turkestan, and he believed that the country could likewise restore peace and order in China's Eastern Turkestan, yet this could happen only at the point of a bayonet:

> Rumours about our pacification of Kokand and Kuldja circulated far and wide; a savage Asiatic begins to understand that Russian power is a guarantee of peace and well-being. <...> Carbine bullets and rifled cannons bring here those elements of civilization which would otherwise be very long in coming to the petrified realms of the Middle Asian khans".[85]

As for Przhevalskii's followers (Pevtsov, Roborovskii and Kozlov), they shared many of his views, though none of them was as bellicose as he was, and their travel accounts lacked his pathos of a European civilizer and enlightener of Asiatic savages. Kozlov, as a Great Gamer, would talk, in the early twentieth century, of a peaceful conquest of Asian peoples by Russian trade and culture, not by arms.

Pevtsov's account of his Tibet expedition (1889–1890) offered quite a positive picture of Kashgarians by noting their sympathetic qualities, such as honesty, hospitality and complaisance, yet with the reservation that these qualities had most likely developed as a result of a long-time oppression of the Kashgarian people, from which they have not yet fully freed themselves. "Still, an unbiased researcher must admit, in all fairness, that the indigenous population of Kashgaria has a rather high moral level", noted Pevtsov. And he added further that the Kashgarians "stand higher in a moral respect perhaps than many civilized peoples of our part of the world".[86]

Pevtsov was also pleased that Kashgarians, being "Sunnit-Mohammedans", "are not imbued with religious fanaticism and are not hostile to adherents of other creeds, especially to Christians". Moreover, they even "entertain sympathy towards us, Russians, thanks to our good name spread all over Kashgaria by our subjects, the Sarts of Fergana, trading in that country".[87] At the same time he spoke rather harshly of the Chinese (Manchu) officials in Kashgaria, in the same vein as did Przhevalskii:

85 Dubrovin 1890, 573 (quoted from Przhevalskii's report on the 'Sovremennoe sostoianie Vostochnogo Turkestana' compiled on 6 June 1877).

86 Pevtsov 1949, 125. (Chapter 4: Etnograficheskii ocherk Kashgarii).

87 Ibid, 126.

> In general, the present-day Chinese regime is strongly disliked by the Kashgarians. The indigenes (*tuzemtsy*) bitterly complain of the Chinese because of their shady requisitions, arrogance, contemptuous attitude to them, and the shameful extortions of the petty Chinese officials… They recall with sadness Yakub Beg … who was cruel and protected too jealously the Muslim traditions, yet he was just and condescending towards the poor.[88]

Pevtsov, however, deemed it necessary to remark that the complaints of local inhabitants might have been strongly exaggerated, and Chinese rule in Kashgaria was in fact not as burdensome as described by the locals.

Roborovskii and Kozlov, having matured as field researchers under Przhevalskii and Pevtsov, shared largely their mentors' negative views of the political regime in China. Yet, when reading their travelogues, one should keep in mind that their harsh criticism of Chinese officialdom was directed mainly at the local representatives of the Manchu Qing administration. To tell the truth, Peking in general, despite its notorious xenophobia, did not prevent Russian (and other foreign) expeditions from travelling in China. The travelers moved around the country with travel permits officially issued to them by the Chamber for Foreign Relations (*Zongli Yamen*) in Peking; however, the officials at the local (provincial, district, city, etc.) level watched closely all movements of the Russian expedition caravans and quite often put obstacles in their way by closing to them access to some territories, especially to Tibet, as Przhevalskii complained in his public speech at the IRGO in 1886.[89] This was a major reason for the travelers to be discontented with the Chinese officials, as evidence their travelogues. According to Chinese historian Rong Xin-jiang, what worried mostly the central government in Peking and provincial authorities, was topographic surveys of the travelers, but that was not until the Xinhai revolution (1911).[90]

A contemporary student of ethnography will find in the travel accounts of both Roborovskii and Kozlov many interesting ethnographical sketches of the

88 Ibid, 146, 147.

89 Gromchevskii, on the contrary, was able to travel from Kashgar to Yarkand and Khotan in 1885 only thanks to the permission issued to him by the Kashgar *taotai* (head of district) who thus wanted to smooth down the incident when Grombchevskii's party was attacked by a group of Chinese soldiers in Kashgar.

90 Rong Xin-jiang (Жун Синь-цзян), *Russkie ekspeditsii i kitaiskie vlasti v kontse XIX – nachale XX veka*, in Popova (ed.) 2008. 225.

Tanguts, the Tsaidam Mongols, the Chantu dwellers of Lukchun (Roborobskii), and the ethnic groups of northeastern Tibet (Kozlov). Even more, Roborovskii supplemented his first-hand observations with excellent pencil drawings – of ethnic types, their dwellings, religious sites, etc. which made 35 black-and-white plates illustrating Przhrvalskii's account of his third Inner Asian expedition (1883). On the other hand, Kozlov, being keen on photography, would take many pictures in the field, and he is known to have brought about 200 glass negatives from his major journey to Mongolia and Eastern Tibet in 1901, a valuable addition to his rich scientific harvest.

These drawings and photographs, the earliest ever taken by travelers in the remote parts of the Asian heartland, serving to illustrate their published narratives, along with ethnographic artefacts, were of great interest to scholars – ethnographers and Orientalists, as well as to the general public, as visual representations of the unknown lands – *nevedomye zemli.* These lands always had a strong attraction for many a Russian, perhaps out of some peculiar innate inquisitiveness of the Russian mind ever bent on exploring all things unknown, wondrous and transcendental, since the time of Afanasii Nikitin's fabulous travels to India in the fifteenth century and the Cossack trailblazers' headlong rush into the vast expanses of uncharted, hence mysterious, Siberia, half as big as the United States. And, of course, as already noted, these same images and narratives also provided a stimulus for all kinds of geopolitical and nationalist visioning of these lands, untouched as yet by the civilizing spirit of the 'superior' European race.

The idea of Russia's civilizing mission in Asia was appealing to many educated Russians of those days, including the famous novelist Fedor Dostoevskii, who strongly emphasized it in his *Writer's Diary* (in the sections penned right after the seizure of the Turkmen fortress of Geok Tepe by Russian troops in 1881 and headlined *Geok Tepe: What is Asia to us?* and *Questions and Answers*). Dostoevskii linked his hopes for Russia's glorious future to the conquest of Asia, a part of the globe he compared to "an undiscovered America", and he spoke of Russia's advance into Asia not only as a civilizing mission but also as a means for national self-realization, with a somewhat messianic undertone:

> The main thing is that our civilizing mission in Asia, from its very first steps, will undoubtedly be accepted and comprehended by people. It will raise our spirit and enhance our dignity and self-knowledge, which we do not have at all or have too little of at the moment. <…> The place where an 'Urus' settles in Asia becomes all at once Russian land. A new Russia could be born that would revive and resurrect old [Russia] with time…[91]

91 See: Dostoevskii, F.M. 1984, 37, 38.

Commenting on these revelations by Dostoevskii, contemporary historian Mark Bassin remarks that the redirection of attention from the Far East, following the annexation of the Amur region in the mid-nineteenth century, to Central Asia "was intended to offer Russia an alternative field for creative work – *an undiscovered America*, in Dostoevskii's telling metaphor – upon which it could exercise its own strengths and independently develop its national qualities". Such a fanciful vision of Asia must have been fully shared by Przhevalskii and men of like mind. In the course of his first expedition to the Ussuri he, in the words of Bassin, "had come to despise his assignment on the Amur" and the region itself and "burned with impatience to get to Turkestan".[92]

Another Russian imperial visionary of note was Petr Badmaev, a prosperous practitioner of Tibetan medicine and entrepreneur, a Buryat by birth, who put forward, in 1893, a project for the annexation of Mongolia, Tibet and China to the Russian Empire.[93] Badmaev's grandiose, if fully adventurous, scheme, despite its approval by Alexander III and the funds allocated for its realization by the Russian government, suffered a predictable fiasco. The only thing Badmaev was able to achieve was sending some of his agents secretly to Lhasa to establish a political link with Tibet's ruling elite. These agents got in touch there with Agvan Dorzhiev whose consequent activities at the court of the youthful 13th Dalai Lama paved the way for a Russo-Tibetan rapprochement, without, however, bringing any substantial political dividends to Russia in the long run.

6 The Final Stage of Explorations: The Academy of Sciences Taking the Lead

The scientific trophies delivered to St. Petersburg by the travelers can be roughly divided into three parts: 1) the route surveys intended for the Military-topographic department of the Main Staff; 2) various instrumental observations (astronomic, barometric, magnetic and meteorological) which went partly to the same MS as supplementary material needed for map-making and partly to the Nicholas Main Physical Observatory (*Nikolaevskaia Glavnaia Fizicheskaia Observatoriia*);[94] 3) naturalist collections (zoological, botanical,

92 See Bassin 1997, 264, 266.

93 On Badmaev and his project, see Semennikov (publ.). 1925 (repr. Badmaev 2001); also Gusev 1995, 3–104; A.I. Andreyev, "P.A. Badmaev i russko-tibetskoe sblizhenie", in Samarukha and al. (eds). 2004, 23–29; Kuz'min 2007.

94 This observatory was established in St. Petersburg in 1849; it was in charge of all meteorological instruments and ran a network of weather stations in Russia, while also collecting

mineralogical and ethnographic). The zoological samples (animals' skins and sculls) were taken to the Zoological Museum of the Imperial Russian Academy of Sciences, the herbariums and seed collections went to the Botanical Gardens;[95] the specimens of rock – to the Geological Room (*Kabinet*) of St. Petersburg University, and ethnographic objects to the Academic Museum (Kunstkamera) and later (in the 20th century) to the newly established Ethnographic museum as well.

The scientific analyses of all these voluminous collections by specialists (zoologists, ornithologists, botanists, geologists and ethnographers), Russian and occasionally European as well, would normally take years and this work is still incomplete. In this way began a lasting cooperation between the IRGO and a host of brilliant scholars, some of whom were full members of the Russian Academy. The travelers themselves maintained close personal contacts with the leading zoologists and botanists, such A.A. Shtraukh, F.D. Pleske, E.A. Bikhner, K.F. Kessler, and K.I. Maksimovich; they consulted with these specialists prior to their expeditions and the same men would later expertly analyze and describe the samples of fauna and flora collected in Inner Asia. Still, being an excellent hunter and having a good knowledge of ornithology, Przhevalskii is known to have classified and described his ornithological trophies himself.

The IRGO exploration activities in Inner Asia continued well into the twentieth century, until they were interrupted by the Great War and the 1917 revolutionary upheaval in Russia. The major expeditions mounted by the Society in the first decade were those of G.E. Grumm-Grzhimailo to Western Mongolia and Tuva (1903), V.A. Obruchev to Dzungaria (1905, 1906, 1909), V.V. Sapozhnikov to Mongolian Altai (1905, 1906, 1908, 1909), and P.K. Kozlov to Mongolia and Sichuan (1907–1909).[96] It is clear to see from the above that the main focus of these journeys was Mongolia, including its northwestern corner known as Tuva or Urianhaiskii Krai, and Eastern Turkestan.

Outer Mongolia assumed major political and economic prominence as well as strategic importance for Imperial Russia in the newly shaped system of international relations in the Far East after Russia's debacle in the war with Japan

---

and expertly processing the results of all available meteorological observations, within and outside Russia.

95 The Imperial Botanical Gardens were established in 1713 by order of Peter I, transferred to the Academy of Medical Surgery (1798) and later (1830) placed under control of the Imperial Ministry of Court; the Botanical Gardens became part of the Academy of Sciences of the USSR in 1930.

96 See Shchukina 1955, 170–181.

in 1904–1905. "The key element" that determined the peculiar character of Outer Mongolia's international status in the post-war period, in the opinion of S.G. Luzianin, was "the global confrontation between Russia and China which, as a special form of East-West interaction, had not only territorial-historic significance, yet also a wider intracivilizational implication".[97] According to the Russian-Japanese agreements of 1907 and 1910, Outer Mongolia and Northern Manchuria became Russia's spheres of influence or 'priority zones', whereas Inner Mongolia, Southern Manchuria and Korea were placed under Japan's sway. However, after the downfall of the Qing dynasty, as a result of the Xinhai revolution of 1911, Mongolia declared independence and began to openly seek recognition by Russia and other Western powers. Yet all too soon the country found itself again in a precarious position balancing between her two neighbour-giants, Russia and China, "still vying for control" over its territory.[98] Tibet too became a virtually independent state, in 1912, and was now also seeking recognition by both Russia and Britain and even more, their joint protectorate.

Kozlov, as an experienced traveler and Great Gamer, retained his interest in Mongolia and Tibet, both geographical and political. His opinion and advise were often sought not only by the military top brass but by diplomats as well, such as the foreign minister, A. Izvolskii, who would discuss with him the cumbersome 'Tibetan issue' in 1906, when the MID contemplated the possibility of sending a diplomatic mission to the Dalai Lama. In 1907 – 1909 Kozlov made another journey to Mongolia and north-eastern Tibet known as the Mongolia-Sichuan expedition. However his initial program approved by the IRGO was radically altered in the course of his travel when he discovered, in early 1908, at the southern fringes of the Gobi, the sand-buried ruins of the medieval Tangut city of Khara-Khoto ('Black City') which he then excavated, with due approval of the IRGO Council.[99] The discovery and sensational excavations of Khara-Khoto distracted Kozlov for a time from Tibet, the more so that Russia and Great Britain agreed, in 1907, not to send their scientific expeditions to the country for a three-year period.

The expeditions of Obruchev, Sapozhnikov and Kozlov virtually marked the end of the era of Russia's geographical exploration of Inner Asia. According to N.M. Shchukina, "The journeys of the [early] twentieth century did not result in the discovery of any major geographical objects, since no large blank spots

97 Luzianin 2000, 39.

98 Baabar 1999, 152–154.

99 See T.I. Yusupova, "Mongolo-Sychuan'skaia ekspeditsiia P.K. Kozlova (1907–1909) i otkrytie Khara-Khoto", in Popova (ed.) 2008, 112–129.

had remained on the maps of Central Asia by the end of the nineteenth century. Yet these journeys allowed [cartographers] to insert a great number of corrections and specific details into the existing maps. Many astronomical points were defined and numerous altitudes were measured in that period, which contributed to the making of the modern physical map of Central Asia".[100]

The geographical *preliminary reconnaissance* of the region was over. The academic circles in Russia fully realized the need for special – single-purpose – explorations, to be focused on one particular type of research – geological, paleontological, botanic, and especially archeological, at one particular locality. This new trend towards specialization was necessitated largely by the needs of the rapidly developing science, its methods and practical applications. Beginning in the late nineteenth century, various specialists were already occasionally attached to the expedition staff – geologists, geodesists, zoologists, ethnographers, and, of course, topographers. For example, the geologist K.I. Bogdanovich travelled together with Pevtsov and published later his geological survey of Eastern Turkestan in a separate volume of the 'Works' of Pevtsov's Tibet Expedition (1892). In this he noted that Kashgaria "deserves serious attention with respect to its gold mines", while Dzungaria "was known, for a long time, as a vast coal basin",[101] a statement which unambiguously called for geological prospecting of the area. Another geologist, V.A. Obruchev, formally on the staff of Potanin's fourth Central Asian expedition (1892–1893), is known to have traveled entirely on his own, dressed in traditional Chinese clothing so as not to arouse the suspicion of the local population.

China's Eastern Turkestan became a great attraction in the twentieth century especially for archeologists, ethnographers and Oriental scholars. In Russia, this interest was continually generated by concurrent exploration of the region by a host of Western travelers, such as A. Grünwedel, A. von Le Coq, A. Stein, P. Pelliot, searching for relics of ancient time, primary manuscripts and books, in the oases towns of Khotan, Turfan, Kucha (Kuqa) and others, on the Silk Road, "one of the world's earliest highways", connecting European lands with Central Asian kingdoms.[102] Remarkably, it was the IRGO expeditions that urged Russian Orientalists to turn their gaze to the oases of Eastern Turkestan. Thus, in 1896, the Society handed over to the Asiatic Museum[103] a bagful of

100 Shchukina, op. cit., 190.

101 Bogdanovich 1892, 38–39.

102 See Wood 2002.

103 The Asiatic Museum was established in 1818 and was the largest depository of manuscripts in Oriental languages in Russia. Since 1930 it has been part of the Oriental Institute of the Academy of Sciences USSR (today's Institute of Oriental Manuscripts of the

fragments of ancient manuscripts in the Old Turkic and Old Uighur languages collected earlier in Turfan by Roborovskii and Kozlov. Their finds spurred the Academy of Sciences to send, in 1898, an archeological expedition to Turfan under D.A. Klements, who thus became the first to start the search for ancient relics in the northern part of Eastern Turkestan.[104] Klements was followed by other travelers – M.M. Berezovskii, who conducted an archeological survey of the oasis of Kucha (1905–1908), and S.F. Oldenburg, the permanent secretary of the Academy of Sciences and Director of the Asiatic museum,[105] who led two expeditions to Eastern Turkestan (1909–1910 and 1914–1915). The second one surveyed the famous site of the cave temples near Dunhuang, one of the 'lost cities' on the northern Silk Road in Gansu province, known as the Mogao (Dunhuang) Caves, or the Caves of the Thousand Buddhas.[106]

These grottoes, with their highly treasured murals and ancient manuscripts, were first visited by Przhevalskii (1879–1880), who dubbed them 'Holy Caves', and then by Roborovskii (1894), shortly before the Caves were discovered by a Chinese Taoist Wang Yuanhu and became a great attraction for Western scholars.[107] On the other hand, the antiques of Eastern Turkestan were also collected by Russian diplomats posted in the region, such as the Consul-General in Kashgar N.F. Petrovskii. His collection alone consisted mainly of items from the Khotan oasis and it was purchased in 1897 by the Imperial Hermitage.[108]

As for Kozlov, he planned to return to Khara-Khoto to continue his excavations. In early 1914 he made plans for another Tibet expedition that was primarily to explore the sources of the three great Asian rivers – Yangtze, Mekong and Salween, originating on the Tibetan plateau. The expedition was con-

---

Russian Academy of Sciences). For the history and collections of the Asiatic Museum, see Baziants, Kuznetsova, Kulagina 1969.

104 See S.G. Kliashtornyi, "Drevnetiurkskie pamiatniki runicheskogo pis'ma Vostochnogo Turkestana", in Popova (ed.), op. cit., 50–64. According to S.G. Klyashtornyi, "the finds of V.I. Roborovskii, P.K. Kozlov and D.A. Klements laid the foundation of the Old Uighur collection at the Asiatic Museum", see Popova 2008, 53.

105 On him see Kaganovich 2006.

106 On these expeditions under S.F. Oldenburg see I.F. Popova, "Pervaia Russkaia Turkestanskaia ekspeditsiia S.F. Ol'denburga (1909–1910)"; "Vtoraia Russkaia Turkestanskaia ekspeditsiia S.F. Ol'denburga (1914–1915)", in Popova (ed.), op. cit., 148–157, 158–175.

107 See Przhevalskii 1948 (2-nd ed.), 91–93. The Caves were described in detail by V.I. Roborovskii, see Roborovskii 1900, 151–220. S.F. Oldenburg, who visited the Caves in 1914, also wrote an article "Peshchery Tysiachi Budd" published with illustrations in the journal *Vostok*, see Oldenburg 1922, 57–66.

108 I.F. Popova, "Rossiiskie ekspeditsii v Tsentral'nuiu Aziiu na rubezhe XIX – XX vekov", in Popova (ed.), op. cit., 25.

ceived as a combination of a reconnaissance trip and a multi-disciplinary field research of the new type and therefore was staffed with specialists in geography, geology, archeology, zoology and botany, as his senior assistants. Yet Kozlov's plans were frustrated by the world war.

It was in 1922 only that Kozlov succeeded in reviving his earlier expedition project, under the aegis of the Russian Geographical Society and with the backing of the Soviet government (*Sovnarkom,* or SNK). The latter attached some political importance to his new Tibet journey, as Kozlov planned to visit Lhasa and the old friend of Russia there, the Dalai Lama. His expedition was fully – and even lavishly – financed by the state, well-kitted and given much public attention, well-deserved as these final exploratory pursuits by Kozlov were crowned again with extraordinary success. Being unable to journey to Tibet (for reasons which will be discussed in more detail in Kozlov's biographical essay), he excavated instead the little-known burial mounds at Noin-Ula, north of Ulan-Batar, in 1924–1925. The excavations were again unplanned and spontaneous, yet, by a streak of unusual luck, they yielded again a great number of antique objects of the extinct Hunn civilization. The event was Kozlov's another great triumph, the finale of his brilliant career of nearly four decades as a field researcher, and that of the whole epoch of the Russian exploration of Inner Asia.

A striking hallmark of the new times was Kozlov's close cooperation with the Academy of Sciences. The latter agency, as soon as it learned about the excavations of the Noin-Ula mounds initiated by the traveler, dispatched a group of academic scholars, including two archeologists, to Mongolia to assist him with this and other field work. And this measure was well-justified within the new system of organization and management of science launched in the country following the Bolshevik revolution. The Academy of Sciences and the People's Commissariat of Enlightenment (*Narkompros*) RSFSR/ USSR became the two major centers of scientific activities, the latter entity being placed in charge of all scientific societies, including the RGO. According to this new system, all projected expeditions, within and outside Russia, were to be 'attached' to concrete scientific institutions, placing thus upon them all responsibility for the implementation of expedition projects.

Having survived the revolution, the RGO was far from thriving as it did before under the patronage of the grand dukes and the tsar himself. As the Academy of Sciences was rising to prominence as the chief organizer of science in the country, the RGO had to considerably curtail its expeditionary activities, especially outside Russia. In 1923, the Society adopted its new charter, to keep in line with the Soviet legislation, and the same year Kozlov's account of his previous journey was published under a catchy title: *Mongolia and Amdo and the Dead-City of Khara-Khoto*. At the same time the RGO Council strongly protested against the plans of the Royal Geographical Society in London to change a number of geographical names on the map of Asia, including those given by Russian travelers. The Society's other issues of much concern in the period were the Franz Josef Land, an uninhabited archipelago in the Arctic Ocean occupied by Russian sailors during the world war (and later annexed by the Soviet Union), and of course Kozlov's politically sensitive expedition to Mongolia and Tibet.

The exploration of Inner Asia in the early Soviet period was substantially limited, being focused mainly on its neighbors, the People's Republic of Mongolia and Tannu-Tuva. Mongolia's final break-up with China in 1924 paved the way to close Soviet-Mongolian political alliance as well as economic, cultural and scientific cooperation, the latter under the patronage of the Academy of Sciences. As for Tannu-Tuva, the former Uriankhai Country, it finally separated from Outer Mongolia in 1921 and became an independent state of its own, the Republic of Tannu-Tuva. It was to these two republics that the Soviet Union would send its specialized exploratory expeditions and individual researchers, their work being coordinated by the *Commission for the Exploration of Mongolia and Tannu-Tuva*. This entity was set up in 1925 at the Sovnarkom USSR under N.P. Gorbunov as its chairman, and later was transferred to the custody of the Academy of Sciences and renamed the Mongolian Commission (*Mongol'skaia Komissiia*).[109]

Along with the Soviet Academy, the field work in Mongolia was organized by the Mongolian Scientific Committee (*Monuchkom*),[110] under Onhodyn Zhamian and Tseben Zhamtsarano.[111] A program of overall field research mapped out by *Monuchkom* included a multifaceted study of Mongolia and Tannu-Tuva, their natural resources and population. The research itself, being of high priority for the national economies of these two republics, was fully financed by the Soviet state, and the researchers were mainly Soviet specialists, some of whom

109 Yusupova 2006.

110 The Scientific Committee of Mongolia (MPR) was established by the government of Autonomous Mongolia in 1921 under the aegis of the Ministry of Enlightenment as the first scientific institution in the country. In 1924 it was subordinated to the Mongolian government and further (in 1930) reorganized and renamed the Committee of Science; in 1961 it was transformed into the Academy of Sciences of Mongolia. See Yusupova, op. cit., 43 and ff.

111 S. Chuluun, R. Tordalai, *Onkhodyn Zhamian i Uchenyi komitet Mongolii: Ot traditsii k nauke (1921–1930)*; T.I. Yusupova, *Vklad Ts. Zhamtsarano v ustanovlenie nauchnogo sotrudnichestva mezhdu Uchenym komitetom Mongolii i Rossiiskoi Akademiei Nauk*, in Chuluun, Yusupova (eds) 2012, 50–63; 32–49.

had been on the staff of Kozlov's last expedition (S.A. Kondratiev, A.D. Simukov, E.V. Kozlova). They were joined by students of the Institute of the Living Oriental Languages in Leningrad, who came to Mongolia and Tannu-Tuva for summer practice and conducted various ethnographic, anthropological and linguistic field investigations. The paucity of funds did not allow the Monuchkom to mount large expeditions; instead the entity dispatched into the field small parties of 3–5 specialists with one or two workmen to assist them. The collected natural specimens were taken to the Committee's head-quarters in Ulan-Batar and thence transferred for proper expertise to the academic institutions in the USSR, namely the Zoological Institute, the Botanical Gardens, and the State Academy of Material Culture, all in Leningrad (renamed St. Petersburg / Petrograd). A large part of the collected materials would remain in the possession of these institutions, while the remainder would be returned to Mongolia to be housed in the newly-established Mongolian national museum in Ulan-Batar.

No exploration in other parts of Inner Asia was planned at that time. In Kozlov's opinion, however, there still remained one particular locality on the Tibetan plateau insufficiently explored – that of the basins of the upper Yangtze, Mekong and Salween, where he wanted so much to go. Yet, Tibet became practically inaccessible to Soviet Russian travelers in the 1920s.

Some Concluding Remarks

Russia's exploration of Inner Asia in the late nineteenth – early twentieth centuries was a unique, multidisciplinary undertaking, namely geographical, topographic (cartographic), and ethnographic, and it also contributed prominently to a number of branches of natural science, such as zoology and botany. The IRGO expeditions can also be seen in a somewhat broader context of Russian – Oriental intercultural and intercivilizational dialogue as well as a manifestation of Russia's geopolitical interests, inspired largely, but not exclusively, by the Great Game.

Still, Inner Asia, as the western periphery of the Manchu-Chinese Empire, was not an object of imperial conquest for Tsarist Russia, like the Middle Asian khanates in Western Turkestan and the Far East, but first and foremost an area of extensive geographical exploration. And although exploration in the imperialist era quite often paved the way for subsequent conquest, this was not the case when Russia started sending her expeditions into Inner Asia, one after another. The travelers themselves, including the 'imperialist-minded' Przhevalskii, perceived their field work primarily as 'scientific reconnaissance' of the unknown lands. As for Przhevalskii's notorious plans for a 'war with China' expounded in his several memos to the General Staff (in the 1880s), they were too unrealistic and adventuresome, such as the annexation of the Ili Region, Northern Mongolia and the Sungari Valley (in Manchuria). These militant plans were in fact his rather emotional response to the anti-Manchu Dungan uprising in the western provinces of China, which led to the temporary occupation of the mutinous Ili region by Russian troops. Yet Przhevalskii's arguments had little appeal for the Tsar's generals who did not want to get involved in any serious hostilities with China, despite the Manchu ruling dynasty's political and military feebleness, and therefore the Russian troops were eventually withdrawn from the area.

There are many Western scholars today, however, who see the Russian exploration of Inner Asia primarily as an integral part of the country's "colonial project", which considerably plays down the huge scientific contribution of the Russian pioneer travelers, turning them into mere "colonial collaborators". But this was not the case as shows the thorough study of the life stories of these pioneers.

The cultural impact of the IRGO-mounted exploratory missions on Russian society in general and particularly its intellectual elite was prodigious, creating in the mass consciousness an appealing image of a scholar-traveler, the intrepid conqueror of nature and discoverer of unknown lands. The public admiration for Przhevalskii and his scientific exploits eventually turned him into a cult figure and a national hero, as evidenced by his obituary penned by Anton Chekhov:

> One Przhevalskii or one Stanley is worth ten scholarly institutions and one hundred good books. Their moral substance, their noble ambition at the bottom of which lay the honor of their native land and science; their persistent aspiration toward the goal they once set for themselves in disregard of all risks and deprivations, and the temptations of their personal welfare; their rich knowledge and industriousness, <...> their fanatical belief in Christian civilization and science, [all these] make them in the eyes of the people devotees [to science], and personifications of the highest moral strength.[112]

To Chekhov, people like Przhevalskii were "men of a different sort – men of action (*podvig*), of faith and of clearly realized purpose". Their country and the

112 First published, without a headline, in the *Novoe Vremia*, 1888, no. 4548. See also Chekov 1985, 442.

entire world should be proud of them. When Checkov wrote these words, he was in effect expressing the *Zeitgeist* of his epoch.

Over a century has elapsed since those days, and the modern heroes look very different from those admired by Chekhov and his generation. Few people remember today that the physical map of Inner (Central) Asia as we know it was largely a product of the strenuous efforts by a whole host of enthusiastic travelers, both Russian and Western, the men who walked many thousands of miles across boundless and lifeless deserts and sky-high mountains just to survey and chart this vast and inhospitable part of the globe. The authors of the book believe that the time has come now to take a fresh look at the great epoch of geographical exploration and to reexamine and reassess the work of these pioneers, their many amazing discoveries and achievements, as well as their failures and idiosyncrasies. The biographical essays about the six most celebrated Russian explorers of the Przhevalskian 'school' included in the volume hopefully will contribute to this end.

CHAPTER 1

Nikolai Mikhailovich Przhevalskii (1839–1888)

Alexandre I. Andreyev

> The wide open space of deserts,
> that's what I am dreaming of day and night
>
> N. PRZHEVALSKII, 1886

•••

> There is a streak of Przhevalsky in all
> who have scorned the multitude and turned to the wilderness
>
> D. RAYFIELD

Nikolai Mikhailovich Przhevalskii is the greatest Russian explorer of Inner Asia of the nineteenth century. His travels, spanning two decades and covering a vast territory of the Asian heartland, marked the onset and the early – 'epic' – stage of Russia's geographical exploration of the region. However, Przhevalskii himself regarded his travels as only a *preliminary scientific reconnaissance* of uncharted lands. This was to be followed in future, he believed, by a more thorough exploration to be conducted by specialists in various branches of learning. He was a prominent geographer and naturalist who brought from his four journeys scores of flora, fauna, rocks etc., along with his most valuable route surveys so much needed by Russian and Western mapmakers. Przhevalskii, who called himself a "traveler by vocation", was a charismatic figure and a born leader. His biographers emphasize his most striking feature – his wanderlust and longing for 'unrestrained wild freedom' (M.A. Engelgardt[1]), yet at the same time he was a sheer pragmatic and a man of undisguised imperial ambitions, an 'empire-builder' and 'conquistador' as portrayed by D. Rayfield and D. Schimmelpenninck van der Oye. Peter Hopkirk called Przhevalskii "the

1 Engelgardt 1891 (repr. in 1995). This and other quotations from the work come from the 2-nd edition: Engelgardt 1995, 392.

 | DOI 10.1163/9789004376267_003

FIGURE 1.1
Major-General Nikolai Mikhailovich Przhevalskii, 1886. NA RGO.

Central Asian Livingstone" and "immortal hero", and the British geographer John Norman Baker especially emphasized Przhevalskii's great contributions to geographical science – he "stands out conspicuously" among the host of other explorers of Central Asia in the nineteenth century and "his career requires separate consideration, and may well mark the turning point in the history of exploration in this area".[2]

Family Name and Ancestors

To tell the story of Przhevalskii we must begin with his ancestry of over four centuries. According to the latest research by Liudmila K. Przhevalskaia, Nikolai M. Przhevalskii, and Valery B. Titov (all three closely related to the Przhevalskii clan), the earliest known ancestor of Nikolai Mikhailovich was Onisim (Anisim) Pereval (Perevalka, Perevalskii), who belonged to 'the horse-owning middle class' (*konnye meshchane*) of Vitebsk.[3] Originally a Russian city on the

---

2 Baker 1931, 280.

3 Przhevalskaia, L.K., Przhevalskii, N.M. 2008, 9–15; same authors, 2014, 6–10; 2016, 209–212; Titov 2012, 27–29. V.B. Titov's work is entirely devoted to the origin of the Przhevalskii clan and in-

MAP 1.1 The routes of N. Przhevalskii's four expeditions in Inner Asia, 1870–1885. Compiled by Alexandre Andreyev on the basis of the map published in: I.P. Magidovich and V.I. Magidovich, *Ocherki po istorii geograficheskikh otkrytii* (Studies in the history of geographical discoveries), vol. 4, Moscow, 1985.

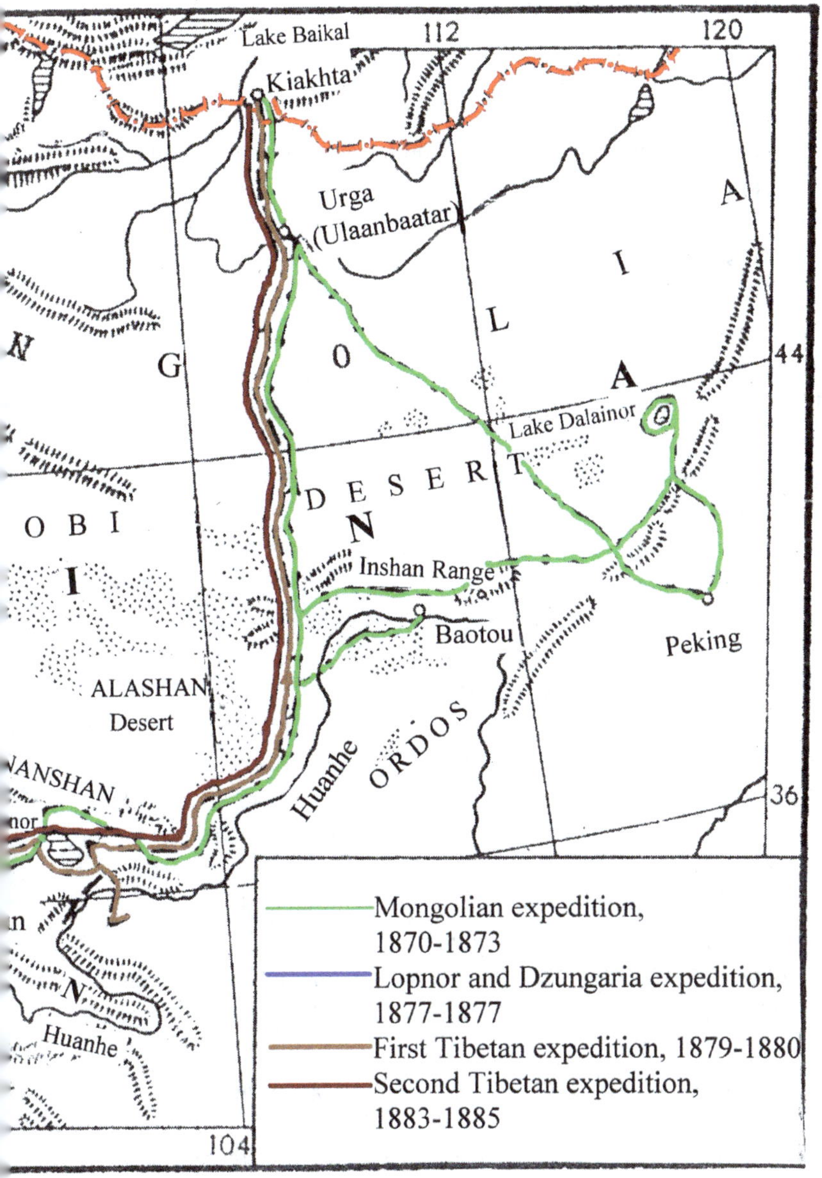
Lake Baikal
112
120
Kiakhta
Urga
(Ulaanbaatar)
A
I
L
O
G
N
A
44
Lake Dalainor
D E S E R T
N
O B I
I
Inshan Range
Baotou
Peking
ALASHAN
Desert
NANSHAN
Huanhe
ORDOS
36
Huanhe
104
Mongolian expedition, 1870-1873
Lopnor and Dzungaria expedition, 1877-1877
First Tibetan expedition, 1879-1880
Second Tibetan expedition, 1883-1885

Western bank of the Dvina River, Vitebsk was the capital of the Vitebsk Princedom, ruled by Russian princes. This state fell under the sway of the Grand Duchy of Lithuania, in the fourteenth century, and later became part of the united Polish-Lithuanian state, Rzcechpospolita. In the days of the first Przhevalskiis, Vitebsk was populated mainly by Orthodox Christians who considered themselves Russians. They spoke old-Russian, similar to the present-day Belorussian language. Nowadays Vitebsk is one of the major cities of Belarus (Republic of Belorussia). Onisim Pereval(ski) is known to have possessed some landed property close to the earth mound of the Castle of Vitebsk, hence his surname which means '[the one who lives] in front of the wall' – *pered valom*).[4]

The fact of Przhevalskii's ancestors originating from Vitebsk was first ascertained by the Vitebsk archivist M.L. Verevkin, in 1890, but was apparently unknown or disregarded by N.F. Dubrovin who published in the same year Przhevalskii's first and still considered to be his best biography. In this Dubrovin claimed that the Przhevalskii's ancestral roots lay with the horse Cossacks of Zaporozhie,[5] not Vitebsk, and he spells Onisim's surname as Parovalski (lit. one 'who 'knocks down' with his mere 'steam' – *parom valit*). Zaporozhie in those days occupied a large territory 'beyond the rapids' (*za porogami* in Russian) of the river Dniper, known as Dikoe Pole (Wild Field), which is today's northern Ukraine. This land hosted Cossacks, runaway serfs, and outlaws from all corners of Muscovite Russia; it was beyond control or jurisdiction of any neighbouring state, forming a country of its own, the 'Cossack Republic' of Zaporozhskaia Sech'. The Zaporozhie Cossacks continuously fought with their neighbours, mainly the Tatars and Turks from the khanate of Crimea and the Ottoman Empire. Dubrovin's version of Przhevalskii's Zaporozhie roots was generally adopted by the traveler's later biographers and it still remains popular with the authors who write about Nikolai Przhevalskii.[6]

Onisim Perevalskii had a son, Kornila (1561–1593), who is much better known to us due to his military exploits. According to the available records, Kornila entered the Polish / Lithuanian military service and hence changed his last name into Przhevalskii. Kornila took part in the later stage of the Livonian War,

---

cludes the Genealogical Table of the Przhevalskiis, from the 16th to the 20th centuries (Appendix, 148 – 156); the list of Przhevalskiis who were persecuted during the Great Purge in the USSR (1936–1938) and earlier, as well as those who perished during the World War II. A brief description of Przhevalskii's ancestry can be found in the collection of documents of Alexandre Iosifovich Przhevalskii (b. 1898), see NA RGO, F. 13, Op. 1, D. 251, ll.1–2 (*Rodoslovnaia Przheval'skikh*), dated 8 April 1842.

4 Przhevalskaia, L.K., Przhevalskii, N.M. 2016, 210.

5 Dubrovin 1890, 1–7.

6 See, for example, Kravklis 2013.

FIGURE 1.2
The Polish coat of arms of Przhevalskii's ancestors, 1581. COPIED FROM V. TITOV'S *VOENNO-SLUZHILYE LIUDI PRZHEVAL'SKIIE* (MOSCOW, 2012).

on the Polish side, and is also known to have been fighting at Polotsk and Velikie Luki against the Russian troops of Tsar Ivan the Terrible. The great courage displayed by the Cossack mercenary was highly appreciated by the Polish-Lithuanian king, Stephen Báthory, who granted him, in 1581, the title of a Polish nobleman together with the 'special crest of arms'. The latter depicted a drawn bow with cocked silver arrow and a helmet decorated with three ostrich feathers against a red shield. This first-named Przhevalskii continued his valiant service; he was promoted as high as the rank of Captain of Cavalry (*rotmistr*), and in 1589 was further granted by the Voevode of Vitebsk, Nikolai Sapega, a landed property of five villages, an act legalized in the same year by King Sigismund III.[7] Thus the Vitebsk horseman Kornila Perevalskii – Przhevalskii became the founder of the noble line of the Przhevalskiis.

Be that as it may, Kornila's activities in Rzcechpospolita allowed contemporary Polish historian Jan Ciechanowicz to conclude that Nikolai Przhevalskii's ancestors were of Polish origin; furthermore, that he himself, as their descendant, represented the exemplary Polish *'rycerz'* (knight or warrior).[8] Yet this conclusion is apparently at variance with the originally Russian / Ukrainian

7 Dubrovin 1890, 1–2.

8 Ciechanowicz 2017 (2nd ed.). This author, however, provides no evidence to support his claims.

surname of the first Przhevalskiis, and the fact that the name Perevalskii was still used by representatives of the clan (Kornila's descendants) in the early eighteenth century.[9] Thus it would be more fair to say that Nikolai Przhevalskii inherited much of his militant spirit and vigor, his hot-blooded recklessness and a strong attraction for free and untrammeled living from his Russian (Belorussian-Ukrainian)-born ancestors, not from Polish knights, some ten generations later. Dressed in his general's uniform, he looks like a "typical Sarmatian", in the words of Ciechanowicz.[10]

Unfortunately, we do not know much about Kornila Przhevalskii's descendants, except their names given in the parish records and first mentioned by N.F. Dubrovin. E.P. Gavrilenkova, having thoroughly researched Kornila Przhevalskii's line of descendants up to the present time, discovered that there were many males in the line, a majority of whom were military men; the others included physicians, chemists, lawyers and building engineers.[11]

Nikolai Przhevalskii's grandfather was Kazimir (Kuz'ma) Fomich who lived on his Skuratovo estate. He was educated in the Jesuit school in Polotsk, though he did not complete his curriculum, having run away from the school. He adopted Orthodoxy and changed his Polish name Kazimir into the shorter Russian Kuz'ma, having thus returned to his Russian roots. Kuz'ma Przhevalskii married a woman called Krasovskaia and had a son Mikhail and two daughters, both of whom died childless.

Mikhail Kuz'mich Przhevalskii (Nikolai's father) (1803–1846) was the one who linked his own destiny with Imperial Russia. He served in the army for two decades, yet, being of poor health, retired in 1835. The only important fact pertaining to his military career is that he took part in the suppression of the Polish revolt in 1831.

Nikolai's mother was Elena Karetnikova, daughter of Aleksei Milhailovich Karetnikov, a well-to-do landowner who lived on his own estate, the village of Kimborovo near Smolensk. Dubrovin portrays Karetnikov as a "very kind man, a philanthropist, and a lover of singing and music". He also had a passion for birds and monkeys. In his house in St. Petersburg he had a separate room where he kept nightingales, goldfinches and some exotic birds in cages; one more room was left for little monkeys, or rather marmosets. Before moving to his country estate, Karetnikov bought ten parrots, one of whom, believe it or not, could sing Russian songs to the great amusement of his numerous guests.[12]

9 Przhevalskaia, L.K., Przhevalskii, N.M. 2016, 211.

10 Ciechanowicz 1997, 273.

11 Gavrilenkova 2011, 39–46, 81–93.

12 Dubrovin 1890, 6, 7.

Karetnikov's daughter was good looking and energetic. Despite Mikhail Przhevalskii's rather unattractive appearance and poor health, she did not reject his wooing and the couple joined in marriage in 1838. A year later, on 1 April (13 April New Style), their first child, Nikolai, was born,[13] to become the famous traveler. They had two more sons – Vladimir (1840) and Evgenii (Eugene) (1844) and a daughter, Elena (1846).

The Przhevalskiis had lived for a few years in a *khutor*, 'a solitary cabin' in the woods outside Kimborovo. They found themselves in rather strained circumstances after their marriage until Elena obtained 2,500 rubles bequeathed by her eldest sister Elizaveta, who had died prematurely. With this money the couple was able to build, in 1843, a new house called Otradnoe ('Joyful').

An interesting thing to note, Nikolai and his brothers knew very little about their family roots, since their father died early. Their information came mainly from their uncle Aleksei Kuzmich and their distant relative, Afanasii F. Przhevalskii, and was based on family legends. Interestingly, it was only two years before his death that he would learn about his earliest ancestor Cossack Karnila Parovalskii and that his clan was not Polish but came from Malorossia (Little Russia), i.e. Ukraine.[14]

Early Years and Education

Nikolai spent his early years in Otradnoye together with his younger brother Vladimir. The life of the Przhevalskiis on their country estate was that of typical Russian landowners of the serfdom era – they owned about 1000 *desyatins* (hectars) of land and 105 serf peasants, which was enough for a comfortable living.

The boy's education started at home, his first teachers being his mother, "a very clever, strict and strong-willed woman", and his uncle on the mother's side, Pavel Karetnikov, who, having squandered his own estate, settled in Otradnoye, in his sister's manor-house. As Przhevalskii would recall years later in his short autobiographical sketch:

13 The date comes from the parish records of the Labkovo church where Mikhail Przhevalskii and Elena Karetnikova were married, as ascertained by E.P. Gavrilenkova. See Gavrilenkova 2011, 46–51. Until then Przhevalskii's birthdate was believed to be 31 March 1839, as given by N.F. Dubrovin.

14 Przhevalskaia, L.K., Przhevalskii, N.M. 2016, 209.

> I grew up in the country like a savage. My education was Spartan. I could go out of the house any time, rain or shine, and quite early got keen on hunting. I first shot acorns from my toy gun, then [shot] from a bow, and when I was 12 or so I've got a real rifle. My hunting teacher was my uncle Pavel Alekseevich Karetnikov, mother's brother, who besides hunting had another inclination – he occasionally indulged in drinking bouts and thence did not go hunting. He also taught me French.[15]

Karetnikov stirred up a love for wildlife in the boy quite early, which eventually turned into a life-long passion, making him a traveler-naturalist. Not surprisingly, one of Nikolai's first books was a collection of fables by the popular Russia fabulist Ivan Krylov, well-illustrated with pictures of animals.

Some two years later his parents began to hire teachers to continue their sons' homeschooling. These were mainly *bursaki* (seminarians) from Smolensk, who proved "bad mentors", in Przhevalskii's own words. The ablest of them, Zeziuminskii, finally prepared the boys for the regular school or gymnasium.

Speaking of Przhevalskii's early childhood, mention should be made of his beloved nanny (*mamka*) Olga Makarova, or Makarievna, as she was commonly called. Makarievna was quite a remarkable person in Przhevalskii's entourage – a short, stout peasant woman, with rather course features, she ran the whole household and kept a tight hand over the domestics, being especially strict and even cruel to young girls who dared to have love affairs before marriage. While being utterly devoted to her masters, she openly tyrannized her equals. Makarievna was passionately fond of Nikolai and his brothers, the landlord's scions; she pampered them, and he remembered her telling him often his favorite fairytale about 'Ivan the great hunter', probably of her own making. He in turn loved her dearly, like a close relative, throughout his life, being unaware of the nasty things she did to other house servants.[16]

As Przhevalskii's English biographer, Donald Rayfield, neatly remarked, this was "a strangely matriarchal life"; Nikolai hardly remembered his father, who died when he was only seven. Therefore all he could say about him in his autobiographical sketch was "I have heard he was a practical man". As for his mother and Makarievna, "these two strong-willed women were allies" and their

15 This sketch was written down stenographically on 2 February 1881 at the editorial office of *Russkaia Starina* journal, and first published after Przhevalskii's death, see Przhevalskii 1888a, 529.

16 Dubrovin 1890, 7, 8.

"formative influence on Przhevalsky – on his attitude to peasants, to morality, to discipline ... – was irreversible".[17]

In 1849 Elena sent her two sons, Nikolai and Vladimir, to Smolensk, where they entered the second class of the local gymnasium. They lived very modestly, in a hired apartment, together with their two servants, under-tutor (*diad'ka*) Ignat and cook Anna, Makarievna's sister. The man would take them to school and back home and accompanied them on their promenades.

Nikolai was generally discontent with his schooling and his teachers in particular. He had no real friends among his classmates, yet enjoyed their great respect. As Dubrovin remarks, he stood "high above them all physically and morally" and struck everyone with his "brilliant mental faculties and phenomenal memory", which soon made him one of the best pupils. Nikolai even became a school celebrity, as he easily memorized whole pages from books during special tests arranged by his admiring mates.[18]

Their long (May to October) summer vacations Nikolai and Vladimir spent in Otradnoe, on their parents' estate, engaged for days in hunting and fishing. This was "a most useful part of the education of the future traveler", in the words of M.A. Engelgardt, – staying out in the woods strengthened him physically, built up his energy, developed his powers of endurance and observation, the things he would need most during his travels.[19]

• • •

The completion of Nikolai's studies in the Smolensk gymnasium in 1855 coincided with the demise of Nicholas I, the Emperor of All Russia, the Tsar of Poland and the Grand Duke of Finland. This put an end to his reign of three decades (1825–1855) known generally as a dark epoch of tsarist absolutism at its height, with its shameful serfdom system and repressive internal policies. Yet this was also the heyday of Russian *intelligentsia,* the status class of educated people engaged in different kinds of intellectual, artistic, cultural, and religious work. The most characteristic feature of the Russian intelligentsia, in the opinion of the late Russian semiotician Iurii M. Lotman, is a 'capacity for self-sacrifice' (*zhertvennost'*), when a person is ready to give up his well-being

17 Rayfield 1976, 3, 4.

18 Dubrovin 1890, 12, 13.

19 Engelgardt 1995, 345.

for the sake of others (his countrymen, *narod*),[20] a trait which is characteristic of a host of Russian pioneer scholar-travelers.

One of Przhevalskii's favourite poets was Nikolai Nekrasov (1821–1878), known for his lyrical poetry imbued with sympathy for the hard lot of Russian peasantry. His adventure novel *Tri Strany Sveta* ('The Three Mainlands of the World' – an allusion to Eurasia and North America washed by the Arctic Ocean) sparked wanderlust in young Przhevalskii.[21] In this novel Nekrasov described vividly the travels of his leading heroes, Kaiutin and Khabarov, to Novaia Zemlia, Kamchatka in the Far East, and Russian America, praising the "spirit of enterprise, courage, and daring". The sources for these descriptions were the land and marine expeditions under S.P. Krasheninnikov[22] and P.K. Pahtusov in the 18th – early 19th centuries,[23] and Przhevalskii might have been particularly impressed with the details of Kaiutin – Khabarov's hunting expedition to the mysterious Novaia Zemlia islands in the Arctic Sea, at the back of beyond.

Przhevalskii's parents belonged to the middle class of landed gentry and they took their privileged social status for granted. They showed no interest in any intellectual or cultural movements of their time, and the only books they had at home were the scripture and popular illustrated magazines. Yet their three sons were different. Their 'semi-savage' upbringing amidst the virgin nature of Smolensk forests and their later education and maturing in the comparatively liberal atmosphere of Alexander I's reign (1855–1881), marked by the abolition of serfdom in 1861 and other sweeping reforms, would eventually shape them into intellectuals of the peculiar Russian type.

The Beginning of a Military Career

In the same year when Nikolai and Vladimir finished the gymnasium they both moved to Moscow. Nikolai joined the Riazan' Infantry Regiment reserve there as a volunteer while his brother enrolled in Moscow University's law faculty. Nikolai wanted to become a military man. What inspired him was an adventure novel that described the exploits of a fearless warrior (*voin bez straha*)

20 Lotman 1999.

21 See Gavrilenkova 2011, 158. The novel was written in 1848–1849 and was co-authored by Avdotia Panaeva, Nekrasov's common-law wife.

22 Krasheninnikov, Stepan Petrovich (1711–1755), Russian botanist, ethnographer, geographer, an explorer of Siberia and Kamchatka, the author of *Opisanie Zemli Kamchatki*, see Krasheninnikov 1949.

23 Pahtusov Petr Kuz'mich (800–1835), Russian seafarer and hydrographer, an explorer of Novaya Zemlia.

FIGURE 1.3
Przhevalskii, aged 17, an officer cadet of the Infantry Regiment, 1856. NA RGO.

which chanced to come into his hands and instantly became his favourite. At the same time he was agitated by current press reports covering the dramatic developments of the Crimean War (1853–1856) that Russia was then waging with Turkey for ascendancy in the Near East. The heroic 11-month defense of Sebastopol by Russian troops inflamed the ambition of the 16-year-old Przhevalskii so much that he wanted to drop everything and travel to the Crimea to join the battle.

The tedium of garrison life, however, damped his ardour all too soon, and his romantic vision of the military calling was ruined finally by the ugly sight of regular drinking bouts and marauding of his fellow cadets. As he recalled this period in his autobiographical sketch:

> I had about 40 comrades belonging to all sorts of rabble; some of them, during our first march, stole somewhere a pair of boots and squandered these on drink in the tavern. This incident strongly upset me and was my first disappointment in the army service – I would occasionally retire to the woods for hunting where I would burst into tears. It was then that I began to look for a way out and [thought about] entering the military academy.[24]

24 Przhevalskii 1888a, 533.

In a letter to his mother dated 3 December 1855, a few months after his joining the army, Nikolai lamented: "There are about sixty of us, but most of them are good-for-nothings, drunks and gamblers. When I see myself with such comrades, I can't help remembering the words 'I'll be a diamond but in a pile of dung'. There are some good men though, but their number is quite small ...".[25]

Przhevalskii remained with his unit for over a year before he was promoted, on 24 November 1856, to the junior officer's rank of ensign (*praporshchik*) and transferred shortly after that to the Polotsk Infantry Regiment stationed in Belyi in his native Smolensk province. The promotion and the new placement, however, did not markedly improve his situation, and he would stay in the field for five years more.

Nevertheless, this time was not in vain. Looking back at it, Przhevalskii recalled with satisfaction that "an enormous change has come about in my ideas and my perception of life – I understood well and studied the community to which I belonged". He finally realized that he could no longer lead his miserable life, wasting his time "in various guard-houses and in the shooting-ground", and must find himself "a broader field of activity", where he would expend his efforts and time more reasonably.[26] Hence he began to prepare for examinations to enter the prestigious Nicholas Academy of the General Staff (Nikolaevskaia Akademiia General'nogo Shtaba, NAGSh for short).

The Academy of the General Staff was founded in St. Petersburg in 1832 by the emperor Nicholas I and Baron A. Jomini 'for the education of officers to serve in the General Staff[27] and for wider dissemination of knowledge in the army'. The foundation of the NAGSh became instrumental in the formation of the corps of officers of the General Staff, and the entity would soon hold the central place in the system of military education in Tsarist Russia.[28] The main subjects in the Academy's two-year curriculum were tactics, military history, administration, statistics, and geodesy, including cartography, land-surveying and technical drawing. Among the graduates of the Nicholas Academy in the nineteenth – early twentieth centuries were some foremost military figures, such as the war ministers D.A. Miliutin and A.N. Kuropatkin; V.G. Skobelev (one of the conquerors of Central Asia), A.E. Snesarev (leading military

25 Dubrovin 1890, 23. The passage is quoted in Rayfield 1976, 8.

26 Ibid., 26.

27 On the General Staff, as an entity distinct from the Main Staff, see the Introduction.

28 See: Glinoetskii 1888. The Appendix to the work includes the list of officers educated in the Academy between 1834–1882, and also the major scholarly publications by the teaching staff of the Academy and officers, who graduated from it, including N.M. Przhevalskii, see pp. 54–55. Some basic information on the Nicholas Academy can also be found in Marshall 2006, 5–6.

orientalist), Grand Duke Nikolai Nikolaevich, and the 'White guard generals' A.I. Denikin and V.O. Kappel, who would fight against Bolshevik (Red) Russia. Therefore education in this privileged institution was the summit of desires for many an ambitious young man pursuing a military career.

In 1865 the General Staff was dissolved, its functions being relayed to the newly established Main Staff (*Glavnyi Shtab*), one of the eight departments of the War Ministry. Four decades later (in 1905) the General Staff would be re-established and henceforth would take responsibility for the mounting of exploratory expeditions to Asia with the participation of army officers.

It took Nikolai about a year to prepare for the entrance exams to the military academy. For this purpose he moved to St. Petersburg in August 1861. Despite the stiff competition, he passed the exams successfully and was admitted to the Academy. His next two years would be a serious trial for him – he was permanently short of money and had to stint himself on everything. His time was generally split between his classes and the intense studies at home, where he buried himself in books. The military sciences were of no particular interest to him, yet unusual photographic memory helped him manage these 'boring' disciplines.

It was during this period that Nikolai decided to try his luck as an essayist. Hoping to earn some money, he wrote an essay titled "Memoirs of a Sportsman", apparently under Ivan Turgenev's influence,[29] that was published in the *Journal of Horse-breeding and Hunting* (*Zhurnal konnozavodstva i okhoty*) in 1861. He did not get a penny for it though, but was happy nonetheless to see his first literary work in print. In fact this was not his first opus – in 1860, while in Kremenets, in Volynskii province, where his regiment had moved, Przhevalskii wrote a paper "On the Essence of Life on the Earth" (*O sushchnosti zhizni na zemle*). This was intended for a discussion within the circle of science-inclined junior officers like himself. The paper, or rather synopsis of his speech, expressed his early and rather muddled views of the evolution of nature in the spirit of materialism and Darwinism, and, paradoxically enough, also dwelt on the idea of afterlife which, he asserted, was inherent in human nature. Yet, an important thing to note, Przhevalskii especially highlighted the holistic idea that nature must be seen as a single unified system: "... every organic body, plant, animal and man are not separate lives, but only fractions of the one uni-

29 The famous Russian novelist and story writer Ivan Turgenev (1818- 1883) began his literary career by publishing a collection of short stories under the title *Zapiski Okhotnika* (A Sportsman's Sketches) in the *Sovremennik* Journal in 1847–1852.

versal life, atoms of that huge whole that we call nature…".[30] This idea accounts well for Przhevalskii's future 'complex method' of field research that he and his many followers used during their expeditions to Inner Asia.

Przhevalskii's first strictly scholarly paper was a "Military-statistic review of the Cis-Amur Country" (*Voenno-statisticheskoe obozrenie Priamurskogo kraia*) that he produced in his second year in the Academy. Such reviews were usually compiled by Academy students in the course of their training. As for Przhevalskii's subject, it was particularly topical, as the Amur region had been recently annexed to the Russian empire and therefore attracted much attention from the military, as well as the general public. Thoroughly researched and well-written, his paper eventually came to the notice of the Imperial Russian Geographical Society. It was highly appreciated by the Society's members, the majority of whom, as already related, were army and naval officers. As a result, Przhevalskii was elected, on 5 February 1864, a full member of the IRGO, a starting point in his career as a military geographer.

•••

A year before, in May 1863, Przhevalskii had volunteered, along with a number of other senior students, to take part in the suppression of a new uprising in Poland against tsarist rule. This was instigated by Polish nobility (*shliakhta*) seeking to restore Rzechpospolita within the borders of 1772. It is most likely that Przhevalskii and his mates did so not so much out of their patriotic feelings, but because the Academy agreed to commission volunteers early and on favourable terms, namely, without final examinations. So he returned to his Polotsk infantry regiment, having been promoted to lieutenant (*poruchik*) and also appointed adjutant-major (*polkovoi ad'iutant*), a post which made him the regiment commander's right-hand man. Przhevalskii remained in this position for about a year; yet, as far as we know, he did not take part in any major punitive operations in Poland. Moreover, after spending some time in the country, he asked for a four-month leave and returned to his family estate, Otradnoe, in the summer of 1864.[31] One of his early biographers, M.A. Engelgardt, noted that while in Poland Przhevalskii showed much more interest in "hunting and book-reading" than in his military duties. Yet his hunting passion once played

30 For the text of the paper see Val'skaia 1967, 94–104. The paper was analyzed by B.A. Val'skii and D. Rayfield (Op. cit, 9), yet none of later Russian biographers has ever mentioned Przhevalskii's manuscript.

31 Gavrilenkov 1989, 17.

FIGURE 1.4
Przhevalskii hunting, painting by A. Bilderling, undated (late 1860?).
COPIED FROM N. DUBROVIN'S *NIKOLAI MIKHAILOVICH PRZHEVAL'SKII. BIBLIOGRAFICHESKII OCHERK* (ST. PETERSBURG, 1890).

a nasty trick on him – while chasing his game, he accidentally found himself on the territory of the insurgents; fortunately, he escaped captivity.[32]

Back in Otradnoe, he again spent most of his time in hunting and reading books, his favourite pastime. Now a member of the Geographical Society, he began to study seriously zoology and botany. His long cherished dream of becoming a pioneer explorer of some distant lands where no European had ever trod had finally matured and fully captivated him by then. Recalling this period, Przhevalskii would say that he was ready to set out for a journey "at the first opportunity".[33] Earlier, in gymnasia, he planned to go to Africa, following the examples of David Livingstone and Samuel Baker, the explorers of the Nile River's source, but later, after Russia began to advance into Central Asia (Western Turkestan) in the early 1860s, he would make the latter region his preference.

However, instead of Central Asia, in late 1864, Przhevalski travelled to Warsaw for his new military posting as "duty officer". There a new military district was established in 1862, in the course of D.A. Miliutin's reforms of the Russian

32 Engelgardt 1995, 349.

33 Przhevalskii 1888a, 535.

army, and consequently the cadet school was set up where Przhevalskii found employment. The reason for moving to Warsaw, as given by Dubrovin, was Przhevalskii's intention to continue his preparations for his future mission "at some scientific center" where he would have access to public library resources. His autobiographical sketch, however, reveals an intriguing fact – having graduated from the Academy, Przhevalskii wanted to be assigned to the General Staff, which afforded excellent opportunities for field research and exploration of Oriental countries, yet his petition to that end was declined because of his Polish name.[34]

In Warsaw

Przhevalskii stayed in Warsaw for over two years (1865–1867), a period he would remember as one of the happiest in his life. Being one of the major cultural centers of Europe, Warsaw was a striking contrast to the provincial backwaters where Przhevalskii had vegetated for five long years before he finally got disappointed with his army service. At the cadet school he taught geography and history – that of geographical discoveries of the last three centuries, and concurrently worked as a librarian. Eloquent, convincing, and startling everyone with his unusual memory, Przhevalskii proved to be an excellent teacher; his charisma made his classes very popular with the cadets. On a personal level, he was very sociable, joyful, and energetic as ever. Not infrequently he invited his cadet-pupils to his home where he entertained them lavishly with all sorts of delicatessen and home-made fruit drinks.

Apart from these improvised small feasts, he led a rather solitary life, being fully concentrated on his studies. Here in Warsaw Przhevalskii finally completed his painstaking self-education, by filling in its many gaps with the help of books and by consulting specialists, such as zoologist and a keeper of the Warsaw Zoological Museum Vladislav K. Tachanovskii (Wladyslaw Taczanowski)[35] and botanist Yu.O. Aleksandrovich from the local Botanical Gardens. Under their competent direction, Przhevalskii did much practical work in zoology, especially ornithology (which was soon to become his favourite subject largely due to his hunting experience), and botany. He mastered, first and foremost, the complicated scientific nomenclature and learned to determine and prepare the faunistic specimens and to make a herbarium. After his first Ussuri journey he would occasionally go to Warsaw to meet Tachanovskii, who helped

34 Ibid., footnote on p. 535. See also Dubrovin 1890, 45.

35 Vladislav Kazimirovich Tachanovskii (Wladyslaw Taczanowski, 1819–1890), Russian zoologist of Polish extraction, specialist in ornithology and arachnology ('spider-expert').

FIGURE 1.5 N. Przhevalskii in Warsaw, 1867–69. NA RGO.

FIGURE 1.6 N. Przhevalskii with comrades in Warsaw. NA RGO.

him define some of the new genera of birds. In this way Przhevalskii finally shaped himself into a genuine 'scholar-traveler', an entirely new profession in the mid-19th century Russia.

His favourite books were Alexander von Humboldt's *Views of Nature* and Karl Ritter's *Asia* (one of the volumes of his masterpiece *Die Erdkunde*), in the Russian translation. He spent nearly all his money on books and was a regular subscriber to the *Izvestiia* (Transactions) of the IRGO, which he received by post from St. Petersburg.

While in Warsaw, he wrote and published the *Notes on General Geography for Cadet Schools,*[36] which subsequently would be widely used in Russia. In this manual Przhevalskii divided geography into three separate disciplines: 1) *physical geography*, basing himself mainly upon Ritter's comparative method; 2) *mathematical geography,* which dealt with the composition of the universe, and 3) *political geography*, that is, regional or area studies (*stranovedenie*). His major focus was physical geography, because "it primarily acquaints one with the structure of the terrestrial globe and the various phenomena of nature". The study of geography, as he underscored in the Introduction, would "give the

36 Przhevalskii 1867; 2nd ed., with additions from I.L. Fateev, 1870.

cadets some idea about nature and man, and this, undoubtedly, will affect beneficially their mental development".[37]

The manual was actually written by Przhevalskii under the influence of his new Warsaw friend, Mikhail Veniukov, who had earlier produced his own manual of physical geography. Veniukov was a military geographer, ethnographer and pamphleteer, a man of progressive views who was dedicated to democratic ideals and openly sympathized with the Polish insurgents. Incidentally, he is known to have translated *La Marseillaise* into Russian. Veniukov was dispatched to Poland by the War Ministry in the capacity of military-regional administrator, yet shortly afterward he was reassigned to sit on the Liublin Committee, which managed the return of the *shliakhta*-seized plots of land to peasants. He was an experienced scholar-traveler and explorer, who had earlier traveled extensively in Siberia, the Far East, including the Ussuri region, Middle Asia and the Caucasus, which naturally attracted Przevalskii to him. In 1869 he would travel to Peking (Beijing) to gather military-statistic data on China and Japan, and also, if possible, "on the neighbouring Middle-Asian lands".[38] Veniukov's geographical studies and spirituality, in the opinion of his biographer V.A. Esakov, drew Przhevalskii close to him, and influenced his fate to a certain degree: Veniukov "inspired Przhevalskii with his plans and his creative work", one result of his influence being Przhevalskii's writing his geographical manual.[39]

In 1866 Przhevalskii petitioned his military superiors for a transfer to the Turkestan region, with a subsequent posting at the General Staff, yet this was not easy, given his Polish surname. Luckily, he had some highly-placed protectors whose mediation proved crucial. One of these was the deputy head of the Warsaw military district Major-General D.I. Chernitskii, who petitioned the General Staff by saying that Przhevalskii, "with his broad knowledge of geography, history and statistics, could be helpful in compiling statistic reviews of the little-known territories in Middle Asia".[40] Chernitskii's petition was further relayed to the Head of the General Staff, Count F.L. Geiden, by Adjutant General N.V. Isakov, with a brief accompanying note of his, which said, inter alia, that Przhevalskii was "a very able and smart officer" who "desired activity".[41] As a result, on 17 November 1866 Przhevalskii was assigned to the General Staff and transferred for further service to the East-Siberian military district, though not to Central Asia, as he had originally wanted.

37 Przhevalskii 1876, [3].

38 Esakov 2002, 41.

39 Esakov 2002, 34.

40 Dubrovin 1890, 45–46.

41 Dubrovin 1890, 46.

In conclusion, I would like to make a brief remark on Przhevalskii's rather controversial ethnic self-identification. There is no evidence whatever that he, as a bearer of the Polish surname and someone whose ancestors served Rzechpospolita, identified himself with Polish nationality. On the contrary, Przhevalskii showed little interest, if any, in the history of Poland, Polish culture, or literature. His library at his House-Museum in Smolensk has only one book dealing with Poland – it is the Russian translation of "The History of the Polish Uprising and the Wars of 1830–1831", written by one F. Smith and published in St. Petersburg in 1865.

At the same time we know that Przhevalskii had some good friends in Poland, one of whom was the above-mentioned Vladislav Tachanovskii. He would correspond with Tachanovskii – a Russian subject with a Polish surname like himself – for many years, and the latter would help him describe many specimens in his zoological collections.[42] What brought them together, however, was not only zoology, but their Poland-related ancestry, as it seems. Curiously, Tachanovskii's first two letters to Przhevalskii were written in Polish, which suggests that Przhevalskii had some knowledge of the Polish language. Yet there is nothing in Przhevalskii's published works and his correspondence indicating that he regarded Poland as his second motherland. He was a great patriot of Russia, ready to sacrifice his life for the cause of the empire of the Romanovs.

Ussuri Trip, 1867–1868

Przhevalskii left Warsaw in late December 1866 in the company of the German zoologist Robert Köcher, who volunteered to accompany him on his journey to Asia. On the way to Irkutsk, the center of the East-Siberian general-governorship, they made a brief stopover in St. Petersburg, where Nikolai called on the Main Staff and the Imperial Russian Geographical Society, in January 1867. At the IRGO he contacted the head of the department of physical geography, Petr Petrovich Semenov, to whom he presented his grandiose project for a large-scale exploration of Inner Asia. This embraced the uncharted continental highlands skirted by the Tian Shan, Pamir and Himalayas; Eastern Asia, from Siberia down to India; and from the Pamir to the Great Chinese wall. Przhevalskii wanted to explore the Gobi desert, the steppes of Mongolia, and to cross through Dzungaria to ascend the world's highest plateau, Tibet. The IRGO Council, however, turned down Przhevalskii's project because it did not want

42 Taczanowskii's letters to Przhevalskii belong to the archive of the Russian Geographical Society (NA RGO), F. 13, Op. 2, D. 243, 40 letters (1870–1887).

FIGURE 1.7
N. Przhevalskii in Irkutsk, 1873. NA RGO.

to take any chances by supporting someone who had no name as yet in geographical science.

Still, Semenov felt a profound sympathy for Przhevalskii, in whom he found a kindred soul. Therefore he suggested that the enthusiastic young officer, keen on geography and travelling, test himself in the field first by exploring some uncharted territory in East Siberia and present the results to the Society. To this end he provided Przhevalskii with several letters of recommendation to his good friends – the general-governor of East Siberia, Mikhail S. Korsakov and the head of staff of the East-Siberian military district and concurrently chairman of the local IRGO department, Major-General Boleslav K. Kukel.[43]

Thus encouraged, Przhevalskii left for Irkutsk, equipped with a shotgun and some books about East Siberia, particularly the Amur – Ussuri region he intended to explore. In Irkutsk he got in touch with local officials, the military and the IRGO administrators. He made a fairly good impression on them, yet his great enthusiasm and frenetic activity struck some as self-conceit and sheer bragging.

43 On Przhevalskii's first meeting with Semenov, see Aldan-Semenov 1965, 209–211; also Dubrovin 1890, 48–49.

His expedition lan took its final shape in May 1867; it included primarily a number of commissions he received from the Main Staff with a view to surveying the Ussuri region, the south eastern corner of the empire squeezed between North-Eastern China and the Sea of Japan. Namely, Przhevalskii was assigned to 1) survey the disposition of the two marching battalions stationed there; 2) collect information on the numbers and condition of the Russian and indigenous (Manchu and Korean) populations in the area; 3) investigate the pathways leading to the borders of Manchuria and Korea and update the Ussuri route chart; and 5) carry out whatever scientific research he would find necessary. On top of that, the Siberian department of the IRGO employed the young officer to investigate the Ussuri flora and fauna and provided him with some funds to that end.[44]

Przhevalskii was delighted and the large scope of his tasks did not scare him at all. On the eve of his journey he wrote in a letter to his Warsaw friend I.L. Fateev:

> In three days, I'll be going to the Amur, thence to the Ussuri, Lake Khanka, and the shores of the Great Ocean, toward the Korean borders. Overall, it's a splendid mission. I am happy beyond words! The main thing is that I can freely dispose of my time, whereabouts, and occupation. Yes, it's an enviable lot, yet also a serious responsibility – to explore the regions, most of which has never been treaded by an educated European.[45]

Unexpectedly for Przhevalskii, his mate Köcher changed his mind at the last moment – he wanted to return to Warsaw where he had left his fiancé, whom he badly missed. So Przhevalskii had to look for another companion, someone of stronger character, and he soon found one – Nikolai Yagunov. The 16-year-old son of an exiled Polish insurgent, he was recently employed as a topographer at the Staff of the East Siberian military district, yet he readily quit his job for the sake of travelling with Przhevalskii to Ussuri's terra incognita. Accompanied by a Cossack, Nikolaev, they set out from Irkutsk on 26 May 1867. Having sailed across Lake Baikal, the party headed further eastward towards the

44 Dubrovin 1890, 51.

45 Engelgardt 1995, 354–355; Gavrilenkov 1989, 20. Iosif L'vovich Fateev was one of Przhevalskii's close friends. He was also educated at the Nicholas Academy and later taught geography at the Warsaw cadet school; Fateev would subsequently hold the post of the deputy head of a department at the Chief Directorate of the Cossack troops. For his correspondence with Przhevalskii see NA RGO, F. 13, Op. 2, D. 264 (1869–1888, 53 letters and newspaper cuttings).

Shilka and the Amur rivers, leading them ultimately to the Ussuri region, their destination. Fully aware of the risks of the journey, Przhevalskii armed himself with a gun, a revolver, and a dagger; he also took along a setter, who proved "an unfailing friend" in imergencies.

Sailing along the Great Amur amidst the impenetrable taiga forests provided some spectacular scenery, which Yagunov depicted in a serious of pencil drawings.[46] It took the travelers about a month before they finally reached the settlement of Khabarovka (today's city of Khabarovsk), the starting point of their Ussuri expedition. There Przhevalskii purchased a boat and hired two Cossack oarsmen to steer it. In this way his small party would sail, for 23 days, up the Ussuri River, from its mouth to the last Cossack settlement (*stanitsa*), Busse. Much of this time Przhevalskii and Yagunov would trudge along the river bank, collecting plants and shooting and stuffing birds, while their boat followed them, their oarsmen being utterly unhappy with such a routine.

Przhevalskii was strongly impressed with "the solemn grandeur of nature undisturbed by man" that he found in the Ussuri, which reminded him of the "primeval forests" and "tropical jungles": "There is not a sign of man's presence around us; everything looks wild, uninhabited, and virgin. One can see here only beasts, which appear for a moment here and there, as if reminding the traveler that these forests are brimming with life, but this life is savage and peculiar…".[47] What struck Przhevalski mostly was the unusual richness of the wildlife, particularly a rather "odd tangle" of the northern and southern types of vegetation and animals. Here one could see a spruce entwined with vine, a cork-oak and walnut standing next to a cedar and a fir-tree. And a hound could track down a bear or a sable, yet at the same time one could meet here a tiger, as big and fierce as in the jungles of Bengal.

Sailing up the river, Przhevalskii surveyed the 28 settlements or *stanitsas* of the Amur Cossack troops, which had recently emerged on the Ussuri's right bank. (These *stanitsas* played the role of frontier posts and post stations along the Sino-Russian border.) The Cossacks migrated here from Transbaikalia in 1858–1862, yet their life in this uninhabited land presented a rather sorry spectacle. In Przhrvalskii's words, "hunger and poverty, together with various human vices which usually accompany them, drove these people into complete moral decay". What made the situation even worse were the typhus and other epidemic diseases which often raged here, and he put much of the blame for

---

46 Yagunov's 'Album of the Amur Views' belongs nowadays to the Przhevalskii Museum (*Dom-muzei N.M. Przheval'skogo*) in his Sloboda estate (renamed Przhevalskoye) in Smolensk Province.

47 Przhevalskii 1937 (2nd ed.), 10.

this on the administrators of the region. (Getting ahead of the story, Przhevalskii's negative description of the life of the Ussuri Cossacks came to public notice right after the completion of his journey, when *Vestnik Evropy* (Herald of Europe) carried his two articles devoted to the subject. This caused much displeasure among Siberian senior officials who accused the traveler of a willful misrepresentation of the condition of the Ussuri migrants. Przhevalskii would squarely deny these accusations in a letter to the editor of the *Peterburgskie Vedomosti* daily published in late 1870, along with a note from General M.P. Tikhmenev, former head of staff of the troops in Primorskii district, which confirmed his first-hand evidence.[48])

From the Ussuri the party moved to Lake Khanka where Przhevalskii and Yagunov spent a month observing and collecting samples of local birds and fish. The sight of the lake, covered with plenty of pink lotus flowers rising above the water surface, fascinated them. Apart from his naturalist studies, Przhevalskii also took a census of the peasants of the three Russian villages located on the Western bank of Khanka. His general opinion was that the Khanka steppes, with their fertile soils and abundance of pastures, were the best place for Russian colonization in the entire Ussuri country. From Lake Khanka, Przhevalskii and Yagunov then journeyed to the coast of the Sea of Japan. Here they surveyed the Posyet Bay (earlier explored by a Russian expedition under Admiral E.B. Putiata), the Olga Harbour and the Vladimir Bay, and they also visited the newly founded port of Vladivostok, the future home base of the Pacific Fleet, which, at that early stage, numbered not more than 500 citizens and troops. Przhevalskii thoroughly examined the coastline, with its many gulfs and bays, and the littoral areas, with a view to using these subsequently for economic and military purposes.

In mid-January 1868 the expedition reached the Cossack village Busse on the right bank of the Ussuri, the last in the chain of Russian frontier posts. Here the travelers stayed for the winter and with the onset of spring they went to Khanka again to watch the migration of birds and replenish their zoological collections. They intended to proceed further to Manchuria, yet the incursion of the Chinese *hunhuzi* bands into the territory of Primorie frustrated these plans. Yet, as an army officer, Przhevalskii could not stay aloof, and engaged in the hostilities, putting aside his research work.

Przhevalskii spent the rest of the year in Nikolaevsk, the Russian post at the mouth of the Amur, sorting out his many collections and writing up his journey. In his leisure hours he was often seen playing cards with naval officers and

48 For the discussion of the conflict, see Dubrovin, op. cit., 99–103; 547–555 (Appendices 1–4). See also Gavrilenkov 1989, 23, 24.

merchants; he turned out a very lucky gambler, which earned him the nickname 'Golden Pheasant'.[49] To account for his compromising passion Przhevalskii would say: "I play cards only to win my freedom". He also tried to earn some money by reselling sable furs that he bought from local merchants and sent to his brother Evgenii in Moscow. Yet again, he did so not for profiteering, but only to secure his living and his scholarly pursuits.

With the arrival of spring of 1869 Przhevalskii was back to Khanka, shooting birds, collecting plants and looking for roads in the western and southern extremities of the Khanka basin. In July, his two-year Ussuri journey was finally over. Its scientific results were quite impressive. Przhevalskii had traversed a territory of over 3000 km, much of which he put on the map; he amassed several zoological and botanic collections, including 310 stuffed birds, 550 bird eggs, 10 skins of mammals, a great multitude of butterflies and spiders, and a herbarium of about 300 species of plants. To these one should add also his detailed descriptions of Russian and non-Russian settlements in the Ussuri region,[50] the meteorological observations, and of course his strategic intelligence. Especially valuable was his information on the communication network in the Primorskii district, namely the pathways to the Manchurian and Korean borders passable for troops and those leading from the Ussuri basin to the harbours on the Pacific Ocean. He proved an excellent trailblazer.

In January 1870 Pezhevalskii was back in St. Petersburg. There he was welcomed cordially by the scholarly communities of the Academy and the IRGO and soon became "one of our circle", in the words of Semenov. On 1 April he reported about his journey at a general meeting of the IRGO members, and towards the end of the year he published, at his own expense, his travel account *Puteshestvie v Ussuriiskom Krae* (Journey in the Ussuri Country), 1867–1869. The book earned Przhevalskii the Society's silver medal (1869), his first award, and an official recognition of his accomplishments as a geographer.

First Journey to Inner Asia: Mongolia, the Tangut Country and Northern Tibet, 1870–1873

The Ussuri trip was a serious test for Przhevalskii, having revealed his excellent abilities as a field researcher and naturalist, as well as his substantial literary talent as shown by his published travelogue. The success of this first expedition foreshadowed his future – in 1870 Przhevalskii submitted to the IRGO a

49 See Dubrovin 1890, 79.

50 See Przhevalskii 1869.

new project, one envisaging exploration of Inner Asia. This time the Society was more benevolent, thanks to the strong backing of the Russian ambassador in Peking, Major-General A.E. Vlangali,[51] then visiting St. Petersburg. As Przhevalskii would later report, "It was mainly due to his [Vlangali's] initiative that my expedition emerged, and he was its warmest supporter from first to last".[52]

Przhevalskii had a chance to discuss with the noted diplomat his preliminary plans before approaching officially the Geographical Society. On 29 March he sent a letter to the then secretary of the IRGO, Baron F.R. Osten-Saken, in which he outlined sketchily the area he wanted to explore:

> I am personally attracted particularly to the northern fringes of China and the eastern tracts of Southern Mongolia, a locality which is practically unexplored by Europeans but is of great interest for geography and natural science; the more so as access with purely scientific purposes to these far-away countries presents no difficulties, according to our ambassador in China, General Vlangali.

Asking for support of his project, Przhevalskii pointed out that the expedition was intended for a period of three years and that he personally could afford 1000 rubles annually to cover his travel expenses. To this he added a very important statement that showed his determination to become an explorer: "Finally, if my service conditions should present any obstacles for the projected trip, I am ready to resign any time and devote myself entirely to the service of science".[53]

Still, before making his final decision, the IRGO's Vice-Chairman Count Litke deemed it necessary to consult Vlangali. In his reply, Vlangali welcomed Przhevalskii's initiative by underscoring the importance of his journey in scientific and geopolitical respects:

> Exploration of the northern fringes of China proper presents such a great scientific interest that one cannot but sympathize with the enterprise and

51 Vlangali Aleksandr Georgievich (1823–1908), of Greek extraction, was a mining engineer who had worked for some time at the Corps of Mining Engineers at St. Petersburg, having taken part in many geological expeditions, including his journey to the Kirghiz steppes in 1849–1851. In 1860–1873 he was head of the Russian diplomatic mission in Peking and later held the post of deputy-minister of foreign affairs. In 1874 Vlangali was elected an IRGO honorary member. Staying in China, he is known to have amassed a valuable collection of Chinese objects of applied arts.

52 Przhevalskii 1875, VI; 2nd. ed. 1946, 42.

53 Quoted from Dubrovin 1890, 90–91; see also NA RGO, F. 1 (1870), Op. 1, D. 10, l. 1 ob., 2.

> wish for its success as much as possible. The upper course of the Yellow River, the Ordos country, [Lake] Kukunor and other places skirting northern Tibet have not been explored so far, whereas geography can enrich itself there with many interesting and scientifically important facts.
>
> We cannot cede the [priority] of these explorations to others and let them be the first in that part of Asia, due to our geographical position in China and also because we have already made some important investigations on the north western outskirts of that country in the Tian Shan Mountains.[54]

At the same time, Vlangali was not certain about accessibility of these regions because of the Dungan rising: "When I was in Peking, the insurgency of the Mohammedans had separated these areas from other provinces, and communication with these became difficult". However, Vlangali believed the situation could change and the obstacles for travelling disappear. The important thing for a Western traveler was not to antagonize the local population by any misconduct. Vlangali also suggested that Przhevalskii should first come to Peking, sometime in the fall, to make preparations for his further inland journey during the winter. There, in Peking, in the lodging of the Lazarist missionaries, the travelers would find the animal and bird collections brought from various parts of China and a vast library. Finally, Vlangali assured Litke that the Russian diplomatic mission in Peking, and he personally, would give their "full assistance and hospitality" to Przhevalskii's expedition. The war minister Miliutin, for his part, also supported the project hence the IRGO willingly gave the traveler a green light. It remained now only to obtain the emperor's supreme approval, which followed on 19 July 1870.

The scientific program of the expedition elaborated by Przhevalskii jointly with P.P. Semenov was rather sketchy, defining only the geographical areas to be explored – Southern Mongolia, Ordos, and Lake Kokonor (Qinghai Lake). As for the types of the projected field work, these included 1) the route surveying, 2) geographical and ethnographic observations, 3) instrumental measurements of various kinds (astronomical for determination of latitudes, barometric, hypsometric, and meteorological), 4) collecting zoological, botanic and partly mineralogical specimens, with a special attention to be given to mammals and birds, which particularly attracted Przhevalskii the hunter. This program would become a standard for Przhevalskii's later journeys. The Main Staff, for its part, commissioned him to gather information about the Dungans

54 See Dubrovin 1890, 92; also the Archive of the RGO, ibid., ll. 3–4 (letter from F.P. Litke to A.E. Vlangali dated 23 April 1870).

and their activities, as their uprising had seriously affected Russia's trade relations with Mongolia and China.

The expedition's main sponsors were the Geographical Society and the War Ministry (Main Staff), which allocated Przhevalskii 1000 rubles each annually (the former in paper currency and the latter in hard money, in addition to officer's salary and travelling allowances between St. Petersburg and Kiakhta),[55] while the Imperial Botanical Gardens provided 300 rubles per year. These sums, however, were not enough for the entire three-year journey, even though the IRGO and the MS would increase their shares to 1500 rubles in the second and third years.

As in the case of the Ussuri trip, the expedition staff consisted of only three men – Przhevalskii himself, his assistant – Lieutenant Mikhail Pyl'tsov, and one Transbaikal Cossack to act as interpreter. Pyl'tsov was Przhevalskii's favourite pupil at Warsaw, so he knew him well and could fully rely on him. (His former companion, Nikolai Yagunov, returned to Warsaw where he entered, on Przhevalskii's advice, the cadet school.)

Przhevalskii and Pyl'tsov arrived in Kiakhta on the Russian-Mongolian border in November 1870, having crossed Siberia in a post-carriage, which was the only means of transportation at a time when the Trans-Siberian railway did not yet exist. From Kiakhta they went to Peking first in order to obtain, with Vlangali's assistance, a Chinese travel passport, as the document could facilitate their journeying in a country where people were generally suspicious of all strangers. While preparing for the journey, Przhevalskii hired a two-wheeled Chinese cart normally used by Russian merchants travelling in China. The cart was a great square wooden box set on two wheels and closed on all sides, drawn by a camel. This was a primitive and extremely uncomfortable conveyance because of constant jolting, even when moving at slow pace. The cart usually accommodated Przhevalskii, his mate Pyl'tsov, and his setter Faust. The party set out from Kiakhta on 17 (29) November in the direction of Urga, the Mongol capital, by the well-trodden 1500 mile-long Kiakhta – Kalgan road, connecting Russia with Mongolia and China.

Urga, known to Mongols as Bogdo-Kuren ('Sacred Encampment'), residence of the Bogdo-Gegen Khutuktu, or the Grand Lama of Urga, the third in the Buddhist hierarchy after the Dalai Lama and the Panchen Lama, was Przhevalskii's first encounter with High Asia beyond Russian borders. The city, seen with the eyes of a European, was an oppressive sight at first glance: "Outwardly,

55 To finance this expedition, the Main Staff had to interrupt Veniukov's journey to China and Japan and transfer the remainder of his sums, amounting to over 2000 rubles, to Przhevalskii, see Esakov 2002, 50–58.

the Mongol part of Urga is disgustingly dirty. All the filth is thrown into the streets and the habits of the people are loathsome. To add to all this, crowds of starving beggars assemble at the market place ... ". More "sickening scenes" awaited the travelers when they passed by a typical Buddhist cemetery outside the city:

> Here the dead bodies, instead of being interred, are flung to the dogs and birds of prey. An awful impression is produced on the mind by such a place as this, littered with heaps of bones, through which packs of dogs prowl, like ghosts, to seek their daily repast of human flesh.[56]

While in Urga, Przhevalskii and his companion were hosted for several days by the Russian consul general, Iakov P. Shishmarev, and his amiable family, and it was from Shishmarev that they heard of the Dungans "who had just plundered Uliassutai and threatened Urga with a similar fate". The ominous news clouded the very start of his journey.

The passage though the eastern Gobi, from Urga to Kalgan, was rather exhausting:

> Our days dragged on with tedious monotony. Following the central caravan route we generally started at midday and marched till midnight, averaging 27 to 33 miles per diem. During the daytime my companion and I generally went on foot ahead of the caravan and shot any bird we saw.[57]

As Przhevalskii penetrated deeper into the Gobi, he meticulously observed and recorded everything unusual that came his way. Apart from hunting, by far the most enjoyable part of the trip, he paid much attention to the soil and vegetation of the desert and took daily meteorological measurements, despite the fact that the Kiahta – Kalgan route had already been surveyed by many a traveler, including the members of the Russian Ecclesiastic missions. The expedition's small caravan was regularly harassed by crows, "our bitter enemies", often perching on the baggage camels and stealing the little skins prepared for the zoological collection. Among other members of the "feathered tribe", peculiarly characteristic for Mongolia, Przheavalskii noticed the sand-grouse and Mongolian larks, "the best songsters of the Central Asian desert". Of the mammals he mentioned the Alpine hares (*lagomys ogotono*) and dseren-antelopes (*Ante-*

56 Quoted from Prejevalsky 1876, vol. 1, 13–14.

57 Ibid., 21.

lope gutturosa), and he described in detail in his diary the modes of antelope-shooting by the Mongols.

Przhevalskii also dwelled at length on the inhabitants of the various parts of Mongolia – the Khalka Mongols, the Sunni Mongols, inhabiting the most barren part of the Gobi, and the Chakhar Mongols, "the frontier police of China Proper". His anthropological descriptions, from the very start, were not devoid of cultural and racial bias. Thus when portraying the Chakhars, he would assert that they "have lost not only the character, but also the [facial] type of *pure Mongols*" by reason of their constant intercourse with the Chinese. "By preserving the native idleness of the past, they adopted from the Chinese only the worst features of their character, and are degenerate mongrels …".[58]

At Kalgan the travelers passed through the Great Wall and soon reached Peking, where they spent two months, until the end of February 1871. By this time, thanks to Vlagali's mediation, they obtained a passport from the Chinese government's *Zongli Yamen* (Foreign Affairs Chamber), permitting them to travel in China, as far as Gansu Province. The party also procured supplies, as well as a large quantity of small articles, as Przhevalskii resolved to pose as a merchant during his journey. His equipment consisted mainly of guns and ammunition for hunting, special measuring instruments some of which he received from the head of the Russian observatory in Peking, Herman (G.A.) Fritsche, and special devices for dissecting animals and drying plants, all packed into four large boxes. He also had some provisions "for immediate wants", that is, a case of French brandy, 36 lbs of sugar and 2 sacks of rice, and he hoped to obtain as much meat as needed by hunting.

Przhevalskii's impression of the ancient Chinese capital was much like that of Urga:

> … indeed a newcomer could hardly be pleased with a city in which cesspools and crowds of naked beggars are the denizens of even the best streets. If we add to this the insolent effrontery of the Chinese themselves and the name of Kwei-tse, i.e. 'foreign devils' with other opprobrious epithets, which they bestowed upon us, it may readily be imagined that Peking is not the pleasantest place in the world for a stroll.[59]

The journey started with a two-month trial foray to the south eastern borderland of the Mongolian plateau, north of Peking, with a visit to Lake Dalai Nor to observe the spring flight of birds of passage. Przhevalskii's party now

58 Ibid., 33.

59 Ibid., 91. The correct transliteration of the Chinese expression is *yang gui zi*.

consisted of four men, including one extra Cossack from the Russian embassy. After this short excursion the travelers returned to Kalgan via Dolonnor, and thence headed westward in the direction of the northern bend of the Yellow River (Huang He), the Ordos Loop, and the Ordos plateau, the first terra incognita on his vast research agenda. From Ordos he would proceed further west to the Alashan desert, another uncharted area, up to the oasis-town of Dyn-yuan-in, the residence of the ruling prince of Alashan. This was the ultimate point of his expedition's route during the first year, from which he would return to Kalgan and thence to Peking, in January 1872.

In the course of the journey Przhevalskii conducted various field investigations, his highest priority being a survey of the entire route and putting it on the map. The route survey was by far the hardest part, especially in areas where there were few landmarks, such as the Mongolian plain, but also because of the suspiciousness of the local inhabitants. "Had the natives, particularly the Chinese, discovered that I was mapping their country", he would later admit in his travelogue, "our difficulties would have been doubled and we should have found it next to impossible to pass through the populous districts. Fortunately, I was never surprised with the map, and no one ever knew that I was sketching my route".[60] The surveying instrument Przhevalskii used was a Schmalkalder compass, which is normally fixed on a tripod stand, but to avoid suspicions he steadied the compass in his hands instead. All distances were measured by the number of hours of travel and "our rate of progress" – 2 ½ to 3 miles per hour. To avoid attracting attention in thickly populated areas Przhevalskii would ride ahead of or remain behind the caravan, and he would also have recourse to many other "stratagems" and "artifices" to fulfil his important tasks "amidst the people who were hostile to us".

Przhevalskii always carried a small field-book in his pocket for noting "all conspicuous objects", and at the end of day's journey he copied his notes to his travel diary. His map, in reeled sheets, was carefully hidden in one of the boxes of his luggage. On this he drew the line of marches, marking all the settled habitations (towns, villages, houses, temples, except nomad encampments), wells, lakes, rivers and streams, however small, and lastly, mountains, hills, and the general outline of the country on both sides of the road. The work of surveying, as Przhevalskii remarked, "simple though it may seem, was one of our most arduous labours ... we often had to travel by day for the sake of our survey, in this way exhausting our strength, as well as that of our camels".[61]

60 Ibid., 111.

61 Ibid., 115.

A trip to Alashan, the southern part of the high plateau of the Gobi to the west of the middle course of the Yellow River, was the final leg of his first year's journey. There, in Dyn-yuan-in (Denyuangying), he met with the Prince of Alashan, Amban, and his three full-grown sons, and succeeded in winning their favour. To all of them he gave presents – a watch and aneroid barometer, unfit for use, to the prince, binoculars to his eldest son, and some "sundry small articles, such as hunting accoutrements and gun powder" to his younger sons. As a result, "Altogether the Amban and his sons, especially, the two younger, were well disposed towards us, and were constantly giving us proof of their goodwill". The Amban, however, made a bad impression on him, being a smoker of opium and "corrupt and despotic to the last degree". In Przhevalskii's words, "his sweet will replaces law and is implicitly obeyed without the slightest opposition from anybody. But the same system prevails throughout the whole of Mongolia and China".[62]

From Dyn-yuan-in the party returned to Kalgan in January, 1872. This final trek was rather troublesome as Pyl'tsov suddenly fell ill with typhoid fever and then the winter set in, with bitter frosts and snow-drifts, which obstructed the caravan's progress. Fortunately, Pyl'tsov's condition improved so he was able to continue as Przhevalskii's assistant. Back in Peking, Przhevalskii dispatched his collections to Kiakhta and reported on his first year's results to the IRGO and the Main Staff. As soon as this was done, he started preparations for a "second act" of his journey – to Gansu province, Lake Kokonor, and Northern Tibet, which in fact constituted an expedition in its own right, to last until October 1873.

While in Kalgan, his Cossacks were replaced by two newcomers from Urga – Dondok Irinchinov (a Buryat) and 19-year-old Panphile Chebaev (a Russian). These two would remain with his party permanently and they proved to be very efficient and hard-working, especially Dondok, whom Przhevalskii nicknamed Didon the Wise (*Didon Mudryi*). Dondok was truly a jack-of-all-trades – he could do any job needed and besides he spoke Mongolian and could understand various Tibetan dialects. His favourite saying was "Don't worry and doubt not".

Przhevalskii started his second journey, again from Kalgan, in March 1872, with a new passport from the Chinese authorities enabling him to travel in Gansu, Kokonor, and Tibet. However, the officials warned him that travelling in these countries was hazardous because of the Dungans' activities and therefore they could not guarantee the safety of his mission. Being fully aware of the danger, Przhevalskii obtained at Peking and Tien-tsin some additional guns

62 Ibid., 243.

(breech-loaders and revolvers), which, he believed, "are the best defense a European can have in travelling in these parts of Asia". (The best of his new guns was a rifle by Berdan). His supplies also included four flat water-barrels, each holding 8 gallons, which he needed for travelling in places where drinking water was scarce.

The travelers' journey inland followed the familiar route back to Dyn-yuan-in. From there Przhevalskii intended to proceed to Gansu and Kokonor, with the assistance of the Alashan prince and his sons, for whom he was carrying many more presents. Upon meeting them, he put on his officer's uniform to make a stronger impression: "They were now more than convinced that I was a high functionary, perhaps the trusted agent of the Emperor himself. ... Henceforward I received the title of the Tsar's officer, by which I was called during the remainder of our journey.[63]

Having sold the merchandize he brought from Peking at a profit in Dyn-yuan-in, Przhevalskii then joined a caravan of Tangut and Mongol 'lama-warriors' heading for the Choibsen monastery (*kumirnia*[64]) in Gansu province and thence to Kokonor, passing through the territories he intended to explore. Thus began his adventurous journey in parts of China which had not been visited before by Europeans.

Gansu was a mountainous, moist and wooded country, making a sharp contract with Alashan, to the north. To the south of Gansu stretched the salty marshland of Tsaidam, the gateway to the Tibetan highlands. The province was the geographical center of China. The inhabitants of Gansu were the Chinese, Tanguts (ethnic Tibetans), Daldis, and Dungans. As the travelers penetrated deeper into the country, they could see everywhere traces of the ravages committed by the Dungan insurgents, which made their Tangut and Mongol companions extremely fearful. As Przhevalskii noted, "The numerous villages were in ruins, human skulls littered the ground, and not a soul was to be seen. Our companions showed symptoms of the greatest cowardice, they refused to make a fire at night, blessed their matchlocks, and begged us to go in front ...".[65]

On the Gansu plateau Przhevalskii found an entirely new world, with a typical mountain climate and nature. As a naturalist, he rejoiced seeing a great variety of plant and animal life, and was particularly impressed with the high mountains he had never seen before, rising to the snow line. The travelers

63 Ibid., vol. II, 37.

64 *Kumirnia* is literally 'an idol temple' in Russian, but here it apparently refers to a small Tangut Buddhist monastery.

65 Prejevalsky, op. cit., vol. II, 61. The English translation reads 'lighted their rifles', obviously a translator's mistake, as the Russian text has *sviatili svoi ruzhia* (lit. blessed their rifles').

would spend the summer months and early fall of 1872 in Gansu, exploring the North-Tatung and South-Tatung ranges, as they were named by Przhevalskii, and known as the Nanshan to the Chinese. These two ranges were separated by the Tatung-gol River, a tributary of the Yellow River.

During his explorations, Przhevalskii ascended the Sodi-Suruksum, one of the peaks in the South-Tatung range, and ascertained its height by measuring the temperature of boiling water, which turned out to be 13,600 ft above the sea. From its summit opened a most spectacular panorama:

> I have never been so high before; at my feet were great mountains covered with wild crags and clothed with forests, through which wound rivers like silver threads. For a long time I could not tear myself away from the spot, but remained as one entranced, and shall remember that day as one of the happiest of my life.[66]

Accompanied by two Mongol guides, Przhevalskii set out on 23 September from Choibsen in the direction of Lake Kokonor. This was the most dangerous part of the trip as his caravan had to traverse a territory controlled by the Dungans. In his travelogue Przhevalskii recounted how they once noticed on the road at some distance a large group of mounted Dungans escorting a flock of sheep. The clash seemed imminent as the Dungans began to fire at them. "The danger was great, but we had full confidence in the excellence of our arms and the well-known cowardice of the Dungans". Undaunted, Przhevalskii determined to force passage and marched on boldly, whereas his guides were paralyzed with fear and implored him to turn back. On observing their forward movement, the Dungans fired a few more shots and then fled to either side of the high road.[67]

Kokonor, which Przhevalskii reached in mid-October, was the high point of his entire journey:

> The dream of my life was accomplished and the object of the Expedition gained! It is true that this success had been purchased at the cost of many hardships and sufferings, but all past trials were forgotten, as we stood in triumph, and gazed with admiration on the beautiful dark blue waves.[68]

66 Ibid., vol. II, 94.

67 Ibid., 104.

68 Ibid., 107–108.

Lake Kokonor, called Tsok-gumbum by Tanguts and Qinghai by the Chinese, was the largest lake in China. Known also for its saline and alkaline water, it lay west of the town of Xining (Sinin) at an elevation of 10,500 ft (3,200 m) above the sea. Here Przhevalskii would spend about two weeks, investigating the lake itself and its environs, and then would strike south, towards Tsaidam and Tibet, following the major Buddhist pilgrims' route.

Staying at Kokonor, Przhevalskii had a chance to meet with a Tibetan envoy, Kambi-nantu. The diplomat had been dispatched by the Dalai Lama to Urga back in 1862, with gifts to the Mongolian Bogdokhan, but Gansu's fall to the Dungans interrupted his journey – for a whole decade. An obliging and amiable man, he offered Przhevalskii his services in Lhasa and told him that the Dalai Lama would be very glad to receive Russians at his palace. This encounter was a strong lure for the traveler, who was keen to be the first Russian to reach the Forbidden City and establish contact with Tibet's mysterious ruler. Therefore Przhevalskii resolved "to advance as far as possible", well aware how important to science is "every additional footstep in these unknown regions of Asia". For this purpose in Kokonor he hired a guide, who was formerly a lama at the famous Kumbum monastery, some 30 miles south of Xining.

Curiously, while at Kokonor, popular report turned the travelers into nearly epic heroes, with Przhevalskii himself being elevated to the rank of 'demi-god' – an invincible Russian *bogatyr*, and even more, a *khutuktu* (great saint) and a prophet. He was also believed to be a doctor and a healer, owing to his plant-gathering and the successful cures he had performed on some fever patients with doses of quinine in the earlier stages of the expedition. Przhevalskii attributed much of his unexpected fame to a variety of incidental factors, such as his party's safe passage through Gansu, their new quick-firing arms which killed animals at unheard-of distances, their mode of preparing skins, and lastly, the secrecy which shrouded the object of their journey.

These circumstances undoubtedly helped ease the way for the travelers. Still, as Przhevalskii would confess,

> it was necessary to keep up my character by dispensing benedictions, prophesying, and every kind of absurdity. Tangutans and Mongols would sometimes come in crowds to pray, not only to us, but to our guns, and the native princes often brought their children, entreating us to lay our hands upon them, and thus confer a lifelong blessing. As we approached Dulan-khit, a crowd of 200 men assembled to worship us, kneeling down on each side of the road.

> I was often consulted in my capacity of prophet, not only as to the future, but in cases of straying cattle or lost pipes, and the like...[69]

From Kokonor the party proceeded southward to the plains of Tsaidam, lying 1,700 ft below and covered with the salty marshes and lakes. This new country was not very attractive to Przhevalskii due to its poor vegetation and the absence of any large animals. There were wolves, foxes and hares, yet the most remarkable he encountered here was the *kulan*, or wild ass. Przhevalskii also found a new species of pheasants, inhabiting the thickets of reeds that he named *Phasianus vlangalii*, in honour of A. Vlangali. A person who was of great help to him throughout the journey, Vlangali would lend him money for the second expedition, when the sums granted by the War Ministry did not reach Peking in time.

The population of Tsaidam consisted of the same Tanguts and Mongols who inhabited Gansu and Kokonor, although Przhevalskii also discovered another branch of Tanguts known as Khara-Tanguts (i.e. black Tanguts). These dwelt in the basin of Kokonor, East Tsaidam and the source of Yellow River and were distinguished by a greater stature, darker complexion, and especially their "predatory instincts". The Tanguts regarded these countries as their own territory, to which they applied the name of Amdo. In general, this people, in Przhevalskii's opinion, presented a marked contrast to the Chinese and Mongols, and their appearance somewhat resembled gypsies.

The southern boundary of Tsaidam was sharply marked by the formidable Burhan-Budda ('God Buddha') range rising boldly "like a giant watchman, who keeps guard over the adjacent cold and sterile highlands of Northern Tibet". The travelers ascended the Tibetan plateau and spent two and half months there, from 4 December 1872 to 21 February 1873. This leg of the journey proved particularly rewarding, notwithstanding the thin atmosphere of the highland and the rigors of the climate, especially the strong wind that seriously impeded their movement.

As pioneer explorers of Northern Tibet, they were amazed at the absence of regular roads anywhere on the plateau; there was nothing but the tracks of wild animals running in all directions. The caravans normally took a straight course, guiding their march by the salient features of the country, but these visitors were not frequent after the outbreak of the Dungan insurrection. The first surprising discovery Przhevalskii and Pyl'tsov made in this no-man's land was its plentiful animal life – immense herds of wild animals roamed from place to place, finding sufficient nourishment, and unaffected by the rarified

69 Ibid., 163.

atmosphere. The most characteristic and numerous mammals of the Tibetan deserts were the wild yak, the white-breasted argali, the kuku-yaman (the latter two of the goat-antelope subfamily), the orongo and ata antelopes, and the kulan (wild ass). Besides these, there were also less rare species, such as the bear, the grey wolf, the steppe fox, the marmot, and the lagomys (rodent). Among the feathered vertebrate, Przhevalskii noticed vultures, lammergeyers, crows, sand-grouses, larks, and linnets.

The travelers' home was a simple yurt (11 ft in diameter and 9 ft high) they brought from Kokonor, with sides and roof covered with three layers of felt and orongo skins. Its interior did not provide much in the way of comfort, accommodating several felts for sleeping upon, two boxes with travel journals and instruments, the rifles ranged round the sides, and an iron-grate in the center where *argals*, whose dry dung fuel continuously burned during the day. Much of the time Przhevalskii and Pyl'tsov spent in "shooting excursions" and conducting various instrumental observations. Their new travel companion and guide was a Mongol, Chutun-dzamba by name, who had been many times to Lhasa with caravans before. Przhevalskii hired him in Tsaidam and he became his chief informant on Tibet.

Unable to trek to Lhasa due to the scarcity of funds, Przhevalski tried to make the best use of the opportunities as they presented themselves – always the guiding principle of his journeys. He moved on, passed the Bayan-hara-ula range, and on 10 January 1873 reached one of the headstreams of the Yangtze-kiang River, known to the Mongols as Mur-usu. His party was actually following the route of the Lazarist missionary Evarist Huc, who as early as 1846 succeeded in getting to Lhasa, a great temptation for Przhevalskii.

The banks of the Mur-usu marked the geographical limit of the expedition, although, as Przhevalskii admitted with a heavy heart, "we were only 27 days' journey from Lhasa". The caravan animals were so exhausted that three out of eleven camels had died, and the rest could scarcely move. Yet the biggest obstacle was the shortage of money. Therefore Przhevalskii decided to return to Kokonor and Gansu, spend the springtime there, and then move homeward, via Alashan and the Central Gobi.

Przhevalskii summed up his short, yet exhausting wanderings in the Tibetan borderland:

> These two and a half month in Northern Tibet were the most arduous of the whole of the expedition. Winter had set in with severe frosts and storms, and the want of even the bare necessaries of life, with other privations, reduced our strength; so that it became a hard struggle for life, and nothing but a consciousness of the scientific importance of our labours

inspired us with strength and energy to carry out the task we had undertaken.[70]

The new round of explorations at Kokonor and Gansu in February – June 1873 proved very productive in many ways, especially the "shooting excursions" on foot, which replenished Przhevalskii's zoological collections with many rare species of birds and mammals. Having left the mountainous Tangut country, the party proceeded to the Alashan desert. While crossing South Alashan, they encountered a caravan of Mongol pilgrims heading for Lhasa in search of a new incarnation of the recently deceased Khutukhtu (Bogdo-Gegen), the first pilgrimage in 11 years. Having met Przhevalskii and his men, the Mongols could hardly believe that they had penetrated Tibet. The travelers made a rather pitiful sight:

> Exhausted with fatigue, half-starved, unkempt, with ragged clothes and boots worn into holes, we were regular tatterdemalions! So completely had we lost the European aspect that when we arrived at Dyn-yuan-in the natives remarked that we were the very image of their own people![71]

But their hardest ordeal lay ahead. After a short respite in Dyn-yuan-in, where Przhevalskii received a thousand lans in money from General Vlangali, along with letters and newspapers from Russia, he started on a return journey to Urga by way of the Central Gobi. This route had never before been traveled by any European.

The final passage across the Gobi, from Dyn-yuan-in to Urga, was accomplished in 44 marches in the burning heat of the desert. This turned out to be the hardest leg of the journey as the travelers, having run out of water, were dying of thirst. "Our situation at the moment was desperate, Przhevaslii wrote in his diary. Only a few glasses of water were left of which we took into our mouths just enough to moisten our parched tongues; our bodies seemed on fire, our heads swam and we were close upon fainting".[72] The problem of water was indeed a matter of life or death. As Przhevalskii recounted, his expedition carried water in wooded casks, yet it evaporated from the heat, so that a cask filled in the morning generally lost the equivalent of several bottles before

70 Ibid., 213.

71 Przhevalskii, 258.

72 Ibid., 269, 270.

evening.[73] On the way, Przhevalskii's 'faithful friend', his setter Faust, died and was lamented bitterly by everyone, and the travelers themselves "nearly perished in the sand".[74]

Yet Przhevalskii's fortune held. On September 5 the party finally reached Urga, where it was welcomed by consul Shishmarev, and two weeks later they were already in Kiakhta. In his short farewell address to his companions upon the completion of the expedition, Przhevalskii would say:

> Fortune had been ever constant, and ensured the success of our undertaking; many a time when it hung on a thread a happy destiny rescued us, and gave us the means of accomplishing ... the exploration of the least known and most inaccessible countries of Inner Asia.[75]

The results of Przhevlskii's first journey were remarkable and absolutely unparalleled, given the vastness of the territories he had explored with his tiny team and his meagre funds. He traversed about 11,600 versts (7656 miles), of which he surveyed and put on the map 5,300 versts (3498 miles), astronomically determined the location of 18 points and made 146 longitudinal measurements. He also discovered and described quite a number of geographical objects, such as six separate mountain ranges in Ordos, Lake Kokonor, the Tatung-gol River with the adjacent North and South Tatung ranges, the Tsaidam plain, several ranges on the Tibetan plateau, the upper reaches of the Yangtze, the Khurkhu mountains in the South Gobi, and many more. His journey filled many blank spots on the map of Inner Asia, as he added 23 mountain ranges, 7 large lakes and 17 smaller ones, 21 rivers, a multitude of springs, wells, and minor settlements. To this, one should also add his valuable meteorological observations, his ethnographic materials and notes, and his vast zoological and botanical collections.[76]

• • •

73 Ibid., 256, fn. 1.

74 See Prejevalsky 1876, 269, 270.

75 Prejevalsky, Op. cit., vol. II, 284.

76 The zoological collections included 42 species of mammals (130 skins) and 238 of birds (about 1000 samples), 84 samples of reptiles and amphibia, about 100 fish (17 species) and more than 3000 insects (these figures slightly vary in Przhevalskii's biography by Dubrovin and the works of zoologists who described these collections); the botanical one – 4200 samples of plants (500 species).

Przhevalskii's pioneering explorations in Inner Asia were an outstanding achievement that instantly made him world famous and a popular hero in his native country. Back in St. Petersburg, he was enthusiastically hailed by the public and invited to endless parties in his honor. On 4 February 1874 he reported on the results of his journey at the general meeting of the RGS members. This was followed four days later by a lecture at the Society attended by its President, Grand Duke Konstantin and War Minister Miliutin. As the latter recorded in his diary,

> Przhevalsky, more than anyone else, can attract listeners and even more, lady listeners; you can see in his whole figure and in his every word an energetic nature. After the lecture I invited him home for a cup of tea and to introduce him to my family.[77]

The following day Przhevalskii showed his zoological collection to the Austrian Emperor, Franz Joseph I, known to be a sportsman, and he was so much impressed that awarded the traveler the prestigious Knight Grand Cross of the Imperial Austrian Order of Leopold. Then, on 21 March the Russian Emperor, Alexander II, visited the IRGO quarters to examine the same collection as well as Przhevalskii's topographic surveys. He was no less impressed and promoted Przhevalskii on the spot to the rank of lieutenant-colonel and granted him a life pension of 600 rubles a year. On the same day Pyl'tsov was promoted to lieutenant and granted a pension of 200 rubles. The emperor also suggested that Przhevalskii's collection, due to its great scientific value, be placed at the disposal of the museum of the Imperial Academy of Sciences, so the explorer sold all the zoological samples he had obtained in Inner Asia and partly in Ussuri to the Academy at a price of 10,000 rubles.

Przhrvalskii's great contribution to geography was not unnoticed by the world scholarly community. The Geographical Society in Berlin elected him a corresponding member; the International Geographical Congress in Paris presented him their diploma, while the French Ministry of Education awarded the explorer the gold order of the *Palme d'Académie*. Finally, on 8 January 1875, one more award came from the IRGO – the Konstantin Gold Medal, its highest honor, given for his outstanding discoveries made in Inner Asia. At the same time Pyl'tsov received the Small Gold Medal, and the Cossacks, Irinchinov and Chebaev, the bronze ones.

Being assigned in the same year to the Main Staff for the purpose of carrying out 'written work', Nikolai Mikhailovich retired to his Otradnoe estate in early

77 Quoted from Rayfield 1976, 84–85.

May 1874 to write up his journey, accompanied by Pyl'tsov. There he was eagerly awaited by his dear ones – his mother and the elderly Makarievna, as well as his new relatives – his mother's second husband, Ivan Tolpygo,[78] "a very nice man", in Przhevskii's words, and their children. Their daughter, the 19 year-old Aleksandra (Sasha), would soon marry Pyl'tsov, much to Przhevalskii's surprise and chagrin.[79] Having become strongly attached to Przhevalskii during their journey, Pyl'tsov had long cherished a desire to be related to his great mentor. However, his marriage and settling in Otradnoe would eventually put an end to his career as a scholar-traveler. As for Przhevalskii, he was 35 now, but he had no intention to settle down to married life. Oddly, he was averse to fair sex and avoided any romantic relations with women, being fully captivated by a much stronger passion for exploration. His two brothers, Vladimir and Evgenii, both residing in Moscow, on the contrary, were married men with children. Vladimir, a lawyer by profession, had been teaching history at various colleges in Moscow, since 1863, and also wrote short stories on historical themes. Evgenii, like his eldest brother, embarked on a military career and held the rank of captain, yet he also had a gift for higher mathematics; he published several manuals and had been employed, since 1866, as an instructor of mathematics at the Alexander military college in Moscow.[80]

Sponsored by the IRGO, Przhevalskii's account of his first Inner Asian journey, *Mongolia and the Tangut Country*, came out in 1875–76 in two volumes, though he originally planned a three-volume edition, The first volume included his own narrative, with a special chapter on the Tangutans and Dungans, while the second one was devoted to the discussion of various scientific subjects, including the climate and the feathered kingdom of Inner Asia in two separate chapters written by Przhevalskii, as well as the region's animal life – that of reptiles and amphibia described by the director of the Zoological Museum, Academician A.A. Straukh; and of fish, by K.F. Kessler. As for the mammals, their short description was given by Przhevalskii in the text of his

78 Ivan Demianovich Tolpygo (1817–1885) was a Smolensk civil servant (an officer of the Smolensk Chamber of the State Property; he married Elena Przhevalskaia in 1854 and they had three children – daughter Aleksandra (b. 1855), and two sons, Nikolai (b. 1856), and Ippolit (b. 1858).

79 Mikhail Pyl'tsov and Sasha Tolpygo first met in Otradnoe on the eve of the expedition and had been corresponding since then. His marriage to Przhevalskii's step-sister, however, was not a happy one. See Gavrilenkova 2011, 102–104.

80 For their short biographies see Gavrilenkova 2011, 81–83, 87–90; also Titov 2012, 161.

travelogue, and he hoped to produce a more substantial account after the completion of his next journey.

The traveler's narrative was subsequently (in 1876–1877) translated into English, French and German (see Bibliography), which evidenced a keen interest it evoked among European scholarship.

Lopnor Expedition (1876–1877)

While writing his travelogue in Otradnoe and St. Petersburg, Przhevalskii began to make plans for his next journey, to start in spring 1876, as soon as he finished the second volume. This was another grand project of his intended to explore several more unchartered lands in Inner Asia on a vast territory from East Tian Shan to the Himalayas.

Przhevalskii presented his program to the IRGO Council and it was approved after a discussion on 31 January of the same year. In this he briefly outlined his tasks which included a) exploration of East Tian Shan, between Kuldja and Hami, during the summer months of 1876; b) survey of Lake Lopnor and deserts to the east of the lake to solve the problem which had long intrigued the naturalists, that of existence of the wild camels and horses, the two endangered species, according to the information he had obtained earlier; c) observation of migrating birds at Lopnor (in the spring of 1877) and thence trekking southward across the Tibetan plateau toward Lhasa; d) from Lhasa Przhevalskii intended to proceed further south to the upper Brahmaputra (Tsangpo) and survey the northern slope of the Himalayas; and finally, to move to Eastern Tibet and South China, or, "if circumstances permit", to cross the plateau, from east to west and return to Russian Turkestan, by way of Khotan and Kashgar.

Przhevalskii also listed the specific types of field work he planned to conduct en route, such as: 1) geographical and ethnographic descriptions, 2) route surveying, 3) astronomical determination of latitudes and also possibly of longitudes, 4) meteorological, psychrometric and hypsometric observations, 5) study of mammals and birds, 6) collecting zoological, botanic and partly mineralogical specimens, 7) taking photographs.

The expedition was planned for a two-year period (1876–1878) at an estimated cost of up to 25, 000 rubles. This sum, apart from the allowances for the expedition staff (its leader, his two assistants, one interpreter, six hired Cossacks and several guides), included the cost of 14 pack and 9 saddle-camels, and an amount of 1,500 rubles for "presents, entertainment and unexpected

FIGURE 1.8
Przhevalskii's famous Lancaster Express carbine displayed at one of the RGO exhibitions in 2009.
PHOTO BY A. ANDREYEV.

expenses". The scientific equipment consisted of a set of various measuring instruments plus a photographic camera to be used for the first time by a scholar-traveler.[81] Indispensable also, though not mentioned in the proposal, was a quantity of fire-arms for each member of the expedition team (10 Berdan rifles and 20 Smith & Wesson revolvers, with ammunition), and its leader, Przhevalskii, personally could boast a very special one – a double-barreled Lancaster Express carbine (called 'stutzer'), presented to him by his former fellow staff officers.[82] No less important was another present, an absolute must for a

81 Dubrovin 1890, 561 (Przhevalskii's note to the RGO, dated 14 January 1876, 560–563). The program was expounded for the first time by Przhevalskii in his letter to V.A. Bel'tsov, dated 25 January 1875, see Dubrovin, op. cit., 204.

82 Dubrovin 1890, 194, 195. The Lancaster Express carbine was ordered in London through a Russian attaché General A.P. Gorlov.

hunter, which Przhevalskii received from Grand Duke Nikolai Mikhailovich Jr. – a setter.

The two assistants of his own choice were an 18-year old, Feodor L. Eklon, whose father was employed in the museum of the Russian Academy, and Evgraph Povalo-Shveikovskii, son of a neighboring squire. Both young men were in the military service – Eklon as volunteer, and Povalo-Shveikovskii as ensign. Originally Przhevalskii wanted very much to take on his team his former Ussuri companion, Nikolai Yagunov, who had finished the Warsaw cadet school by then, yet he died tragically, having drowned while swimming in the Visla in the summer of 1875.

Przhevalskii's project was received favourably by both the war minister D.A. Miliutin and the minister of foreign affairs N.K. Giers. The former spoke of his mission as "a rather useful undertaking", and the latter thought that the exploration of the territories adjacent to East Tian Shan and the Lopnor basin would be very important in a political respect. "As for Tibet, which is of importance for Russia as a center of the Lamaist teachings", wrote Giers in a letter to Semenov, "the Foreign Ministry cannot but regard with special sympathy M. Przhevalskii's resolution to penetrate [into the country] up to Lhasa. It is highly desirable that he spend as much time as possible on the study of the peculiar ways of life there".[83]

The expedition started in May 1876. Przhevalskii and his two assistants took a train from Moscow to Perm and thence they proceeded, on 13 postal horses, two *tarantases* (springless four-wheeled carriages) and two carts carrying their rather bulky luggage of 130 poods (about 2,000 kg), to Kuldja in northwestern Xinjiang. (Przhevalskii, however, refused to take along the camera with auxiliary appliances, offered by the head of the Military-Topographic Department Otto von Shtubendorf, as altogether these weighed some 17 poods (272 kg), too heavy to carry.) On the way to Kuldja, in Semipalatinsk, the party was joined by Irinchinov and Chebaev, and later five more hired Cossacks arrived.

Kuldja was the chief commercial center of the Ili River Valley, a vast territory occupied by Russian troops in 1871, in the wake of the Dungan uprising. From here the expedition route led straight into the region Przhevalskii wanted so much to explore, with Lake Lopnor and adjacent deserts, situated within the domains of the new Kashgarian ruler Yakub Beg Badaulet ('Lucky One'), known as Yettishaar (Seven Cities Kingdom).[84] While in Kuldja, making preparations for the journey, Przhvalskii obtained a Chinese passport which allowed him to travel to Tibet via Hami, an oasis town in eastern Xinjiang.

83 Ibid., 207, 208.

84 On him, see Introduction, 35.

On 24 August his caravan left Kuldja in the direction of the Tarim River and Lake Lopnor. Moving along the fertile Ili Valley, Przhevalskii traversed a most picturesque land buried in verdure of gardens alternated with pastures abounding in livestock, a land that he called 'Asiatic Lombardy'. The travelers then ascended the Little Yulduz high plateau, in the center of the Tian Shan, another scenic spot ("a paradise for cattle-breeding"), and it was here that Przhevalskii suffered his first setback. One of his assistants, Povalo-Shveikovskii, proved totally unfit for field work – he could not stuff birds, failed to make even the simplest topographic surveys, and besides was a poor hunter. So Przhevalskii had to send him back home, together with one of the interpreters, who proved a dishonest and lazy man.

Beyond the Tian Shan lay Yakub Beg's kingdom, Yettishaar. Having approached the first of its "Seven Cities", Korla, the party was not allowed in by the local authorities. The travelers had stayed outside Korla for two weeks before Przhevalskii, with the mediation of the governor-general of Russian Turkestan, Baron K.P. Kaufman, obtained Yakub Beg's permission to travel in his country. Moreover, trying to please the Russians, who had earlier acknowledged his rule[85] (concurrently with the British), Yakub Beg sent to Przhevalskii one of his trusted agents, Zaman Beg, with an escort. Zaman Beg was instructed to take Przhevalskii to Tarim and Lopnor, but in fact he and his men were attached to keep an eye on the Russians and obstruct their movements. Still, outwardly Zaman Beg was quite friendly and helpful to Przhevalskii in many ways. Formerly a subject of the Russian empire, he posed as a Russophile and probably was one indeed. (At the same time Yakub Beg was expecting another important Russian visitor – Captain of the General Staff A.N. Kuropatkin, dispatched from the Turkestan military district, in May 1876, at the head of a mission to demarcate the borders between Fergana Oblast' and Yettishaar. Kuropatkin was well-aware of Przhevalskii's journey and offered him assistance in resending his correspondence to Russia as well as in exerting influence on the Kashgarian ruler, if need be.[86])

On 4 September Przhevalskii set out from Korla to the Tarim River, the principle waterway of the Takla Makan desert, being carefully shepherded by Zaman Beg and his men. The lower reaches of Tarim, flanked along its right bank for miles by quicksands, made a rather cheerless picture. The flora and fauna were poor, yet it was known to be the habitat of the wild Bactrian camels Przhevalskii was looking for, the only endemic species in the area. He would later

85 The Russians sent several diplomatic and military missions to Yakub Beg (in 1868–1875) under Captain Reital (1868, 1875) and Baron A.V. Kaulbars (1872).

86 Dubrovin, 565–570, letter from Kuropatkin to Przhevalskii, of 15 January 1877, Korla.

lament in his travelogue that never in his previous journeys had he visited such a place absolutely unfit for shooting ("For a whole month the six of us did not shoot anything, and I have not even seen a single animal around".[87]) From Tarim Przhevalskii failed to proceed directly to Lopnor, as he had expected using the old maps. His guides told him there was no direct pathway to the lake (which was not true). So he stationed his caravan with all the luggage and the collected materials for the winter at the oases-village Chargalyk, not far from Lopnor, and set out southward, toward the Altyntag mountains, in the company of Eklon and three Cossacks, to ascertain the northern frontier of the Tibetan plateau. He also hoped to find the wild camels at Altyntag, their dwelling place, as he was told by the Lopnor inhabitants.

The excursion to the Altyntag lasted for 40 days (from 26 December 1876 to 5 February 1877). During this time the party traversed over 500 km in a most extreme condition of winter frosts, rarified atmosphere and sterile terrain. The result of this foray was most sensational – the discovery of an enormous range unknown until then to Western geographers.

> We succeeded in exploring these mountains, that is to say their northern slopes, over an extent of 300 versts (198 miles) east of Chargalyk. Throughout the whole of this distance the Altyn-Tag serves to buttress a lofty plateau overhanging the Lop-Nor desert and most probably forming the northern limit of the Tibetan highlands; at least that is what the inhabitants gave us to understand ...[88]

As Przhevalskii would subsequently ascertain, there existed a continuous "gigantic mountain ridge" stretching from the Upper Huang He to the Pamir.

> This wall encloses from the north the highest elevation of Central Asia and divides it into two distinct parts: the Mongolian desert, in the north, and the Tibetan highlands, in the south. Nowhere on the planet will one find on such a vast territory two adjacent worlds so strikingly different in their geological formation and topographic relief, their absolute altitude and climate, flora and fauna, and, finally, their origin and the historical destiny of the indwelling peoples.[89]

87 Przhevalskii 1947, 45.
88 Prejevalsky 1876, 80–81.
89 Quoted from Gavrilenkov 1989, 58.

Because of the wintertime the party was unable to cross over the Altyntag to get onto the Tibetan plateau. Later on, already at Lopnor, Przhevalskii found a road leading to Tibet, yet a journey there became impossible due to the lack of camels and guides, as well as the rough condition of the road itself.

At the foot of the Altyntag, Przhrvalskii met, incidentally, only one Bactrian camel, yet he failed to shoot the animal. This occurred, as he recorded in his diary, on 15 January 1877, a memorable day, that of the tenth anniversary of his "wayfaring life", which he never regretted.

Przhevalskii and his companions finally reached the mysterious Lopnor in the early February. Lopnor was an old appellation unknown to the present inhabitants of the area, who used other names – Kara-koshun or Kara-kurchin. The lake was located between two deserts – Takla Makan and Kumtag. In fact Przhevalskii found two successive lakes at the Tarim's end, Kara-buran-kul and Kara-kurchin (i.e. Lopnor). (As he would explain, Tarim, at its mouth, formed a shallow lake, Kara-buran-kul, from which it reemerged as a full-flowing river, yet soon diminished in size, being squeezed by the sands of the Kumtag, and it ultimately overflowed to produce a large reedy marsh, Lopnor. This was a freshwater lake of elliptical shape, 2–3 ft deep.) Przhevalskii was certain that he discovered the historical Lopnor as shown on old Chinese maps, and thus he became the first Western traveler to explore it. (The famous Marco Polo, who passed through the land of Lop in the 13th century, did not know about the existence of the lake.) He surveyed the entire hydrographic system of the Lower Tarim, including lakes Kara-buran-kul and Lopnor, which had never been done before by travelers at any river basin of Inner Asia. In particular, he put on the map the northern bank of Kara-buran-kul, the western bank of Lopnor, and besides, discovered the large river Cherchen-darya (Cherchen), falling into Kara-buran-kul.

At the same time Przhevalskii and Eklon conducted some important ethnographic research among the local inhabitants of the Tarim area, the Karakules, and the Karakurchins. These were two primitive tribes of the "Arian type", Mohammedans by creed, yet with clear signs of physical and mental degradation, in Przhevalskii's view, totally isolated from the rest of the world. At Lopnor Przhevalskii also collected some fascinating information about the Russian Staroveri (Old Believers), seekers of the promised land of Belovodie (White Waters, the Russian counterpart of Shambhala), who came to Lopnor in two large groups as early as 1861–1862.[90]

Having learned from the natives that the wild camels dwell in the Kumtag desert, Przhevalskii sent there some local hunters who shot and skinned three

90 Przhevalskii 1947, 57.

samples of the much coveted animal. "I have no words to say how happy I was to obtain finally the skins of the animal which was reported by Marco Polo long ago and which has not been seen by a single European since then".[91]

The party spent about two months (February and March 1877) on the shores of Lopnor and then moved back to Korla. They arrived there on 25 April, and five days later Przhevalskii was received in audience by Yakub Beg. The latter assured the traveler of his sympathy for the Russians in general and him personally, though "facts testified to the opposite".[92] In the memo Przhevalskii would later forward to the Main Staff titled 'On the current state of Eastern Turkestan', he spoke of the Kashgarian king rather negatively as a typical Eastern tyrant, "suspicious, cruel, and perfidious", whose rule in Yettishaar was one of "murderous terror". Hence, Przhevalskii suggested that Russia should take advantage of the situation by moving its border south from the Ili Valley and occupy both the Great and the Little Yulduz, being no man's lands, with their superb pastures "where many thousands of Kalmucks can be settled, who are more inclined towards us than toward the fanatical Moslems".[93] Yet a month later Yakub Beg was dead.

Despite the isolation during his long travels, Przhevlskii was well aware of current world affairs through his correspondence with St. Petersburg (IRGO and the Main Staff) and the newspapers he occasionally received. So when he learned that Russia had declared war on the Ottoman Empire (in April 1877) in support of the liberation movement of the Slavic peoples in the Balkans suffering badly under the Ottoman yoke, he immediately volunteered to join the acting Russian army. Yet his military superiors thought he should continue his journey, which would be of greater use to Russia.

The return to Kuldja in early July marked the end of the first stage of the expedition. There Przhevalskii began at once to make preparations for his next journey – to Tibet. Since Eastern Turkestan was seized by civil strife, he intended to reach the highlands via Dzungaria. In his travel diary he wrote enthusiastically on 3 July: "The first act of the expedition is over! A complete success! Lopnor has become an established fact of science! As in the earlier expeditions, such a success was attained at the cost of many misfortunes, physical and moral". Yet his forthcoming journey, with forbidden Lhasa as his destination, was fraught with even greater trials. "Ahead of us are multifarious

91 Ibid., 67.

92 Ibid., 90 (a brief description of Przhevalskii's audience with Yakub Beg).

93 Quoted from Dubrovin, op. cit., 576–577. The memo *O sovremennom sostoianii Vostochnogo Turkestana*, dated 6 June 1877 (Appendix 12, 570–577).

enemies – both nature and people; we are in for a hard struggle. May God grant [us] enough health and vigour".[94]

Yet this time fortune turned away from Przhevalskii. His new journey, which he started with a reformed team from Kuldja on 28 August, was interrupted at Guchen at the end of November, largely because of his illness, a severe itch. This began already in Kuldja and was diagnosed by a local doctor as *pruritis scrofi*, yet the medications he proscribed were of little help. Being tormented by the itch day and night and unable to walk or stay in the saddle, Przhevalskii realized he could not properly conduct his regular field work, particularly to do the surveys. So he had to interrupt the expedition and return to Russia (Zaisanskii Post), for a course of medical treatment. As soon as his condition improved, he set off bravely from Zaisan to Tibet in March 1878, but his ill fate was still after him. Soon enough he would receive a telegram from the head of the General Staff, Count L.F. Geiden, telling him that his further travelling in China was unwelcome "under the present circumstances"; then came the shocking news from Otradnoe about his mother's death. And finally, his illness recurred which made Przhevalskii to give up on his plans and postpone the Tibet expedition for a year.

On 31 March, his 39th birthday, Przhevalskii recorded in the diary that the day marked the end of his expedition:

> Now only one half of the work has been done: Lopnor is explored, yet Tibet remains untouched. For the fourth time I cannot reach it – the first time I returned from the Blue River, the second time from Lopnor, the third time from Guchen, and the fourth time the expedition was terminated at its very start. Still I don't give up. As soon as my condition improves, I will be on the road again next spring.[95]

Despite Przhevalskii's modest appraisal of his achievements, the Lopnor expedition made some outstanding contributions to the knowledge of Inner Asia, the foremost being the discovery of the Altyntag range and Lake Lopnor. Thus it turned out that the northern frontier of Tibet lay three degrees farther to the north in latitude than it was believed until then. At the same time the discovery of Lopnor gave rise to a lively scholarly polemic between Ferdinand von Richthofen and Przhevalskii. Richthofen contested Przhevalskii's claim for two main reasons – firstly, because the lake was positioned on his map some 150 km further north than on old Chinese maps, and, secondly, because Chinese

94 Przhevalskii 1947, 99.

95 Ibid., 127.

sources referred to Lopnor as a salt lake, whereas Przhevalskii described it as a freshwater basin. The discussion of the Lopnor problem would last for several decades before it was established in the early 20th century that Lopnor was a 'wandering lake' due to the Tarim River changing its course, so that its terminal lake altered its location. Two other remarkable travelers joined the polemic, Sven Hedin (on Richthofen's side) and Petr Kozlov (on Przhevalskii's side), with the result that it was determined that both parties were right in their own way. The clue to the solution of the Lopnor's enigma, according to Soviet geographer E.I. Murzaev, lay in the changes in the hydrographic systems of the Tarim river and Konche-darya, both feeding the lake.[96]

Przhevalskii's last and most valuable scientific trophy came as a surprise, in early 1878, when he stayed in the hospital at Zaisanskii Post. This was a skin and a scull of a wild horse (*tarpan*) shot by the Kirghiz hunters in the sands of southern Dzungaria and presented to him by the former head of the local police, A.K. Tikhonov. Przhevalskii was delighted and he immediately reported the lucky event to A.A. Strauckh at the Zoological Museum.[97] On the basis of these unique samples zoologist I.S. Poliakov would later give a description of the wild horse and the name of *Equus Przewalskii.*

The Lopnor expedition assured firmly Pezhevalskii's reputation as an outstanding explorer of Inner Asia and earned him more prestigious awards, in 1878–1879. The French Geographical Society sent him their gold medal (which had been conferred on him earlier); the German Geographical Society in Berlin awarded Przhevalskii the Humboldt gold medal, instituted in the same year to commemorate the society's 50th anniversary, and actually made him its first recipient. The Russian Academy of Sciences and the Imperial Botanical Gardens each elected Przhevalskii an honorary member (in November 1878). Finally, the Royal Geographical Society, in April 1879, awarded him its Founder's gold medal "for the great additions he has made to our knowledge of Central and Eastern High Asia, by his successive expeditions into the unexplored parts of the great plateau of Mongolia and the lofty deserts of Western Tibet, and for the admirable way in which he has described the regions in the published narratives of his journeys".[98]

His faunistic collections from Lopnor and Western Dzungaria (those of mammals, birds, reptiles and fish) Przhevalskii donated to the Zoological Museum of the Academy of Sciences (in November 1878).

---

96 Murzaev 1947, 17. On the Lopnor problem / Richthofen – Przhevalskii polemics, see also Przhevalskii 1880, 1–6; Hedin 1886, 305–361; Kozlov 1898, 85–98; Horner, Chen, 145–166.

97 Letter to A.A. Strauckh, dated 4 February 1878, SPbF ARAN, F. 50, op. 2, D. 208, ll. 6, 6 ob., 7.

98 See "Sketch of N. Prejevalskii", in *Popular Science Monthly*, vol. 30 (January 1887).

Towards the Forbidden Land (1879–1880)

Przhevalskii returned to St. Petersburg on 23 May 1878 and soon after retreated hastily to Otradnoe. There he was able to fully recover during the summer months thanks to the curative effects of nature in his native Smolenshchina. Back to his writing-desk, he wrote up his travel account to be printed later that year under the title *Ot Kul'dzi za Tian-Shan' i na Lob-nor* (From Kuldja across the Tian Shan to Lopnor). The book was translated into German and English in 1878–1879.[99] One more important work he produced in Otradnoe was a paper, "Apropos of the remarks of Bar. Richthofen on the article 'From Kuldja across the Tian Shan to Lopnor'",[100] which started a lively polemic between the two geographers concerning the location of Lake Lopnor.

Fully fit now to resume his journey, Przhevalskii submitted to the Geographical Society, on 24 August, a fresh proposal for a Tibet expedition. In this he highlighted, from the very start, the political importance of exploration of Tibet for Russia. The "spiritual dominion" of the Dalai Lama, he argued, extended widely over the entire Inner, East and South Asia, and the 250 million Buddhist followers there constituted "a frightful force". Therefore Russia should try to win over to its side these Buddhists by taking advantage of a rather favourable situation – the isolated Tibet had no friends as yet and "the name of Russia" enjoyed much sympathy of Asian natives. Finally, he pointed out far-sightedly that the holy city of Mongolian Buddhists, Urga, was under "our strong influence", and that the Mongols in general were hostile to the Chinese, hence they could serve Russian interests.

Tibet was also important as an object of varied exploration. Przhevalskii particularly emphasized the fact that his previous journeys showed that the geographical maps of many parts of Asia, including Northern Tibet, based largely on Chinese sources were "absolutely incorrect". At the same time, continued Przhevalskii, we know very little of the "natural products" of Tibet, clearly a reference to the mineral resources of the country. "Finally, the study of the indigenous population, its way of life, relations with its neighbours, as well as the acquaintance with Buddhism at its very heart – all this is of so great interest that one may take risks, without much consideration, disregarding all hardships of the journey". Moreover, added Przhevalskii, "scientific

99 The English translation was prefaced by British traveler Thomas Douglas Forsyth, who had led two missions to Yakub Beg in 1870 and 1873–1874, the translator being the same E.D. Delmar-Morgan who translated *Mongolia and the Tangut Country*. The volume also included F. Richthofen's "Remarks on the results of Col. Prjevalsky's journey".

100 See ft 96.

investigations ... will camouflage the political goals of the expedition and will [allow us] to ward off all suspicions of our adversaries".[101]

At this preliminary stage Przhevalskii considered two alternative routes for his expedition, both leading to the Tibetan plateau: 1) from Kiakhta, via Urga and Kokonor, and) from Zaisanslii Post, via Hami, Shazhou and further to the upper Blue River. Each of these had certain advantages, yet the easiest and shortest to get to Tibet was the former, as he had tried it before. As for the routes from Russian Turkestan passing through Khotan and Lopnor, he declined both as "impassable", as there had been no caravan communication between Eastern Turkestan and Tibet proper for a decade and a half under Yakub Beg. Besides, it was impossible to obtain at Lopnor any fresh animals – pack camels or yaks, as only these were fit for traveling on the Tibetan plateau, and it was also too difficult to hire guides there. He personally preferred the Zaisanskii Post – Guchen – Hami – Shazhou route: in that case, provided the expedition started in February / March 1879, it would be able to reach Lhasa in October / November. If so, the winter of 1879/80 and the spring – summer of 1880 would be entirely devoted to the exploration of Tibet proper, after which the expedition would return home.

Przhevalskii's main assistants in his revived Tibet expedition were to be Feodor L. Eklon and Vsevolod I. Roborovskii, Eklon's gymnasium mate. Roborovskii fully met Przhevalskii's strict requirements – he was a junior officer (ensign), educated at the infantry cadet school in Helsingfors (today's Helsinki); could draw skilfully from life and do surveys, and showed keen interest in natural sciences. On a personal level, Roborovskii was quite sensible and easygoing.

The staff also included a preparator (A. Kolomiitsev), two interpreters (one speaking Turkic and the other Tibetan), two guides (from the natives), and a seven-man Cossack escort, making fifteen men altogether, including the leader, Przhevalskii. The cost of the expedition was estimated at about 28, 000 rubles for a two-year period, a sum Przhevalskii received from the Treasury. (The latter actually paid him only 20, 000 rubles, the remaining 8, 000 being left from his previous Lopnor journey).

Przhevalskii's project was readily approved by both the War Ministry and the Ministry for Foreign Affairs. In his report to the Emperor Alexander II, A.N. Kuropatkin (now head of the Asiatic Section of the Main Staff) underscored:

> The object of this research will be Tibet. Along with scientific aims, it is suggested that, as much as may be possible, the expedition will also

101 Dubrovin, Appendix 14, 580–581.

> gather intelligence about the political situation in Tibet ... Such an effort .... may open the way to our influence over all of Inner Asia, right through to the Himalayas".[102]

An expedition to Tibet, with Lhasa as its destination, as projected by Przhevalskii, coincided in time with the aggravation of Anglo-Russian relations following the Berlin Congress of 1878. By giving support to his most audacious project, St. Petersburg apparently wanted to challenge British interests in the strategically sensitive region of Asia.

What served as a mighty incentive for Przhevalskii and his military superiors was the concurrent exploration of the Tibetan Highlands started by the British from India, in the mid-1860s, with the help of specially trained spies, the so-called pundits ('wise men'), selected by the Great Trigonometrical Bureau at Dehra-Dun among the north Indian tribes. Disguised as Buddhist pilgrims or merchants, the pundits, for nearly two decades, were secretly penetrating into Tibet, the most outstanding achievements from the point of geography being attained by Nain Singh, Kishen Singh, Lalu and Kinthup. Incidentally, Nain Singh was able to visit Lhasa (in 1865 and 1874), determine its location and altitude, and collect much valuable information on Tibet and its 'forbidden' capital, still inaccessible to Western travelers.

The heads of the Imperial Russian Geographical Society and the Main Staff were well aware of these British initiatives. As early as 1869, they also considered the possibility of sending to Lhasa their secret agent, a person of Buryat extraction, yet these plans were frustrated by the Dungan uprising.[103] But now, a decade later, thanks to Przhevalskii, the IRGO resolved to send, quite openly, a whole mission to Lhasa, with both scientific and political goals. This was St. Petersburg's earliest attempt to establish contacts with the xenophobic leaders of Tibet.

•••

The journey, known in the IRGO annals as Przhevalskii's First Tibet Expedition, started on 21 March (3 April) 1879 from Zaisan and was terminated 17 months later, on 29 October 1880 in Kiakhta, that is, it was shorter than originally planned. On the way to Tibet the party passed through the very heartland of Inner Asia, having traversed a vast uncharted territory between 91° and 97 °

102 Quoted from Schimmelpenninck van der Oye 2001, 224 (fn 39).

103 See Introduction, 36. See also Andreyev, A. 2001, 163–180.

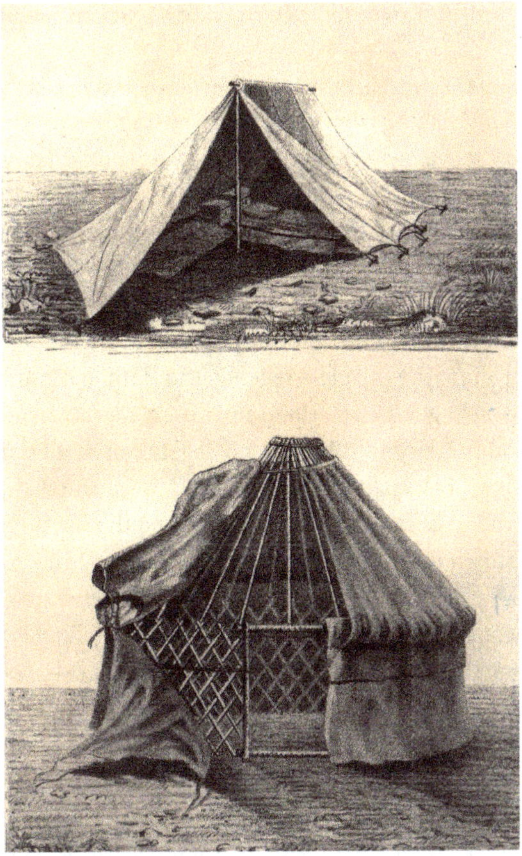

FIGURE 1.9
The expedition tents, drawing by V. Roborovskii, from Przhevalskii's account of his 3rd Inner Asian expedition: *Iz Zaisana cherez Khami v Tibet*, St. Petersburg, 1883.

FIGURE 1.10 Kitchen utensils, drawing by V. Roborovskii, from Przhevalskii's account of his 3rd Inner Asian expedition *Iz Zaisana cherez Khami v Tibet* (St. Petersburg, 1883).

east longitude, namely between the Urungu River in the north and the Tangla range in the south.

Trekking across Dzungaria, up the Urungu River, Przhevalskii and his assistants frequently went hunting and fishing, yet their catches were meagre. The animal kingdom in the area was rather scanty, still Przhevalskii made one important discovery – he caught sight of two herds of wild horses, known to the locals as *kertag*, roaming in the desert, the only place in the world where the animal could be found. Unfortunately, Przhevalskii failed to steal up close enough to shoot any as the *kertags* proved very timid. In the same desert he also observed some wild camels, whose homeland was Lopnor.

Having passed over the Tian Shan, the travelers reached the Hami oasis and thence moved to Shazhou. The passage through the Hami desert was extremely exhausting, as the area was absolutely sterile, with no vegetation at all and no animal life, not even lizards or insects to be seen around. Because of the scorching heat, it was impossible to do any route surveys during the daytime, so the caravan would march at night and early in the morning. In mid-June the travelers finally came to Shazhou, also known as Dunhuang, one of the oasis towns surrounding the Takla Makan and Lop deserts, on the ancient Silk Road. Some 12 miles outside Dunhuang they discovered a unique Buddhist site – 'The Caves of the Thousand Buddhas', which would later become a strong attraction for Western orientalists. Przhevalskii used the occasion to make a tour of the place and visit some caves that he would later describe in his travelogue.

Beyond Shazhou, while passing over the formidable Tian Shan Mountains, Przhevalskii discovered two ranges to which he gave, as their first discoverer, the names of the German geographers, Humboldt and Ritter. The Humboldt range stretched to the west of the upper Huang He and consisted of several chains which formed an Alpine country, while the Ritter range ran perpendicular to it.

Finally, in the beginning of September, the party reached the encampment of the Mongolian petty prince Dsun-dzasak in south Tsaidam, situated near the major Lhasa-bound route. Caravans from Mongolia and China usually stopped here for a short respite and to replenish their supplies before ascending the formidable Tibetan plateau. Przhevalskii left a part of his collections in the encampment, and in the middle of September his caravan of 34 camels and 5 riding horses set out in the direction of Tibet.

Having ascended the plateau, the party traversed a number of inland ranges, one after another, between the Burhan Buddha and the Tangla. The travelers were now in a wild, solitary, and mysterious land – the Land of Snows, with its sullen-looking siliceous mountains and Alpine plains. Yet the river valleys were

FIGURE 1.11 The animal kingdom at the Tibetan Plateau, drawing by V. Roborovskii, from Przhevalskii's *Iz Zaisana cherez Khami v Tibet*, St. Petersburg, 1883.

brimming with life, as one could see here numerous herds of exotic animals Przhevalskii had already seen on his first journey – ada and orongo antelopes, wild asses (kulans), and yaks, feeling no fear of man and gazing lazily upon the passing caravans. This picture of "fabulous abundance of herbivores" grazing peacefully on the fine pastures of the Shuga valley made him say: "It seemed that we found ourselves in the primeval paradise, where men and animals have not yet known of the evil and sin".[104] Roborovskii made a pencil drawing of the place to be later reproduced in Przhevalskii's travel account.

Still the unfavourable climatic and physical conditions of the Northern Tibetan plateau rendered the country uninhabited for hundreds of miles; even nomadic tribes, noticed Przhevalskii, found it difficult to dwell here due to the insufficient fodder for their cattle which "cannot roam here the way the antelopes, kulans and yaks do". That is why the Mongols spoke of the country as *guresu gadzir* – the 'savage land'.

Trekking across the plateau known to the Tibetans as Chang Tang, Przhevalskii, accompanied by Eklon and Roborovskii, gave himself up to his great passion of hunting. The only serious obstacle for it was the blinding snow; as soon as it fell down and covered the ground in a thick layer, Przhevalskii had to put on dark glasses, yet these were of little help as the light, being reflected by the

104 Przhevalskii 1948, 172.

snow, hurt his eyes from the sides all the same. One of his best trophies was a new species of bear, *Ursus lagomyiarius n. sp.*, feeding on small rodents like the Alpine pika. The local Tibetans regarded the animal as sacred and called it 'God's dog'.

The first humans the expedition met was a group of the *egrai*, a nomadic tribe of the Tangut extraction, roaming on the Tangla range and earning their living by robbing the passing caravans of pilgrims and traders, those commuting between Tsaidam and Lhasa. These brigands attacked Pzhevalskii's caravan at once but were repelled and hastily retreated, having lost some of their comrades in the skirmish. Having descended from the Tangla, the travelers reached the first nomads' camps at the banks of a little river, San-chu, with their peculiar black tents and large flocks of sheep and yaks browsing nearby. Moving further south, Przhevalskii met three Mongols returning from Lhasa – one of them was his old acquaintance from Tsaidam, and two others were lamas from Karchin in Inner Mongolia. From them he learned that the dwellers of Lhasa were agitated by rumours about the Russians coming to "steal the Dalai Lama" and "destroy the Buddhist creed" and that the Tibetan authorities sent out pickets, up to the Tangla, to prevent the foreign trespassers from coming to the holy city.

A few marches after, in November 1879, the expedition, while approaching the main frontier post of Nagchu (spelled Napchu by Przhevalskii), some 250 versts (about 160 miles) north of Lhasa, was halted by a Tibetan cordon. Nikolai Mikhailovich decided not to force his way to the 'forbidden city' (which he could easily have done with his well-armed party) and turned back, having put down in his diary: "For us to act against the fanaticism of the entire [Tibetan] nation was pointless and impossible – so we had to resign to the inevitable".[105] Still, he demanded that the Lhasa emissaries, who visited his camp at the Bumza Mountain, give him a special paper stating the reason for not admitting him to the Dalai Lama's capital, which they reluctantly did.[106]

Thus Przhevalskii's fourth attempt to reach Lhasa ended in a fiasco.[107] However, the two and half weeks he spent at Bumza, waiting for a reply from Lhasa, were not in vain. He succeeded in gathering whatever information he could about Tibet, its population, the Dalai Lama and Lhasa, his informants being

105 Przhevalskii 1948, 224.

106 Upon his return to St. Petersburg, Przhevalskii handed over this document to Semenov. It was reproduced in Przhevalskii 1948, 225; Gavrilenkov, op. cit., 80. See also NA RGO, f. 13, op. 1, d. 116 (a draft copy of the Russian translation).

107 According to Przhevalskii, his first abortive attempt was made in 1873, two more in 1877, and the final one in 1879, during his first Tibet expedition. See Przhevalskii 1883.

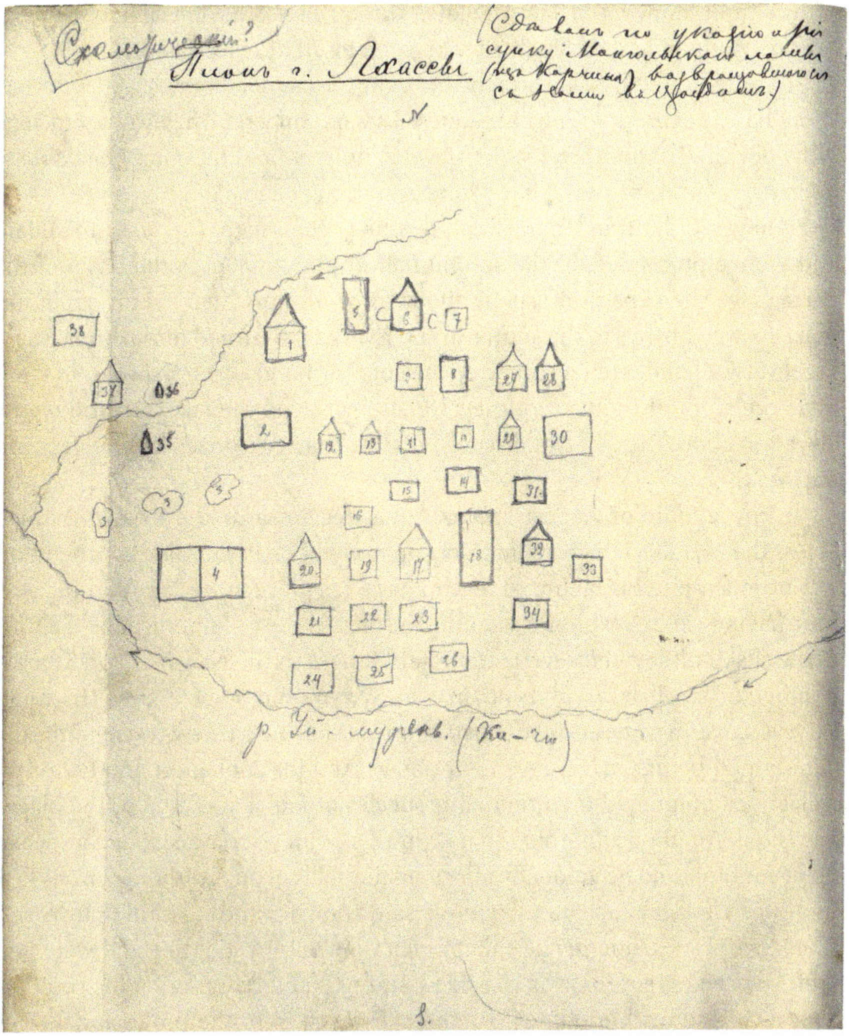

FIGURE 1.12 Plan of Lhasa, hand-drawn, from Przhevalskii's journal of his 3rd expedition to Inner Asia. NA RGO, F. 13, OP. 1, D. 68, L. 122 OB.

mainly the three Mongolian lamas. Even more, one of them, the Karchin lama, drew at his request a plan of Lhasa which he then copied into his diary. This was a rather crude sketch titled "The schematic plan of the city of Lhasa" followed by a detailed set of explanations which included 38 topographical sites.

(This simple drawing in fact predates the well-known Plan of Lhasa (1884) drawn by pundit Kishen Singh after his secret visit of Tibet's capital in 1878.[108])

In a letter to the Russian attaché in Peking, Koiander, Przhevalskii reported: "I can hardly express with what sadness I set out on a return journey. But such is my destiny. Let another, luckier traveler, finish what I have not been able to do in Asia. As for my part, I have done everything I could".[109]

Przhevalskii returned to the Dsun-dzasak encampment (*hyrma*) in Tsaidam and thence proceeded, in the second half of March 1880, to the basin of the Huang He. The expedition spent about three months there, yet it could not make its way further to the source of the river and returned to Kokonor. There Przhevalskii conducted a survey of the southern and eastern shores, to complete the work of the first journey. He also sounded the lake, ascertained its elevation above the sea, and discovered more than two dozen small rivers falling into it.

In the middle of August the party headed towards the Russian border, across the Middle Gobi, reaching Urga on 19 October. Przhevalskii's expedition was over. Despite his failure to reach Lhasa, his journey was another big success, having yielded plentiful scientific results. In the geographical field alone, Przhevalskii discovered several mountain ranges in the Nanshan and Tibet, a number of small lakes in Tsaidam and Syrtyn, and some rivers. His route surveys covering about 4000 versts were based on 23 new, astronomically determined points, an achievement per se. And his zoological and botanical collections numbered many endemic species of fauna and flora of the places he visited. An important thing to note at this point, Przhevalskii had a special liking for birds and he gradually mastered the skills of an ornithologist. His first tutor was Vladislav Tachanovskii in Warsaw, known as the father of Polish ornithology, and subsequently he spent much time in bird shooting and watching their seasonal migrations at lakesides, and was able to describe his ornithological collections himself, with some assistance from the professional ornithologist, F.D. Pleske, from the Zoological Museum.[110] (One of Przhevalskii's favourite sayings was *'byt' kak vol'naia ptitsa'* – lit. 'to be free like a bird'.)

Upon his return to Russia, Przhevalskii was warmly welcomed as almost a national hero by the general public and scholarly circles. On 10 January 1881 – three days after his arrival in St. Petersburg – he was granted an audience by the emperor who showed genuine interest in his journey and asked him many

108 See Andreyev, A. 2014, 41–55.

109 Quoted from Gavrilenkov, op. cit., 81.

110 See Przhevalskii's article on the new species of birds he discovered in Inner Asia, Przhevalskii 1887.

questions. In the same year Przhevalskii was elected an honorary citizen of St. Petersburg and his native Smolensk, and Moscow University bestowed on him the title of the honorary doctor of zoology. Several more geographical and other scholarly societies in Europe and China also elected him as their honorary member, in 1880–1886.[111] He was now at the pinnacle of his fame in his most unusual career as a scholar-traveler.

His military career was also progressing. Holding the rank of Colonel, Przhevalskii was appointed, in 1882, a 'supernumerary member' of the Military-Scientific Committee (VUK) at the Main Staff, and he would remain in that capacity until his death. His duties as a VUK member, according to M.K. Baskhanov, consisted mainly of counselling his superiors on various issues, including those related to the military-geographical and military-statistical study of China and Mongolia and mounting expeditions to these countries.

On 15 March 1881 an exhibition of Przhevalskii's zoological collections[112] was inaugurated in the Main conference hall of the Academy of Sciences. This attracted a large number of visitors, including the imperial couple. The Empress Mariia Fedorovna was so impressed that she begged Przhevalskii to tutor her son, the teenage tsarevich Nikolai in geography of Asia in a series of informal talks about his travels there, which he willingly agreed to do. The occasion allowed the traveler to befriend his pupil, the future tsar of Russia Nicholas II, and certainly strengthened his monarchic feelings.

Back in Otradnoe after the course of tutoring, Przhevalskii was not happy with his life there anymore. His dear mother and uncle Karetnikov were away from the estate, and the place itself was not a solitary refuge as it used to be, being utterly spoiled by 'civilization', with its *kabaki*-taverns in the villages around, the cut down forests, and the railway built in the vicinity, breaking the silence with train whistles. Also annoying were his many neighbours, trying to arrange a match for him with their daughters. For this reason Przhevalskii resolved to find himself a better home in some out-of-the-way spot – a '*medvezhii ugol'*' (lit. 'bears' place'), and he found one with the help of his father-in-law, Ivan Tolpygo. This was an estate called Sloboda situated in the Porechskii uezd of the same Smolensk province which he purchased from a retired artillery officer

111 These included the Dutch geographical society (1880), the North-Chinese department of the Royal Asiatic society (1880), the Swedish anthropological and geographical society (1881), the geographical societies in Viena (1884), Leipzig and Dresden (1886), as well as the German Academy of Naturalists Leopoldina (Deutsche Akademie der Naturforscher Leopoldina) in Halle (1886).

112 A collection of vertebrates obtained in the Tibet expedition Przhevalskii donated to the Zoological Museum, and the herbarium to the Botanical Gardens.

FIGURE 1.13 Przhevalskii's zoological collections exhibited at the conference hall of the Academy of Sciences, 1881. *Vsemirnaia illustratsiia*, 1881, T. XXV (24), no. 648, 6 June 1881.

Leonid Glinka. Sloboda stood in a most picturesque area, amidst virgin forests and many lovely lakes, one of which was the beautiful Lake Sapsho, reminding him of Lake Baikal 'in miniature'. To this new refuge of his Przhevalskii moved in July 1881; he regarded it as his "nest" from which he would "fly into the heart of Asian deserts".[113] In Sloboda, he would build a new manor house, a few years later, right on the shore of the lake. But to begin with, he converted an old cabin – known as *khatka* (little hut) – in the garden into his study, which was destined to become his most 'intimate nook'.

Here he completed, by the end of 1882, the account of his Tibet journey titled '*Iz Zaisana cherez Khami v Tibet i na istoki Zheltoi reki*' (From Zaisan across Hami to Tibet and the source of the Yellow River). The book was lavishly illustrated by Roborovskii's excellent drawings. It is worth mentioning here that Przhevalskii's erlier published accounts of his first three journeys (to Ussuri and Inner Asia) had no illustrations at all.

113 Dubrovin 1890, 365.

FIGURE 1.14
The Sloboda estate. COURTESY OF THE HOUSE-MUSEUM (DOM-MUZEI) OF N. PRZHEVALSKII, SMOLENSK DISTRICT.

FIGURE 1.15 Przhevalskii's *khatka*, his village asylum. COURTESY OF THE HOUSE-MUSEUM (DOM-MUZEI) OF N. PRZHEVALSKII, SMOLENSK DISTRICT.

FIGURE 1.16 Lake Sapsho, 2010s. PHOTO BY TATIANA GNATYUK (2013).

Final Explorations and Death at Karakol

While his book was being printed in St. Petersburg, Przhevalskii approached the Geographical society, on 9 February 1883, with a proposal for his next two-year journey to Inner Asia to do a more thorough exploration of Tibet (primarily) and Eastern Turkestan. In this he again made a special point to say that Tibet, with its vast territory of over 20,000 sq. miles, still remained an "almost entirely unknown land", despite the success of his three previous journeys and the explorations of other travelers. He further named the Europeans who had earlier visited 'some places' on the plateau – Oderico di Pordenone, the missionaries Grüber and Dorville, Samuel van de Putte, Huc and Gabet, "who nearly followed the path of Mongolian pilgrims, yet none of them left any records of their journeys"; and finally, the Indian pundits secretly dispatched to Tibet, since the mid-1860s, by the 'Ost-Indian geodetic bureau'.[114] Przhevalskii was apparently familiar with the travel accounts of these scouts published in the 'Proceedings of the Royal Geographical Society' in London, and the pundits' tales served as a strong irritant. If pundits, disguised as Buddhist pilgrims,

114 Dubrovin 1890, 588–591 (Appendix 16).

managed to get to Lhasa from India, why not try to do the same from the opposite direction? Przhevalskii was ready to meet the challenge of these obscure agents of the Raj, whose real names were not even given in their published records.

Lhasa again was on Przhevalskii's agenda, although, after the setback he had suffered in his previous journey, he was not so certain that he would reach it ("If they [the Tibetans] permit, we'll go there; if not, we'll bypass it"). Yet the place he certainly wanted to visit this time was Lake Tengri-nor (Namtso), the largest in Tibet, situated north-west of Lhasa. From Tengri-nor, "if circumstances permit", Przhevalskii intended to proceed to South Tibet (the province of Tsang) and further up the Brahmaputra River (Tsangpo), following the route of pundit Nain Singh.

The second stage of the expedition was to be devoted to an additional exploration of Eastern Turkestan. Przhevalskii wanted to get there from a base in Tsaidam in two separate parties – one bound directly for Lopnor and another for the large oasis town of Keriya and thence via Cherchen to Lopnor again. From Lopnor the expedition was to proceed to the Tibetan plateau, towards Karakoram, and, after completing a new series of explorations there, it was to trek down the Khotan River to Lake Issyk-kul, back to Russia.

This new expedition was conceived by Przhevalskii on 'a large scale', judging by the length of his routes, his voluminous kit (which included this time a photographic camera and 200 kg of Chinese silver bullion, so-called *yambs,* as their money resource), and the staff numbering 21 men. The funds for the journey were readily allocated by the government and amounted to 43 thousand and a half rubles. And indeed the mission, known as his Second Tibet Expedition, was another ambitious enterprise and yet a grueling trial for Przhevalskii and his companions. His main assistants on the journey were Vsevolod Roborovskii and Petr Kozlov, a 20-year old volunteer of the Sophia infantry regiment.[115] Feodor Eklon, who originally had been taken on the staff, however, changed his mind; like Pyl'tsov, Eklon wanted to get married, which strongly upset Przhevalskii. (Commenting on the incident, M.A. Engelgardt, one of Przhevalskii's early biographers, noted ironically that "by that time his (Przevalskii's) love for traveling had turned into a fanatical passion; he could not even imagine that

115 There is a romantic story of how Przhevalskii met Kozlov (1863–1935), then working in the winery not far from his new Sloboda estate, by mere chance in his garden. The traveller got into a friendly chat with the trespasser, in the course of which he sensed in the youth a 'kindred soul'. Przhevalskii then invited him to take part in his 4th expedition and personally trained him as a field surveyor and collector. For more detail, see P. Kozlov's biographical essay in this volume.

someone, who had known the sweet living amidst deserts, could aspire for something better".[116]) The rest of the staff included an amateur-entomologist, Mikhail Protopopov, from Troitsosavsk, his two old companions and personal friends, Irinchinov and Teleshov (the latter had participated in his first two journeys), an interpreter, Abdul Yusupov, and the usual armed escort of Cossacks and soldiers.

The expedition set out from Kiakhta – after a parting church service – in the direction of Urga, on 21 October 1883. Its caravan consisted of over 50 camels (40 of which were harnessed) and 7 horses, marching in six echelons, followed by a herd of 30 sheep intended as 'provisions in the flesh'. At the head of the party were Przhevalskii, Teleshov and a Mongol guide, whose service was indispensable. From Urga the party traversed the Great Gobi, already for the third time, heading for the Alpine country of Gansu. This was a familiar route which allowed Przhevalskii to halt again near the Chertynton and Choibsen monasteries, where his expedition had earlier been received in a rather friendly fashion by their abbots and monks.

From Choibsen the party moved to Kokonor and thence to Tsaidam, where it arrived in May 1884. There Przhevalskii unexpectedly found himself in a desperate situation as his old acquaintance, the chieftain of Dzun-dzasak, refused point blank to sell him fresh camels and sheep and to provide a guide to lead him to Tibet. This threatened to wreck the expedition, at the outset, so Przhevalskii had to resort to extreme measures – he arrested Dzun-dzasak, put him in a camp tent, with a sentry posted outside, while one of Dzun-dzasak's close assistants was chained, and another one flogged. These violent acts proved effective – the camels, sheep and a guide were soon placed at Przhevalskii's disposal. To proceed on his journey, Przhevalskii set up a depot at the *hyrma* (wattle and daub fortress) of Barun-dzasak (Dzun-dzasak's neighbour), where he left some of his luggage, the collected materials, and camels, under a surveillance of six Cossacks, so as to be able to travel light on his forthcoming Tibet journey.

Having passed over the formidable Burhan Budda range, the travelers headed for the Odon-tala boggy salt marshes ('Starry Steppeland'), known to the Chinese as Sin-su-hai ('Starry Sea'), where the springhead of the Huang He lay. "Our long-cherished desire was crowned with success – we have seen, with our own eyes, the mysterious cradle of the great Chinese river and drank water from its source", rejoiced Przhevalskii.[117] More than that, he determined the precise geographical coordinates of the place and put it on the map. This was

116 Engelgardt, op. cit., 384.

117 Przhevalskii 1888, 158.

the first big success of his fourth expedition. Also, he discovered two lakes in the area (known to Mongols as Djarin-nor and Orin-nor), which he named Lake Russian and Expedition Lake in memory of Russian pioneer explorations of Tibet. Przhevalskii's major trophy for his zoological collection consisted of skins of a new species of the Tibetan bear (*Ursus lagomyiarius n. sp.*).

The party then crossed the watershed of the Huang He and Yangtze, having entered the basin of the upper Blue River (Tib. *Dji-chu*), another unknown land to be explored. The explorations, however, were not fully safe, as the travelers were unexpectedly attacked twice by the groups of roaming Tanguts and Goloks. The second attack occurred when the expedition bivouacked on the shore of the newly-discovered Lake Russian and was particularly dangerous as the 14 expedition members had to fight off no fewer than three hundred mounted Tanguts, yet they managed to repel their attack. Fortunately, none of the Russian travelers were killed or wounded in the course of this skirmish. The expedition faced more trouble during this Tibet 'excursion' when Roborovskii nearly drowned while crossing a mountain river.

By the end of July, 1884, the party was back to their temporary depot at Barun-dzasak in eastern Tsaidam. It was then that Przhevalskii gave up on his plans of going to Lhasa, feeling that he had little chance of being admitted to the holy city by the Tibetans after he had twice 'beaten' the Tanguts. Instead, he decided to explore more thoroughly Tsaidam (the uncharted southern and western parts of the plain) and the Kunlun Mountains, which he had already crossed twice before.

On 26 August the second stage of his journey began – the expedition trekked westward across the width of Tsaidam to Lake Gas, where Przhevalskii set up a new depot, and thence proceeded south of Tsaidam and north of a series of unknown ranges, along the southern edge of Altyntag. This was his second excursion to an absolutely uninhabited area, with no wild animals or birds to be seen. During 54 days the travelers covered about 800 km, having again ascended the Tibetan plateau through the 'Valley of Winds'. The result of the trip was a discovery of some new ranges, to which Przhevalskii gave names of his own – Tsaidam, Columbus, Moscow (the peak of which he called Mount Kremlin) and Mysterious (later renamed Przhevalski range by the IRGO), crowned by the highest peak in the entire Kunlun (7723 km) he called the Hat of Monomakh.[118] Putting these ranges on the map was of paramount geographical importance. "The famous Kunlun", Przhevalskii wrote, "that backbone of Asia, as Baron Richthofen calls it, before our last journey was entirely unknown through 12°

118 The Hat of Monomakh is the legendary cone-shaped and lavishly decorated headgear of the Russian tsars and grand dukes.

of longitude, from the meridian of Naijin-gol in Tsaidam to that of Keriya in Eastern Turkestan. We have now passed along this unknown belt of the most ancient ranges of Asia, and have in some measure elucidated its topography".[119]

On 11 (24) January 1885 Przhevalskii returned to his depot at Gas. The third and final stage of his expedition now was exploration of Lopnor and visiting some of the major oases in Eastern Turkestana. He reached there by passing over the Altyntag, following the track discovered earlier by a Cossack patrol under Irinchinov. His party camped at the village Abdal close to the lake, a place which he had already visited in 1877. The local dwellers remembered him from those days and received rather warmly. This made things easy for Przhevalskii and he spent about two months at Lopnor, collecting naturalistic and ethnographic materials and observing the springtime migration of birds, a real delight for a keen ornithologist like himself. From Lopnor, moving south-westward along the rim of Takla Makan, the travelers went to Cherchen, where they were caught by a severe sandstorm, thence to the oases-settlements of Niya, Yasulgun, Keriya, Polu, Chira, and finally to the thickly populated Khotan, which they reached in September 1885. They were moving home now.

On the way Przhevalskii discovered one more unknown mountain chain, which he named Russian Range. The inhabitants of all the places his expedition visited were friendly and sociable. Curiously, they often visited the expedition bivouac in the evening to listen to the Cossacks playing the Russian harmonica, which left them spellbound.

Having ascended the Bedel pass in the Tian Shan on 29 October, the party crossed the Chinese-Russian border, an occasion celebrated by firing a salvo at the nearest rock. The expedition ended shortly after at Karakol, on the shore of Lake Issyk-kul (in today's Kyrgyzstan).

•••

The scientific results of Przhevalskii's second Tibet expedition were many, but the most important were his new geographical discoveries in Northern Tibet, Tsaidam, Kunlun and Eastern Turkestan. They were incorporated in the two maps produced by the Main Staff's Military-Topographic department – that of his last expedition and the summary map of his four journeys, both scaled 100 versts to the inch (1 verst = 1.07 kilometres). The latter one was based on the 100-verst Map of Asian Russia. Comparing them with any of the maps made before Przhevalskii's first expedition shows how radically the cartographic

119 Quoted from: Royal Geographical Society, Supplementary Papers, Vol. III, 58.

image of Inner Asia had changed as a result of his four pioneer journeys covering a period of 15 years, the 'Przhevalskiian Era'.

The explorer's return to St. Petersburg in January 1886 was as triumphant as before. While the media and the scholarly community hailed his new geographical exploits, Przhevalskii was quickly promoted to the rank of major-general by order of the Emperor, thus joining the brilliant cohort of Russian 'geographical generals'. However, as someone averse to public attention, he soon left for his estate at Sloboda. There he started the construction of his new manor house with the top story facing Lake Sapsho, according to his own design. He also had a pond dug nearby for breeding fish, and planted a little fruit garden featuring apple, plum and cherry-trees as well as some exotic plants, such as the wild rhubarb from Gansu and the melons and water-melons from Khotan and Aksu, which were grown in a special greenhouse. His cherished dream was to enclose his abode with greenery, so as to have nightingales singing in the trees throughout the spring. And the dream came true – almost every evening in spring 1887 Przhevalskii would hear the melodious warble of these song-birds, which reminded him so much of his happy days in Gansu, "the most luxurious corner of Inner Asia". Enchanted, he would put aside his work and sit by the window lost in memories.[120]

In early June the Main Staff summoned Przhevalskii to take part in a discussion of a potential Russo-Chinese military confrontation at a 'special council', presided over by General N.N. Obruchev, the head of the General Staff and the Chairman of the Military-Scientific Committee (VUK). For this occasion the explorer wrote a 100-page memorandum, "New considerations regarding a war with China", which he submitted to Obruchev.[121] "My plan for the march on Peking from Transbaikalia has been finally accepted", Przhevalskii gleefully informed Roborovskii. "The Amur [region] and Turkestan will be the secondary theatres of war. A flying squad will be needed from Lopnor to Sinin. Of course, I'll take command of the squad, and you'll be my assistant. <…> In general, my triumph was complete. There are opponents, of course, but they are weak due to their ignorance". The only "bad thing", in Przhevalskii's opinion, was that Russia was not willing to start the war, yet he was certain that the Chinese themselves would "spark it off" in a year or two.[122]

Interestingly, while writing his memo, Przhevalskii approached the consul-general in Kashgar, N.F. Petrovskii, for his advice who readily shared his own

120 Dubrovin 1890, 435.

121 NA RGO, F. 13, Op. 1, D. 43, 101 pp. (*Novye soobrazheniia o voine s Kitaem*, 25 May 1886).

122 NA RGO, F. 13, Op. 1, D. 139, L. 16, Przhevalskii's letter to Roborovskii, dated 6 June [1886].

ideas.[123] At the same time he drew on his latest observations, made during the recent tour of the oasis-settlements in southern Kashgaria.

Przhevalskii penned another analytical paper later that year, "An outline of the present situation in Central Asia". Originally published in *Russkii Vestnik* (Russian Herald) and then reprinted as a separate brochure,[124] this rather lengthy essay expressed fully his political and ethno-cultural views of the countries and peoples he had explored for nearly two decades, much in a Eurocentric and even chauvinistic vein.[125] Przhevalskii particularly stressed two points. First, the instability of the Qing dynasty, especially its precarious position in Mongolia, Eastern Turkestan and Tibet; and second, the strong Russian sympathies of the Inner Asian Moslem and Buddhist peoples who had come under Manchu rule in the mid-seventeenth century who openly expressed their desire to become "subjects of the White Tsar". While there was a great deal of truth to the first claim, the latter was largely an exaggeration. Przhevalskii saw China's leaders as arrogant and xenophobic. As for the 200-year "notorious friendship" between Russia and China he regarded it as nothing but the former's "fawning" on her eastern neighbor, as evidenced by its retrocession of the Kuldja (Ili) region, temporarily occupied by tsarist troops, to Peking in 1881. The relations between the two empires were far from being neighbourly. Indeed they were "hanging by a thread" that could break "any day now", since China, encouraged by her recent victory over the Dungans and the return of Kuldja, and instigated by "our opponents" (i.e. the British), had become more arrogant and intractable and hence might declare war on Russia.

Przhevalskii's essay appalled Russian liberals and Orientalists (sinologists), who retorted with a critical review carried by *Vestnik Evropy* (Herald of Europe) in 1887. To this Przhevalskii responded with his own retort, although he thought the polemic was meaningless due to the absolute incompatibility of

123 NA RGO, F. 13, Op. 2, D. 185, ll. 9–11, dated 19 April 1886. Nikolai Feodorovich Petrovskii (1837–1908) was a Russian diplomat, the consul in Kashgar in 1882–1903. He is known also as a collector of historical and archeological artifacts in the region as well as an active Great Gamer. Petrovskii's views, expounded in his letter to Przhevalskii, were summed up in four points, his main arguments being: a) China's ally in the war would be England, and the country actually could be the instigator of the conflict; b) the most important "strong points" for Russia, hence "the weakest" for China and England, would be Ili and Kashgar; c) the purpose of the Kashgarian war (to be waged in the interests of Russia) would be to drive away the Chinese from Kashgaria. This could be achieved either by inciting a popular uprising in the area or sending Russian troops there.

124 See Przhevalskii 1886, 472–524; as a separate brochure under the same title: Moscow, Universitetskaia Tipografiia, 1887, 65 ps.

125 See Introduction, 46, 48, 49.

his views with those of his critics. As he put it in a letter to his old friend I.L. Fateev: "They have armchair humanitarianism, whereas I have conclusions drawn from harsh, practical life. They have soap bubbles they call ideals, I have might as the only recognized criterion of right".[126]

In early January 1887 Przhevalskii went to Petersburg again to attend the annual general meeting of the Imperial Academy of Sciences, where he was presented with a gold medal in his name as a token of his great scholarly achievements over the years. The medal had his portrait and name and the reverse was inscribed: "To the first explorer of nature in Central Asia", encircled by a laurel wreath. The initiative came from a group of leading Russian academicians, K. Veselovskii (secretary of the Academy), A. Shtraukh, F. Ovchinnikov and others, who pointed out in their proposal to the Academy of 3 May 1886 that Przhevalskii's expeditions into Inner Asia "constitute one of the most outstanding enterprises in the history of scholarly journeys". He was "the first European to have penetrated into the very heart of High Asia and acquainted us with the countries, known until then only from scarce Chinese sources". However, he was not content with his geographical discoveries alone, like the majority of other travelers, but also obtained valuable natural-historical collections. "His amazingly rich zoological and botanical materials allowed us to determine the composition and character of fauna and flora of these countries, practically unknown until the travels of Przhevalskii".[127]

Shortly thereafter, in February 1887, the Academy organized an exhibition of Przhevalskii's major zoological collections amassed during his four expeditions. These included the rare specimens of vertebrae, birds, reptiles, amphibia and fish of Inner Asia (279 items altogether). The most exotic exhibits were the clothes of the Lopnor dwellers made of the skins of ducks and swans.[128] The exhibition gathered a large number of visitors and was attended on 2 February (two days before its formal inauguration), by Przhevalskii's monarchic patrons – the Tsar, Tsarina, and Tsarevich, as well as the Grand Dukes, Georgii Aleksandrovich and Vladimir Aleksandrovich, who were shown around by Przhevlskii himself and his two assistants, Roborovskii and Kozlov.[129]

Apart from these visits to St. Petersburg, Przhevalskii spent most of his time after the return from the expedition (until March 1888) on his Sloboda estate.

126 Quoted in Dubrovin, op. cit., 428 (fn. 1).

127 See Dubrovin, op. cit., 427.

128 See Katalog 1887.

129 The exhibition was described in great detail as "the most unique, not to be found in any West-European zoological museum", in *Novoe Vremia* (daily), 9 and 10 March, 1887, p. 2 (*Zoologicheskaiia kollektsia N.M. Przheval'skogo*).

He was not alone there; his dear nanny Makarievna who ran the household was close by, as were his estate manager, E.S. Denisov, and his servants whose services were absolutely indispensable, the cook Archip and the coachman Zakhar. A group of carpenters were building a new house for him into which he would move in June 1887. In the meantime Przhevalskii stayed in the old one where he hosted the members of his large expedition family – Teleshov, Nefedov, Bessonov,[130] Egorov, and, of course, Roborovskii and Kozlov, his right-hand men, nicknamed Volia and Kizo, who occasionally came over from St. Petersburg.[131] They would have a good time together, their favourite recreation being, as usual, shooting and fishing and camping out in the woods. More visitors would come to Sloboda to catch up with Przhevalskii, including his old Smolensk schoolmates, his close friends from the military (F.A. Feldman, I.L. Fateev and A.A. Bilderling), and his relatives. Finally, local peasant villagers he befriended often accompanied him whenever he went hunting or fishing.

There was yet another person to whom Przhevalskii paid particular attention at this time – a teenage son of the neighboring squire, Kostia Voevodskii, who had lost both of his parents. Przhevalskii volunteered to take care of the boy, having sent him to the cadet school in St. Petersburg at his own expense, and he eventually became Kostya's second father. In general, Nikolai Mikhailovich had a most sensitive and generous nature; he showed much concern for the welfare of his close ones, such as the elderly Makarievna and his travel companions, especially young men, whom he always treated in a fatherly manner, when journeying together and after.

In March 1888, having finished the account of his fourth journey, Przhevalskii went to St. Petersburg again. The book, titled *Ot Kiakhty na istoki Zheltoi reki. Issledovanie severnoi okrainy Tibeta i put' cherez Lob-nor po basseinu Tarima* (From Kiakhta to the Sources of Yellow River; the Exploration of the Northern Outskirts of Tibet and the Way across Lopnor along the Tarim Baisin), had been written like his previous travelogue in his little 'hut' amidst the garden, his *sancta sanctorum* in Sloboda, to which no one had access.

In fall 1887, while writing the book, Przhevalskii came up with a project for a new, – fifth, – journey to Inner Asia. Its main object was again Tibet, with Lhasa as final destination. In the introductory chapter "How to travel in Central Asia", Przhevalskii sketched a program of future explorations, having listed the regions to be "reconnoitered" primarily by Russian travelers. His priority

130 Nefedov and Bessonov stayed at Sloboda with Pzhevalskii permanently from 1886 through 1888.

131 Roborovskii was preparing for entrance examinations to the military Academy and Kozlov attended the cadet school, both in St. Petersburg.

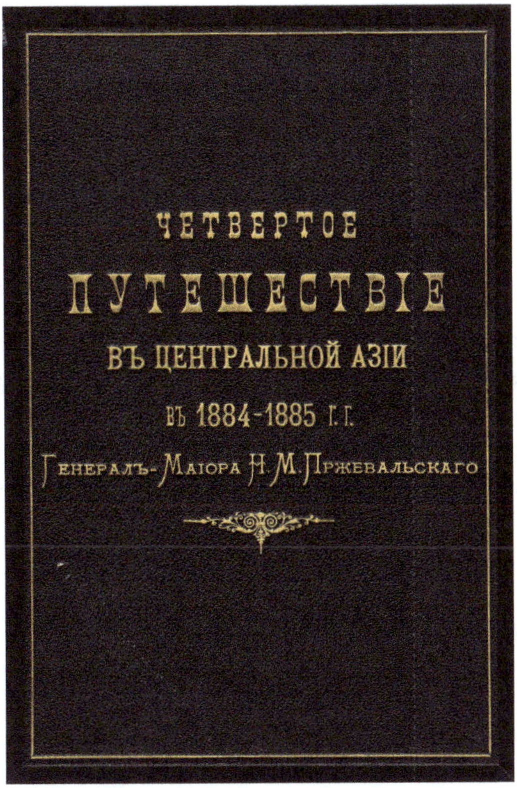

FIGURE 1.17
The front cover of N. Przhevalskii's last travelogue: *Chetvertoe puteshestvie v Tsentral'noi Azii*, St. Petersburg, 1885.

again was Tibet and its uncharted areas – the eastern part of the plateau (Amdo and Kham) and the territory "from Lhasa to Gartok" in the south. The new expedition was to start from Karakol on the shore of Issyk-kul and proceed from there by a familiar route, via Khotan and across the Altyntag, to Lake Gas in Western Tsaidam where Przhevalskii planned to set up a base with a depot, in the manner of his previous journey. From Gas he intended to ascend the Tibetan highlands with his main caravan and head for Lhasa by the pilgrim's road he had already trod before.

Przhevalskii's project was approved unhesitatingly by the IRGO, the Main Staff and finally the Emperor himself, on 15 March 1887. This time the government allocated him some 80,000 rubles, 15 times as much as it did for his first journey in 1870. The expedition staff was to consist of 27 men, and Roborovskii and Kozlov were to act again as his main assistants.

Along with the purely scientific program, Przhevalskii also intended to conduct some political reconnaissance. He anxiously watched the developments on the Indo-Tibetan border as the British gradually advanced toward Tibet. The border clash between the English and Tibetans at the checkpoint Lungthur in

March 1888 did not escape his attention. Hence while expounding the program of his fifth expedition in a letter to the new war minister, P.S. Vannovskii, he wrote, apparently hinting at this incident: " Aside from the scientific results, I can possibly collect information concerning the current activities of the English toward Tibet via Sikkim and the sentiments of the Tibetans".[132]

He also took care to equip his party with sufficient arms, as he did on previous journeys. However this time he requested the military to provide him with one light machine-gun (so-called *mitrailleuse* or the Maxim gun – the first recoil-operated machine gun, invented by H.S. Maxim in 1883), a model that the Russian army was testing at that time. He justified his request by saying that he would need this new gun in case his expedition caravan was attacked by a large group of hostile Tibetans, as had happened before. As a result, on the order of the war minister Vannovskii, one of the Maxim guns was specially adapted to Przhevalskii's requirements at the Nobel arms factory at Izhevsk and dispatched to Karakol.

•••

Przhevalskii left Sloboda for his new expedition on 5 August 1888, accompanied by Roborovskii, Kozlov, Teleshov and Nefedov. He departed with a heavy heart and full of misgivings, leaving behind his dear ones – his ailing nanny Makarievna (who would soon die) and his ward, Kostia Voevodskii, whom he entrusted to the care of his old friend, General F.A. Feldman.

The party first went to St. Petersburg to collect the necessary supplies. The highlight of the stay was an audience, on 10 August, with Emperor Alexander III at his Gatchina Palace. The reception was unusually cordial – Przhevalskii presented the Tsar a copy of the freshly published account of his last journey, with a dedication to the heir-apparent, Nikolai Aleksandrovich, as "the high patron of my latest explorations in Central Asia". At parting the Tsar addressed Przhevalskii with many kind words and treated him as "one of his family".[133] The Tsarevich, on his part, asked his former tutor to keep him posted on the progress of his expedition. Both the emperor and his heir felt a great sympathy for Przhevalskii, being proud of his achievements as the Russian pioneer explorer. In turn, Przhevalskii revered the ruling dynasty, as evidenced by the names he gave to some of the highest mountain peaks in the Kunlun, such as

132 NA RGO, F. 13, Op. 3, D.5, l. 7 ob. (draft of a letter to P.S. Vannovskii, undated).

133 See Dubrovin, op. cit., 451 (quoted from Przhevalskii's letter to E.S. Denisov of 17 August 1888).

FIGURE 1.18 Seated: P. Kozlov, N. Przhevalskii and V. Roborovskii, with the expedition team at the background, 1888. NA RGO.

Tsar-Liberator (*Tsar'-Osvoboditel'*), Hat of Monomakh (*Shapka Monomakha*), and Kreml.

Przhevalskii and his companions left Petersburg a week later, on August 18. Their passage to Karakol was a long, yet exciting, trip across a large tract of Russia and its new domain, Russian Turkestan. They travelled first by train from Moscow to Nizhnii Novgorod, thence by a steamer, down the Volga and across the Caspian Sea, and proceeded further on to Samarkand on the recently constructed Transcaspian railway. Arriving in Tamerlane's former capital on September 20, they changed onto postal *troikas* which brought them to Pishkek and finally to Karakol on the shore of Lake Issyk-kul.

Now Bishkek, Kyrgyzstan's capital, Pishkek then was a sleepy village of about hundred adobe houses, with narrow dirty streets. Przhevalskii stopped here to purchase provisions and camels. Together with Roborovskii, he went to Vernyi (Alma-Ata, or Almaty), the center of the Semirechensky military governorship, both to obtain Chinese silver and to select a group of soldiers and Cossacks for his escort. The explorer was very choosy; he admitted no "married men" and "town-dwellers," whom he found totally unfit for long journeys. The best candidates for the job were sportsmen.

Shortly after their return to Karakol, Przhevalskii suddenly began to feel sick and his illness progressed quickly. He suffered blood loss, high temperature and stomach pains, caused most likely by typhoid fever he caught by drinking unboiled water while shooting pheasants at Pishpek. A doctor was summoned and Przhevalskii was moved from his camp tent to the infirmary. It was to no avail, and he expired a few days later, on October 20 (2 November, N. St.). Przhevalskii's sudden demise, at the outset of his new journey, came as a great shock to his fellow-travelers; everyone sobbed, bitterly mourning the loss of their leader.

The Russian historian A.A. Kolesnikov, however, questions the general consensus that Przheval'skii succumbed to typhoid, arguing that his disease was not properly diagnosed by doctors (which is not true[134]) and no post-mortem medical examination was carried out. If so, Kolesnikov believes the explorer could have been poisoned by some "slow acting poison" by his foes, who opposed any Russo-Tibetan rapprochement. They knew that the main object of his expedition was Lhasa, and they wanted to prevent Przhevalskii from reaching it, as "the establishment of contacts between Russia and Tibet could substantially alter the geopolitical situation in the region".[135] However, there is no evidence of any conspiracy against Przhevalskii, from either Russian or British side.

Przhevalskii was buried with military honors outside Karakol, on the high shore of Issyk-kul overlooking the majestic Tian Shan, as was his dying wish. The following year the town Karakol was renamed Przhevalsk, and in 1894 an impressive monument was erected on the spot, designed by Przhevalaskii's friend, General A.A. Bilderling, and executed by architect I.N. Shreder. The nine-meter high monument consisted of a massive rock of Tian Shan granite, surmounted by a bronze eagle extending its wings wide. In its beak the eagle held an olive branch while its claws rested on a bronze plate depicting the map of Asia with all of Przhevalskii's expedition routes carved on it. In the center of the rock one could see a round bas-relief with a profile of the traveler, an enlarged replica of his IRGO gold medal.

Two years earlier the explorer's many admirers sponsored a more famous tribute in Alexander Gardens, in the center of St. Petersburg, a short distance from the Geographical Society and the Main Staff – the two institutions to which Przhevalskii owed his career of military geographer – and the Winter Palace, residence of the Romanovs, his patrons and sponsors. Visitors today can still see the bronze bust on a high pedestal of granite, with a pack-cam-

134 Przhevalskii was treated by several doctors who diagnosed the cause of his death as typhoid fever, according to Roborovskii; see Roborovskii 1888, 277–280.

135 Kolesnikov 2008, 67.

FIGURE 1.19 Przhevalskii's burial place with the memorial tombstone at Issyk-kul, 1888. NA RGO.

FIGURE 1.20 The consecration of the Przhevalskii's monument at Issyk-kul, 1894. NA RGO.

el resting at the base, bearing the dedication *To Przhevalskii, first explorer of Central Asia.*

•••

On 8 November of 1888 the IRGO held an extraordinary meeting in Przhevalskii's memory. This provided an occasion for leading scholars, including members of the Russian Academy of Sciences, to assess the traveler's important contributions to the fields of geography, zoology and botany.[136] Przhevalskii's scientific legacy is indeed enormous and varied. It includes his route surveys, various field measurements and observations (astronomical, barometric, magnetic, meteorological), and huge collections of animal and plant specimens. Less well known are his first-rate political and military intelligence on China, particularly its western borderland (Xinjiang), and Tibet, the parts of Inner Asia which were most attractive geopolitically to Tsarist Russia. The explorer gathered these details en route by means of questioning the locals and they were included in his special memos to the Main Staff and the General Staff. Overall, Przhevalskii's four expeditions to Inner Asia considerably expanded our knowledge of this region, having yielded priceless nuggets of information about the least known lands and peoples in the 'Heart of Asia'.

His route surveys of over 16,000 km – out of the total distance of 31,500 km he traversed – cover a vast territory, from Pamir to the Greater Khingan Mountains and from Altai to Central Tibet. These surveys radically changed the map of Inner Asia, with its basic oro- and hydrographic systems, its deserts and steppe-lands, filling in many of the previous blanks.

Przhevalskii was a great pioneer explorer, in the same league as David Livingsone. But unlike the Scot, he was also a brilliant naturalist. Despite the fact that he regarded his expeditions as no more than a reconnaissance of uncharted lands of Inner Asia, he sought to produce a detailed record of the main natural features of the places he visited, namely, each locality's relief, climate, soil, minerals, fauna and flora, on the one hand, and, on the other hand, of its various ethnic groups. The resulting picture portrayed the unity of nature and man into one whole, according to Przhevalskii's all-embracing "synthetic" formula – "one Earth, one mankind and one ethical law" (that of justice), which guided his wordview.[137] However, his ethnographic achievements are somewhat more modest, and comprise a series of lively and detailed sketches of the

136 See *Izvestiia IRGO*, 1888, vol. XXIV (4), 233–272 (Speeches made by P.P. Semenov, A.A. Strauch, K.I. Maksimovich, and A.I. Voeikov.)

137 See Fateev 1945, 3, 143.

many Asiatic peoples he came across during his travels, such as Khalka-Mongols, Torgouts, Tanguts, and Dungans, which he incorporated into his published travelogues.

His zoological collections consist of over 7,500 specimens of mammals, birds, reptiles, amphibians, fish, and insects, which were real treasure-trove for the Zoological Museum of the Academy of Sciences. Some of the new species of bird and mammal in it were named after Przhevalskii, such as Przhevalskii's horse (*equus przewalskii* Poliakoff, 1881), Thorold's (White-lipped) dear (*cervus albirostris* Przewalskii, 1883), Przhevalskii's gerbil (*brachiones przewalskii* Buchner, 1889), Przhevalskii's steppe lemming (*eolagurus przewalskii* Buchner, 1889), Przhevalskii's gazelle (*procarpa przewalskii* Buchner, 1891), crested tit-warbler (*leptopoecele elegans* Przhewalskii, 1887), as well as after his assistants, Pyl'tsov, Roborovskii, Kozlov, Teleshov, and Eklon.

His botanical collection numbers about 16,000 plants making 1,700 species, 218 of which and 7 genera were completely unknown to botanists of the day. Apart from the collected herbarium, Przhevalskii recorded in his diary and described minutely in his books the vegetation of many parts of Inner Asia. Thus in the account of his first journey the reader will find a detailed description of many native plants and the way these are utilized by locals. One of his rarest finds was a new genus and species of the Solanaceae family, *Przhewalskia tangutica Maxim* (1881), a tropane alcoloid-containing plant he found in the Burkhan Budda Mountains and in the upper Blue River of Northern Tibet. The two most remarkable plants which grow everywhere in Inner Asia, especially in Dzungaria, Gobi and Tsaidam, according to Przhevalskii, are saxaul (*Haloxylon ammodendron*) and dyresun (*Lasiagrostis splendedens*). Saxaul called *zak* by the Mongols, is "most precious for the nomads of deserts", being used as fodder for camels, and even for goats and rams in Alashan, and it is also a perfect fuel.[138] Meanwhile, Przhevalskii donated a mineralogical collection of several hundred samples to the Geology Room of St. Petersburg University, headed by Prof. A.A. Inostrantsev.[139]

It took many years for scholars to study these unique materials, which resulted in the publication of the "Scientific Results of the Journeys of N.M. Przhevalskii into Central Asia" (*Nauchnye rezul'taty puteshestvii N.M. Przheval'skogo po Tsentral'noi Azii*) by the Academy of Sciences, between 1888 and 1912.[140] Published

138 Przhevalskii 1948, 42–46.

139 This collection was exhibited at the Geology Room but has never been properly described by mineralogists; to this day only few sections of the basalt rocks from it have survived. On Przhevalskii's geological collections see Andreyev, A.I. (ed.) 2013, 265–279, 286.

140 See Nauchnye rezul'taty, 1888–1912.

in three large-size volumes of many separate installments, they featured descriptions of his zoological collectionsby E.A. Bikhner, V.V. Zelenskii, F.D. Pleske, V.L. Bianki, Ya.V. Bedriaga, S.M. Gertsenstein, while K.I. Maksimovich studied the plants, and the geographer and climatologist A.I. Voeikov analysed his meteorological observations. The heir to the throne, Nikolas Alexandrovich, enthusiastically subsidized the publication's cost of 25,000 rubles. In 1891, the IRGO began to award a silver medal in the explorer's name, as prestigious as the Society's small golden medal, as well as the Przhevalskii prize for travelers. The first to receive the Przhevalskii medal that year, were K.I. Bogdanovich, Roborovskii and Kozlov, all of whom had taken part in the Tibet expedition of 1889–1891 under M.V. Pevtsov to complete Przhevalskii's aborted fifth Inner Asian journey. Mikhail E. Grum-Grzhimailo, who participated in his brother Grigorii E. Grumm-Grzhimailo's Inner Asian expedition of 1889–1890, also received the medal. Half a century later, in 1946, when the RGO was celebrating its 100th anniversary, this medal was replaced by a gold one to be awarded to travelers once in two years, at the initiative of the Council of Ministers of the USSR.

Przhevalskii enjoyed wide popularity in the USSR after the end of the World War II, when all his travelogues, including the account of his first Ussuri journey, were republished in 1946–1948. And in 1952, Sergei I. Yutkevich shot a color film about Przhevalskii that portrayed him as a great hero – 'the Russian bogatyr' – fully within the framework of Soviet ideology.[141] E.M. Murzaev's preface to his edition of Przheval'skii's first travelogue was in a very similar vein:

> Przheval'skii's work, his heroism, audacity, fearlessness, firmness of purpose and his selfless devotion to science are fully consonant with the Soviet epoch. Nikolai Mikhailovich was a truly Russian man, and wherever his fate took him – the Gobi desert, the steppes of Mongolia, the mountains of Tibet ..., he always remained an advert patriot of Russia, his motherland. His whole life serves as an example for us, one that is calling for persistent labour and the lofty principles of service of one's country and one's people. The names of many Russians of the XIX century are

141 Six decades later (in 2015), now in post-Soviet Russia, the TV Company Istra-Art shot a documentary *Nikolai Przhevalskii: Ekspeditsiia dlinoiu v zhizn* (Nikolai Przhevalskii: A Life-long Expedition), to revive the image of the great Russian pioneer explorer. The film was produced by Vladimir Shuvannikov and Tatiana Borshch and was sponsored by the RGO. Istra-Art also made a film (end of 2017) about another great explorer, Petr Kozlov, the man who discovered Khara-Khoto, *Petr Kozlov. Taina Zateriannogo Goroda* (Petr Kozlov, the Mystery of the Lost City).

FIGURE 1.21
The front cover of P. Kozlov's book about Przhevalskii's travels *V Serdtse Azii* (In the Heart of Asia), St. Petersburg, 1914.

> well known and respected around the world. In this galaxy, geographical science is brilliantly represented by Nikolai Mikhailovich Przheval'skii – a Russian *bogatyr*.[142]

Today, Nikolai Przheval'skii is remembered as one of the greatest explorers of all times, the 'Great Trailblazer' (as Gavrilenkova called him), an object of Russia's national pride; a naturalist and nature-lover, with a strong romantic streak in his heart. His name is borne by a mountain range in the central part of Kunlun (the one separating Xinkiang-Uigur autonomous region and Tibet); a glacier in Altai, a cape and peninsula in one of the Kuril Islands, Iturup, a small settlement Pristan'-Przhevalsk on the shore of Issyk-kul, some 12 km north of Karakol, where his burial place is (the town Przhevalsk reverted to its old name

142 Murzaev 1946, 37.

Karakol under the Kirzhiz authorities in 1992); and many animals and plants. Still, Przhevalskii is little known outside the scholarly community – most people first hear of Przhevalsky because of the wild horse (*equus przewalskii*) that has his name, as zoologists Eugene and Roald Potapov ironically noted.[143]

Sloboda, Przhevalskii's favourite nook and refuge from civilization, suffered badly during World War II. His mansion was completely destroyed in 1941 by German Nazi troops and his exotic garden and birch grove were cut down. Sloboda was renamed Przheval'skoe in 1964, and Przhevalskii's house was restored in the 1970s and turned into a public museum (*Dom-Muzei N.M. Przheval'skogo*), much owing to the efforts of two enthusiasts and Przhevalskii's long-time biographers, the married couple of V.M. Gavrilenkov and E.P. Gavrilenkova. The Museum accommodated Przhevalskii's library of about 1000 volumes, which was fortunately rescued from the Germans, who removed it from the mansion, manuscripts, some of Roborovskii's drawings, stuffed birds and animals as well as various memorabilia. Przheval'skoe today is situated in the very heart of the Smolensk Lakeland National Park – *Smolenskoe Poozerie*, one of the most scenic areas of Central Russia.[144]

Another museum in the explorer's honour is the N.M. Przhevalskii Issyk-Kul Memorial Complex. It was established in 1957 on the shore of Lake Issyk-Kul, in Kyrgyzstan, close to his grave and his monument. The museum displays a small collection of Przhevalskii's personal items, the most valuable being his rifle, manufactured in 1883 by St. Petersburg armourer V.V. Lezhin; his photographs; and books published by him and other outstanding Russian travelers of his epoch – Roborovskii, Kozlov, Potanin and Grumm-Grzhimailo.

The Unknown Przhevalskii

In this final section we will address some intriguing issues relating to Przhevalskii's intimate life, namely: 1) his overt misogyny – a strong aversion to women and married life, which suggests that his sexual orientation was not traditional; 2) the myth of Przhevalskii's being the biological father of Joseph Stalin as claimed by some authors in today's Russia on the basis of their facial resemblance; 3) Przhevalskii's attitude to religion. His psychological portrait would be incomplete without shedding light on these matters.

143 Potapov, E., Potapov, R. 2006, 86–91.

144 Re the Przhevalskii museum in Sloboda, its creation and exhibits, see Gavrilenkova 2011, 8–18, 21–24 et al.

1) *Przhevalskii's Misogyny*

Przhevalskii's biographers, M. Engelgardt and A. Fateev, sketched an insightful psychological portrait of the traveler. The most remarkable feature of his character, according to Engelgardt, was a "love for wandering". Przhevalskii was a "hardened vagabond" and his aversion to settled life only grew with time and finally turned into "a morbid passion". The Asian "wilderness" attracted him mightily and more than once he expressed a desire to die in a desert, while traveling with his companions.

A. Fateev, on the other hand, noted that travelling instilled in Przhevalskii a feeling of profound joy and admiration for nature, an esthetic feeling, which markedly distinguished him from many other travelers and naturalists of his time. He was a special kind of scholar-traveler, a "traveler by vocation", in his own definition, one who "organically merged his life's mission with his profession".[145]

Having grown up on his parent's estate, amidst virgin 'wild nature', Przhevalskii developed a strong passion for the outdoors and eventually came to believe, much in the spirit of J.-J. Rousseau, that the pure 'state of nature' was opposed to the corrupting influence of civilization. He would surely have subscribed to Rousseau's statement "Nature made man happy and good, but society depraves him and makes him miserable".[146] Hence his idealization of 'natural living' and 'natural men', such as 'savage Asiatics', and his hostility to the civilized 'settled life' of cultured Europeans, especially urban life with its permanent bustle, which often made him physically sick. However, as Engelgardt remarked, Przhevalskii had no reason to complain, since his merits were judged at their true worth by his contemporaries – the scholarly community and public in and outside of Russia.

Here lies the greatest paradox, Przhevalskii's duality. When at home, in Russia, he would speak harshly of Western civilization and its "perverted" influence on humans, but, as soon as he found himself in Asia, he would castigate "Asiatics" for their many vices, both moral and physical. Therefore one should not take at face value all of his derogatory statements, such as when he calls the Chinese in Peking "rogues", the Europeans "abandoned wretches", and the whole of Asia "a rotten place", which naturally made him a misanthrope in the eyes of his contemporaries. These utterances, in the words of Engelgardt, were the result of Przhevalskii's "sanguine, fervid" character, going on to add,

145 For a discussion of this aspect see Fateev 1945, 3–18.

146 Rousseau 1995, 934.

> having noticed some evil in a certain milieu, he would attack it and smash it to smithereens, without too much thinking. <...> He did so partly because of his self-assurance, typical of strong-willed men, partly because of his want of logic and abstract thinking, but mainly because he had no need to investigate properly the sort of life from which he tried to escape. A tramp is always opposed to settled life.[147]

Przhevalskii's other striking oddity was his open and strong aversion to the fair sex and married life. Unlike his married brothers, Vladimir and Evgenii, he strongly opposed marriage in general, and never took a married person on his expedition's staff. He believed that family life presented one of the greatest obstacles for a traveler in far-off lands, although there are examples when married couples traveled together in Mongolia, China and Tibet and successfully conducted joint explorations, such as Grigorii and Alexandra Potanin, Petr and Elizaveta Kozlova, George and Teresa Littledale, etc., to name a few. Some researchers claim that this hints at a homosexual predilection on Przhevalskii's part,[148] yet one can also see in it something different. Placed within the framework of Russian traditional mentality, Przhevalskii was a typical *podvizhnik,* the Russian word which can be translated as a devotee – in his case a devotee to geographical exploration, basically a quest for the unknown, persistent and unending, and also an act of personal self-sacrifice – for the sake of something he viewed as a lofty goal, like scientific knowledge or the benefit of his native land.

In a letter to his Kiakhta friend Lushnikov, written in 1886, he would say: "... what you said about a general's lady will probably not come true. I am too old [46] and my profession is not right for marriage. It's in Central Asia that my posterity lie ... in a figurative sense of course: Lob Nor, Kuku Nor, Tibet etc. – these are my children".[149] Yet in the same year Przhevalskii seems to have had a romance with a 28-year old married woman, Xenia (Aksiniia) Mel'nikova. This fact came to light only recently, after the publication of the recollections of Marfa Batseva (Mel'nikova by birth), his alleged daughter.[150]

147 Engelgardt, op. cit., 397, 398.

148 One of them, L.S. Klein, having thoroughly examined Przhevalskii's biography, concludes that he was a typical homosexual, although, Klein admits, the available evidence is only indirect, and "there are no direct proofs whatever of his homosexuality", see Klein 2002, 157, 178 (this work includes a whole chapter on Przhevalskii).

149 Durbrovin 1890, 422; quoted in Rayfield, op. cit., 184.

150 See Gavrilenkova 2011, 159–213. Two photos of Marfa, taken at the ages of 20 and 45, are reproduced on p. 160 and 162. Marfa Mel'nikova's memoirs were written in 1952–54 and a copy of these was sent to the IRGO.

What we learn from Marfa's unsophisticated narrative is this: Przhevalskii came to know Xenia in the summer of 1882 through her husband, a landless peasant, Kirill Grigorievich Mel'nikov, from Borovskoe village, not far from Przhevalskii's Otradnoe estate. He was a skillful carpenter who made wardrobes as well as fine fiddles for local squires. Mel'nikov was often away from home during the summer time. Przhevalskii met Kirill accidentally in the forest when shooting, as the peasant was also a good hunter, and they became friends. Przhevalskii would occasionally come to Kirill's home, alone or together with Pyl'tsov, and sometimes stayed there overnight. Gradually a mutual sympathy arose also between him and Xenia, yet it was not until he returned from his fourth journey in 1886 that they became lovers.

Przhevalskii found in Xenia that special type of a Russian peasant woman so much admired by the poet Nikolai Nekrasov – simple, unassuming, sincere, quiet, yet fervent in spirit like himself. As for Xenia, she was strongly impressed from the first by Przhevalskii's striking appearance, his masculine aura – he was tall, broad-shouldered and good-looking – and his attractive personal qualities – his kindness, sensitivity and gallantry. Xenia has never met a man like him before. As a teenage girl, she had been married by her parents to a man she did not love, as was the usual practice in the Russian countryside of those days.

Naturally enough, she and Przhevalskii kept their love secret from Kirill and other villagers, and the rest of the world. As time went on, Xenia got pregnant, and in 1887 a baby was born to her, a girl named Marfa by Kirill, who earnestly believed it was his daughter. Przhevalskii often visited Xenia (when Kirill was away from home, busy with his carpenter's work), yet finally he had to leave his beloved to go on his fifth expedition.

Przhevalskii's sudden death was a severe shock for Xenia when she learnt about it. Being a religious person, she felt badly about her liaison and made a pilgrimage to the Beloberezhnaia Pustyn' Monastery, near Briansk, to repent of her great sin of adultery.

This romantic, yet sorrowful, tale was brought to public notice in 1999, when E.P. Gavrilenkova published Marfa Mel'nikova's memoirs in her collection of essays, *Neizvestnye stranitsy zhizni N.M. Przhevalskogo* (The Unknown Life of N.M. Przhevalskii). The author fully believes Marfa's account, yet Przhevalskii's other biographers, such as N.N. Kravklis and L.S. Klein, are skeptical about its authenticity. Kravklis, for example, puts forward several counter arguments. First, Przhevalskii never mentioned Xenia or his romance to his brothers or his close friends; second, if Marfa was indeed Przhevalskii's daughter, he must have left her something by his last will, as he is known to have split up his fortune between his relatives not long before his death, but he did not. And finally, Przhevalskii could not come to the village Borovskoe at the time of Marfa's conception, as it was 147 km from Sloboda. However, another person might well have been there, Tsohar, one of Przhevalskii's companions in the expedition. He was expected by Przhevalskii at Sloboda at that time, but was slow in coming, for some unknown reason.[151]

The only way to ascertain the truth, according to Kravklis, is a DNA test of the living descendants of Przhevalskii's brothers and those of Marfa Mel'nikova, who died in 1975. Until then, the woman's claim will remain the subject of speculation.

2) *Stalin's Father?*

A claim that Przhevalskii was Stalin's father was first made in 1939, when two jubilees were widely celebrated in Soviet Russia – Przhevalskii's centenary and Stalin's 60th birthday. It was then that the popular Polish daily *Zycie Warszawy* (Life of Warsaw) revealed this most sensational fact, largely on the basis of Nikolai Przhevalskii's and Joseph Dzhugashvili's facial resemblance. The necessary details were elaborated later to make the claim plausible, such as Przhevalskii's alleged visit to Gori in Osetiia (Transcaucasia), Stalin's birth-place, where he had a love affair with Ekaterina Geladze, Stalin's future mother. No serious evidence was provided to support the fact.[152] The myth or simply a fib of the explorer's blood tie with the Soviet leader has enjoyed a great longevity and survives to this date. As recently as 2013 one could see, right in the center of St. Petersburg, an advertising pylon with two large portraits, those of Nikolai Przhevalskii and Joseph Stalin, with a provocative caption below: "Stalin – the son of Przhevalskii?".[153]

---

151 Kravklis, 2013, 198; see also an article by the same author, "O general'she ne mozhet byt' i rechi! Pochemu Nikolai Przheval'skii tak i ne zhenilsia" in *Smolenskaia Gazeta*, no. 117 (845), 20 October 2001. For L.S. Klein's arguments, see Klein 2002, 178–182. As for Western authors who addressed the subject, see, for example, Simon Karlynsky, "Gay Life before the Soviets", *Advocate*, Apr. 1, 1982, 31–34.

152 Kravklis, 2013, 195–199; see also Gavrilenkova, op. cit., 51–55.

153 This was an advertisement of the pseudo-scientific magazine *Zagadki Istorii* (Enigmas of History), which in 2012 carried an article with the same title authored by one Igor Radionov (2012, no. 12, pp. 12–13). In this article Radionov adduced several arguments in support of the myth, the strongest being the fact that Nikolai Przhevalskii used to transfer money by post to "some unknown person" in the Caucasus in 1881–82, and the person could well have been Ekaterina Geladze. Moreover, Radionov gives support to another myth – that Nikolai Przhevalskii himself was in fact an extramarital son of Grand Duke Aleksandr Nikolaevich, the future emperor Alexander II, who had visited Smolensk in

FIGURE 1.22 An advertising pylon in St. Petersburg, showing the portraits of Nikolai Przhevalskii and Joseph Stalin, 2013. The caption below reads: Stalin – the son of Przhevalskii? PHOTO BY A. ANDREYEV (2013).

The facts do not support this legend. Joseph Dzhugashvili, the future Stalin, was born on 6 (18) December 1878.[154] Meanwhile, nine months earlier, Przhevalskii had been staying at Zaisanskii Post with his Lopnor expedition, making preparations for another journey to Tibet. The expedition, as we know, was interrupted by his ailing condition, and he returned to St. Petersburg in late May 1878, which excludes any possibility of his visiting Gori at the time of Ekaterina Geladze's conception. More to the point, Przhevalskii had never been to Transcaucasia. The myth was conclusively laid to rest by the DNA tests taken by N.M. Przhevalskii, a nephew of Nikolai Mikhailovich's great granddaughter, and Stalin's own nephew, Aleksandr Vasilievich Burdonskii. The tests evidenced beyond any doubt that the famous traveler and the Soviet leader were not genetically related.[155]

1837 and had a fleeting romance with "a local beauty", Elena Karetnikova. If so, "Stalin was the nephew of the tsar and had the right to ascend the throne of the [Russian] monarch".

154 The date was changed in official records to 9 (21) December 1879 in the mid-1920s.

155 Kravklis, op. cit.

Oddly, the myth remains alive and there are still people who believe, or rather want to believe, that Przhevalskii was Stalin's biological father, just because they look so much alike on photographs.

3) *An Agnostic and Darwinist*

Przhevalskii's parents, as we know, were both Orthodox Christians and he himself was christened at birth. As a Christian he must have attended the religious services at the local church in the village Lobkovo, close to the Otradnoe estate, together with his parents, and thus grew up as a believer in God. Yet his attitude to Christianity and religion in general changed radically in later years, apparently under the influence of Charles Darwin's theory of evolution.

One can already feel it when reading Przhevalskii's early essay, 'On the essence of life on Earth', which he penned in 1860, when he was only 21, and discussed with his fellow officers in Kremenetsk. According to B.A. Val'skaia, the subject of the essay was probably prompted by Darwin's book – *On the Origin of Species by Means of Natural Selection*, which had only been published a year earlier, in 1859. Przhevalskii's essay also mentioned several more names of outstanding Western naturalists and zoologists he called "the great geniuses" – A. Humboldt, G. Cuvier and J. Audubon, who apparently had influenced his views of the world order. In his own essay Przhevalskii emphasized the role of nature in the creation and evolution of plants, animals and humans on the planet, saying, for example, that "man is the crown of the creations of nature", that is, not of God's creation.

Having thoroughly analyzed Przhevalskii's paper, B.A. Val'skaia, maintains that it clearly reveals him as an advocate of 'natural-scientific materialism' and Darwin's teaching. Moreover, the author refers to Prof. P.V. Terentiev, who thinks that Przhevalskii's ideas "are consonant with the theory of biosphere" developed decades later by the great Russian naturalist V.I. Vernadskii.[156] Hence, Val'skaia concludes, "the question of the essence of life on Earth was investigated by Przhevalskii in order to scientifically substantiate and reject the Christian dogma of the immortality of soul".[157]

Indeed, from Przhevalskii himself and his biographers we learn that, as an adult, he never went to church. He spoke very harshly of the Orthodox clergy in general and was known to be on bad terms with the priest of the Lobkovo church, Father Mikhail. Marfa Mel'nikova also evidenced in her narrative that Przhevalskii never prayed and never crossed himself before icons and over the table before taking meals, as most Christian Russians did at the time, which

156 Val'skaia 1967, 95–97.

157 Ibid., 97.

surprised the peasants from Borovskoye who knew him. At the same time, keeping in line with the age-old tradition, Przhevalskii would celebrate the Christmas and Easter holidays on his journeys and received congratulations on these days from relatives and friends with the traditional *Khristos voskres*! ('Christ is risen!') on Easter Day; and he arranged a church service for a group of Cossacks in his party before their departure from Russia with his expeditions. We also know that Przhevalskii wore a little cross his mother hung on his neck with a blessing when she sent him to the gymnasium in Smolensk. He never spoke openly against Christianity and occasionally mentioned God's name in colloquial clichés, such as 'God grant' (*dai Bog*).[158]

Przhevalskii's attitude to Eastern religions, Buddhism and Islam alike, was strongly negative since he regarded both as major impediments to the "progress" of the civilization in Asia. The "pernicious impact on the masses" of these religions, in his opinion, "root too deep and put up barriers too strong for the successful dissemination of Christianity".[159] One can only guess after reading this statement whether Przhevalskii wanted Christianity to replace Buddhism and Islam on the Asian continent as a prime mover of civilization.

All this clearly suggests that Przhevalskii was an agnostic, who, however, prudently refrained from making any critical remarks in public on Christianity, the faith of his forefathers, and a Darwinist, as many naturalists of his day, a believer in the evolution of the human race and social progress. A convincing proof of this is that a new species of argali that he discovered in the Kharku Mountains of South Gobi on the third journey was named, on his suggestion, *Ovis darwinii*, after the recently deceased British naturalist.

Przhevalskii often spoke of the 'good luck' that rescued him and his companions more than once in the most dramatic situations. This leads one to think that the explorer was a believer in providence – that of omnipotent Nature, the source of organic life on Earth, rather that of God and His beneficent care. Indeed, the assumption goes well with Przhevalskii's life-long passionate admiration for nature and the task which he set himself as a naturalist of making a profound study of it.

158 See, for example, ft 82 with a quotation from Przhevalskii's diary.

159 Przhevalskii 1887, 16, 17.

CHAPTER 2

Mikhail Vasil'ievich Pevtsov (1843–1902)

Mikhail K. Baskhanov

Mikhail Vasil'evich Pevtsov belonged to a group of Russian 'geographical generals' – these were influential military geographers who made a career and name for themselves exploring the vastness that is Inner Asia. Unlike N.M. Przhevalskii, Pevtsov's name is less known in the sphere of geographical science and details of his academic and personal life still remain relatively unknown. At the same time, Pevtsov's scientific activity and his contribution to the study of geography and ethnography of Inner Asia quite rightly places him in the same category as other distinguished travellers and explorers of this remote region of Asia.

Early Years

Pevtsov was born on May 21, 1843 into a noble family on an ancestral estate in the Ustiuzhinskii *uezd* (district) of the Novgorod *guberniia* (province).[1] The *Pustosh Belousova* estate belonged to Pevtsov's father, which included a total area of about 300 hectares of land. At the young age of seven, Pevtsov became an orphan and was brought up by his relatives in St. Petersburg. He received his education at the prestigious First St. Petersburg High School. During his summer vacations, Pevtsov would spend time at the ancestral estate, in the company of local hunters. He would roam the remote villages of the Novgorod *guberniia*, with either a gun or a fishing rod over his shoulder, satisfied by the small things in life and getting used to various deprivations of camp life.

Pevtsov dreamt of studying at the St. Petersburg University, however financially this was not an option. Pevtsov, therefore, decided to join the military and to continue his education in the cadet school at the Fourth Infantry Corps in city of Voronezh.[2] Colonel I.A. Kienskii, a schoolmate of Pevtsov, remembers that at the school 'they thrived and there was never a concern of how to get

1 See *Rossiiskii gosudarstvennyi voenno-istoricheskii arkhiv* (RGVIA), Moscow, F. 400, Op. 21, D. 2997: A full record of service of General Staff Major-General M.V. Pevtsov.

2 For the details on the school at Voronezh see Forrestt A. Miller, *Dmitrii Miluitin and the Reform Era in Russia*, Charlotte: Vanderbilt University, 1968, pp. 110–111.

 | DOI 10.1163/9789004376267_004

FIGURE 2.1
Russian General Staff Colonel Mikhail Vasilievich Pevtsov, early 1880's.
THE COLLECTIONS OF THE OMSK STATE MUSEUM OF LOCAL LORE (OSMLL, 13073/14).

bread on a daily basis'.[3] Kienskii mentions that Pevtsov was 'an extremely honest and gentle young man, who never quarrelled with anyone <…> he had a good soul and a sympathetic heart; willing to give up his last shirt to help a fellow cadet, in spite of his limited means <…> he was small and frighteningly frail, yet his health was good'.[4]

Pevtsov was a capable and diligent student; he showed great interest in history, geography and mathematics. His excellent mathematical knowledge earned him the nickname 'Pythagorean pants' amongst his fellow cadets. The large school library was his favourite space and he spent most of his spare time within its walls. Of the small amount of money that was allocated to each cadet, Pevtsov spent his share purchasing books, which he read avidly. Pevtsov's academic success in the sciences was noticed by the Chief of Staff of the Fourth Infantry Corps, General Baron Delvig, who persistently advised Pevtsov to enter the Nicholas General Staff Academy after graduation.

In December 1862, Pevtsov was assigned as warrant officer to the 39th Infantry Tomsky Regiment, which was stationed in Tula *guberniia*. Pevtsov's start date coincided with the Polish uprising (1863–64). The regiment in which he

3 Linda 1902, p. 1.
4 Ibid., pp. 2–3.

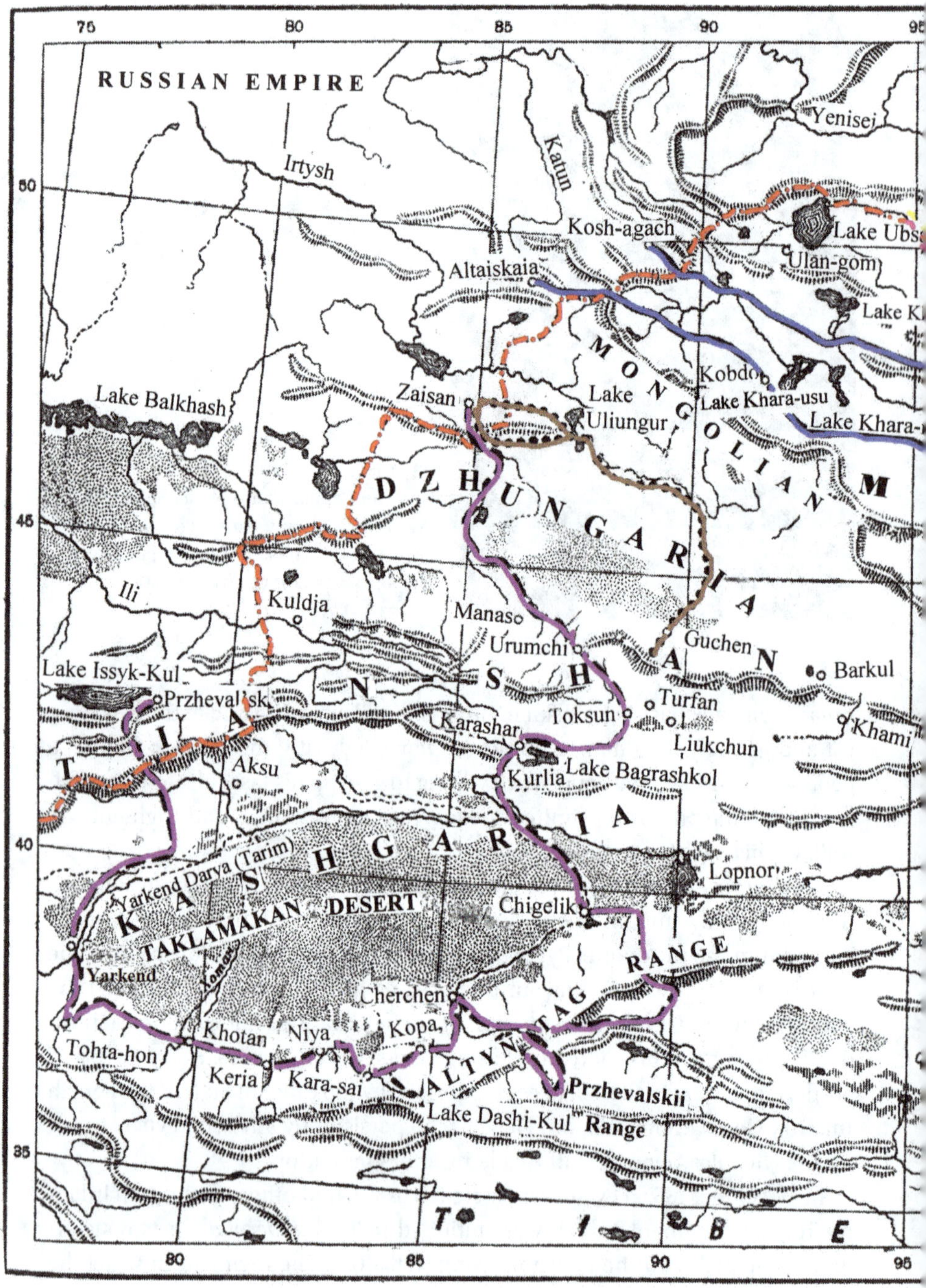

MAP 2.1 M. Pevtsov's expedition routes. Compiled by Alexandre Andreyev on the basis of the map of Pevtsov's travels enclosed in Pevtsov's *Puteshestvie v Kashgariiu i Kunlun'* (Moscow, 1949).

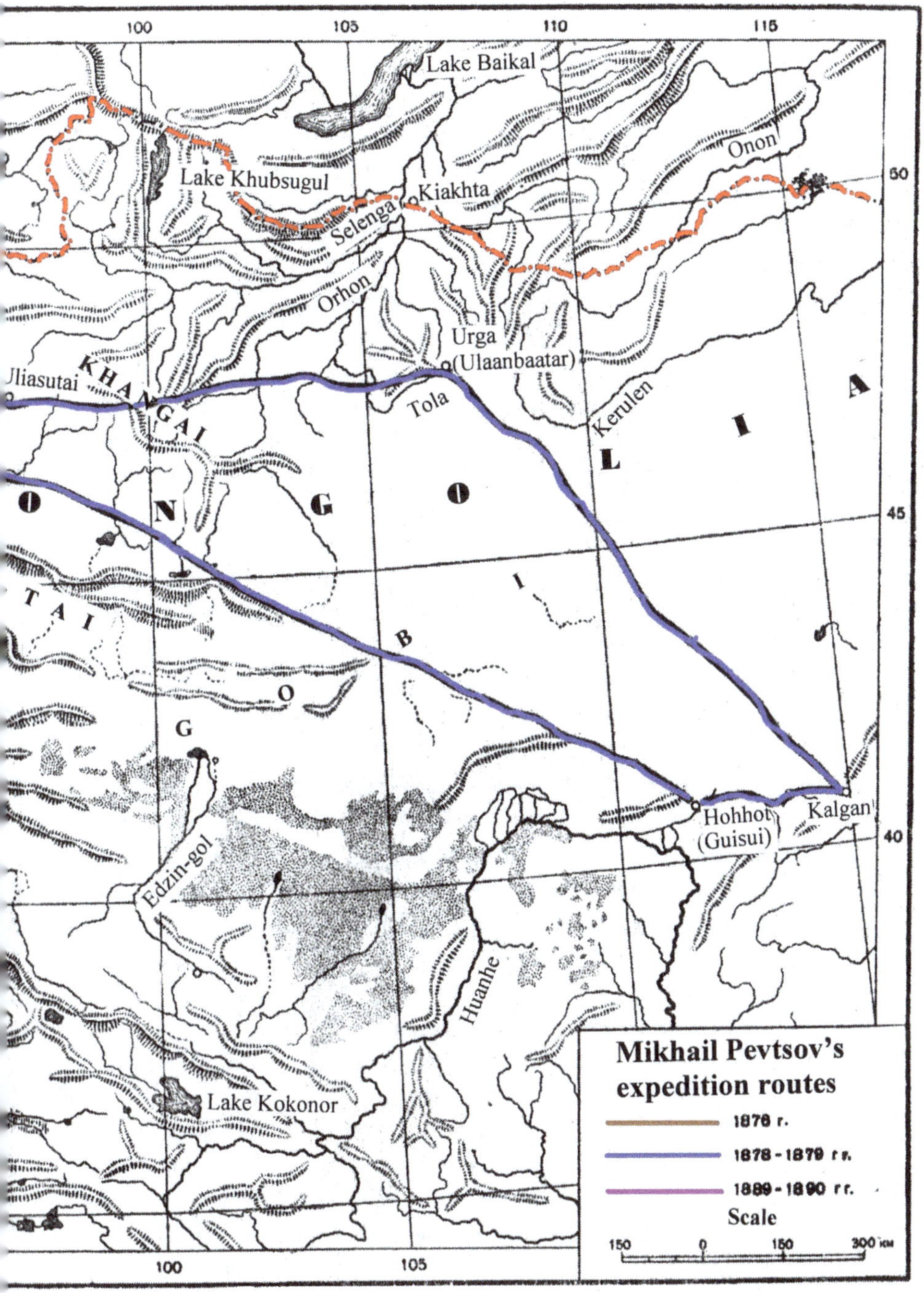

Lake Baikal
Lake Khubsugul
Kiakhta
Selenga
Onon
Orhon
Urga
(Ulaanbaatar)
Uliasutai
KHANGAI
Tola
Kerulen
MONGOLIA
ALTAI
GOBI
Hohhot
(Guisui)
Kalgan
Edzin-gol
Huanhe
Lake Kokonor
Mikhail Pevtsov's
expedition routes
1876 г.
1878 - 1879 гг.
1889 - 1890 гг.
Scale
150 0 150 300 км

served was transferred to Poland to strengthen the Russian troops. In the period between September 1863 and May 1864, Pevtsov took part in combat operations against Polish insurgents, for which he was awarded the bronze medal in memory of the suppression of the Polish Insurrection. After the uprising suppression, the regiment remained in the Kingdom of Poland (*Tsarstvo Pol'skoe*) and garrisoned in Lenchitsa in the Kalishskaia *guberniia*. During Pevtsov's service in the Warsaw Military District he was introduced to Maria Renast, who would later become his wife.

Whilst in Poland, Pevtsov was fascinated with military history and geography and dreamt of entering the Nicholas General Staff Academy. In 1868, at the Headquarters of Warsaw Military District, Pevtsov successfully passed the Academy entrance examinations. He was set on applying to the prestigious geodesics department at the Academy, but in 1868, there were no admissions to the department (it seems admissions would only take place the following year). In August 1869, he successfully passed the Academy exams and enrolled as a student. Pevtsov's time at the academy is mentioned in the memoirs of General Iu.A. Schmidt, then head of the Military-Topographical Department of the Omsk Military District, a friend and colleague of Pevtsov during his time in Omsk.

During Pevtsov's sojourn in the Academy, he had great tendencies for mathematical sciences, regardless of being a student on the general course he nonetheless maintained close relationships with outstanding students in the Geodesics Department including N.Ia. Tsinger, P.P. Kulberg and others. Through them, he made ties with Professor A.N. Savich and the astronomers from the Pulkovo observatory. Even though he was on the general course at the Academy, he successfully kept track of the geodesics course. He would organise private trips to Pulkovo to see his comrades, he practised observing and assimilated various methods of observing the heavenly bodies.[5] In spite of his successes in the sciences, Pevtsov graduated the Academy on the 2nd lowest discharge, due to lower marks achieved for foreign languages.

During his studies in St. Petersburg, Pevtsov also developed his knowledge in the field of natural sciences. He often visited the library and the museum of the St. Petersburg University, where he furthered his skills as an adept preparator of birds and beasts; he studied taxidermy and the basics of conservation and storage of zoological collections. In St. Petersburg, Pevtsov consciously prepared himself for the role of future traveller and explorer, passionately developing the qualities and skills required of a natural scientist. During his study at the Academy Pevtsov had attended meetings at the Imperial Russian

5 Ibid., p. 4.

Geographical Society (*IRGO*), and in May 1867 he had become a full member of the Society.

Service in Western Siberia

Captain Pevtsov graduated from the Academy in 1871. The following year he was appointed to the Staff of Troops of Semipalatinsk *oblast'*(region), which was part of *Zapadno-Sibirskoe general-gubernatorstvo* (Governor-Generalship of Western Siberia); its south-eastern region had a common border with China. The troops based in the Semipalatinsk region played a crucial role in Russian foreign policy during the Muslim Rebellion in China (1864–77). In 1868, a vast area of western China was enveloped in the Muslim rebellion. The Qing dynasty lost all control of Kashgar, Dzhungaria and part of Gansu Province. Kashgar was seized by Kokand chancer Yakub Beg, the Ili region was now ruled by the Sultan of Alakhan, and the rest of the territory of Dzhungaria and Gansu was now in the hands of Dungans insurgents. In 1871 Russian troops defeated the army of Sultan of Alakhan and occupied the Ili region. The troops in the Semipalatinsk region played an active role in the campaign against Kuldja, the capital of the Sultanate. Whilst in the Semipalatinsk region, Pevtsov immersed himself into life by the border. Part of his work at this point included reconnoitring the areas bordering China, inspecting troops, participating in cartographic work and preparing military-statistical accounts. Pevtsov's activity suggested that most of his duty time was spent in the saddle, travelling to remote areas of the region, this had of course greatly contributed to the development of his passion for long distance travel and research.

In Semipalatinsk Pevtsov was part of a small but highly professional group of General Staff officers. The area commander was Major-General V.A. Poltoratskii, who prior to his appointment in Semipalatinsk served as Head of the Asiatic Department of the Main Staff, and as such was actively engaged in the study of Asia. Poltoratskii was a member of the IRGO and had made a name for himself in the circle of Russian geographical science by exploring Tian Shan and the Trans-Ili region. Poltoratskii himself would even encourage his subordinates to undertake military-geographical explorations.

Pevtsov's other colleague was General Staff Captain Julian Sosnovskii. By the time Pevtsov had arrived at Semipalatinsk, Sosnovskii, a member of the IRGO, had already undertaken a number of important reconnaissances along the Russo-Chinese border. In 1872 he led the Bulun Tokhoi expedition to study the valley of the Black Irtysh River in Dzhungaria. Between 1874 and 1875, Sosnovskii led the Russian science and trading expedition to China, which was of

FIGURE 2.2 The residence of Pevtsov during his service in Omsk. OSMLL, 7301/3–6.

great scientific importance as it mapped the unknown regions of Western China. Working with outstanding officers such as Poltoratskii and Sosnovskii could not but had an effect on Pevtsov's future as an explorer of Inner Asia.

In 1875, Captain Pevtsov was transferred to Omsk and appointed as Assistant to Senior Adjutant of the Headquarters of the West Siberia Military District, which was later, renamed the Omsk Military District in 1882. Pevtsov's immediate supervisor was General I.F. Babkov, under whose leadership Pevtsov would end up serving for 15 years. Babkov was an active member of the IRGO and made his name by researching the geography of Altai and Southern Siberia as well as his numerous efforts in organising Russian expeditions to Inner Asia. Pevtsov's new position opened up great opportunities for geographical exploration. In Omsk, Pevtsov lived in a private house on the outskirts of the city in the suburb of Kadyshevskii. A temporary observatory was set up in the Pevtsov's courtyard, where he would carry out daily observations. In addition to his official duties, Pevtsov also taught geography at the Omsk cadet school. Amongst his pupils was young cadet Lavr Kornilov, future Commander-in-Chief of the Russian Army and well-known military orientologist. Pevtsov also independently studied Kazakh and Arabic whilst in Omsk in order to better read the writings of Arab geographers and use these newly learned oriental languages on official duty and during his travels.

During Pevtsov's service in Western Siberia, he enjoyed the full support of the region's two most influential people – the Governor-General of Western Siberia N.G. Kaznakov and Chief of Staff of the West Siberia Military District Lieutenant-General I.F. Babkov. Major-General G.E. Katanaev, Pevtsov's colleague, notes in his memoirs that Pevtsov

> ... being a scientist by nature rather than a clerk, would more willingly undertake not the routine duties, but rather the more interesting ones which were assigned to him by Babkov and Kaznakov. Both of whom often entrusted Pevtsov with matters of special importance.[6]

Pevtsov's First Dzhungarian Expedition (1876)

Pevtsov's first independent geographical exploration had come about as a result of an unusual set of circumstances. In 1874 the Qing Empire started to regain control of the lost territories of Western China, which had been lost in the Muslim uprising. The Chinese army was forced to operate at a considerable distance from the main supply bases in Inner China and as a result, they were in desperate need of food supplies. At the request of the commander of the Chinese Expeditionary Force, Zuo Zongtang, the Russian government in October 1874 allowed the delivery of food to the Chinese authorities. At the beginning of January 1875, the Chinese Expeditionary Corps were sent almost 18,000 *poods*[7] of flour.[8] Due to the increased number of attacks on grain caravans by Dungan gangs of robbers the Russian government ensured the caravans with armed escorts. A Cossack force was sent with a caravan on its way to the town of Gucheng, under the command of the General Staff Captain Pevtsov.

In July 1876 Pevtsov reported on the placement of its Russian garrison in Bulun Tokhoi and on the movement of the main unit in Gucheng.[9] Pevtsov was assigned with gathering as much detailed information on the area the caravan covered, especially between Bulun Tohkoi and Gucheng, where until that time no other traveller had done so.[10] Pevtsov's activity in Dzhungaria was part of a

6 Ibid., p. 45.

7 *Pood* – a unit of weight of the Imperial Russian weight measurement equal to 16.38 kilograms.

8 RGVIA, F. 846, Voenno-uchenyi arkhiv, D. 6887, l. 16: Register of the events in near-border region.

9 Ibid., ll. 85–87: Report of General Staff Captain Pevtsov to the Chief of Staff of West Siberia Military District, 6 July 1876.

10 Pevtsov 1879, p. 1.

larger program of the Russian General Staff to study the area upon which soon hostilities between Zuo Zongtang's expeditionary force and Yakub Beg's army would play out. Pevtsov's expedition initiated extensive strategic reconnaissance of the area. He was followed by officers of the General Staff including Captain A.N. Kuropatkin (October 1876) and Lieutenant-Colonel N.M. Przhevalskii (August 1876), who were set out with the task of reconnaissance and assessing the combat potential of Yakub Beg's army.

Pevtsov's first Dzhungaria expedition took place in 1876 from May to September that year; he was to guard the grain caravans. He and his hundred Cossacks were accompanied by class topographer, A.V. Skopin, who was instructed to conduct a route survey. The Dzhungaria basin which Pevtsov had to cross was located between enormous mountain systems – Tian Shan to the south and Altai to the north, it consisted of many semi-deserts, desert plains, and high plateaus. The first Russian to visit this part of Dzhungaria was the Russian Ambassador Fedor Bankov in 1654. Geographical studies and descriptions of this part of Inner Asia were conducted much later when in 1864 the southern mountainous part was surveyed by O.V. Struve and G.N. Potanin. Lieutenant (*poruchik*) L. Matusovskii, a military topographer, led the topographic work in Dzhungaria between 1871 and 1872. Matusovskii conducted work between Zaisan (Russian frontier post) and Manas, through Ulungur Lake, whilst Captain J. Sosnovskii undertook the return journey in 1875 from China to Zaisan through Lake Ulungur and the Chiliktinsk Valley. Their route surveys helped compile the preliminary map of Northern Dzhungaria, which required additions and refinement as parts of this territory remained unknown.

On May 16, 1876, Pevtsov's detachment set out to Zaisan along the Koba Valley to the Chinese town of Bulun Tokhoi. The caravan route ran through the eastern, least studied part of Dzhungaria. They first came across the salt lakes of Ulungur and Baga Nur, which were previously visited by Russian travellers. Pevtsov then carried on through the desert with its sparse vegetation and headed to the foothills of the Mongolian Altai along the valley of the Urungur River. This section of the road remained completely unknown to Europeans, moreover, the locals themselves knew very little about it. Pevtsov was able to survey the Urungur Valley as well as Lakes Ulungur and Baga Nur. He was able to better establish a more accurate position of Lake Baga Nur and measure the depth of Lake Ulungur. On the basis of barometric determinations, he found that the two lakes occupied a large part of the basin. Pevtsov avidly surveyed the Ulungur River, which was previously undertaken by Captain Sosnovskii just over 70 *verst*[11] upstream from the town of Bulun Tokhoi. Pevtsov conducted the survey to the outskirts of the Southern Altai Mountains after passing 255 *versts* in the region, all of which was unmarked and unknown territory. Pevtsov then made his way south through the most south-eastern point of Dzhungarian Gobi, before reaching his final destination – the town of Gucheng, located in an oasis on the northern slopes of Tian Shan.

The entire journey between Zaisan and Gucheng took 47 days with the caravan safely reaching its destination proving the first task of the journey was a success. Pevtsov allowed his detachment to rest after what was a difficult journey. From Gucheng he undertook a short trip of 40 *versts* south into the Tian Shan Mountains to determine the general direction of the main ridge and the height of its snow line. Pevtsov determined the approximate height of the Bogdo Ula mountain peak to be 15,500 feet.

The return journey took Pevtsov over a month having gone along a somewhat different route only to become familiar with the eastern mountainous part of Northern Dzhungaria. However, he did not come across the spur of the Mongolian Altai mountain range, as shown on the map of the Military-Topographical Department of the Main Staff (1859). At the border mountainous area south of Zaisan Pevtsov repeated the error of the *Survey Map of the Region of the Black Irtysh River*, which was put together by Lieutenant Matusovskii and by General Staff Captain Sosnovskii, which indicated a non-existent mountain junction linking ranges of Tarbagatai, Saur, Urkashar, and Semistai. After inspecting the mountainous part of Northern Dzhungaria the Pevtsov expedition turned east to Lake Ulungur and returned to Zaisan on September 10, 1876.

Whilst travelling Pevtsov produced astronomical observations in six various points and carried out barometric determinations in 17 points (the height was defined twice, both on the journey there and on the journey back). The expedition marked up the city plans of Gucheng and Bulun Tokhoi, as well as neighbouring astronomic points. Pevtsov reported the results of the geographical research to the Headquarters of the West Siberia Military District in a separate report.[12] The report was Pevtsov's first scientific work. Upon his return from the expedition, Pevtsov had put together a geographical outline of his visit, to which he added astronomical, magnetic and barometric notes concerning

11 *Versta* (*verst*, pl.) – is an obsolete Russian unit of length, one *versta* equals to 1.0668 kilometres (0.6629 miles).

12 RGVIA, F. 846, D. 6887, ll. 153–158: Report of General Staff Captain Pevtsov on the works in Western China.

Dzhungaria, as well as materials relating to the zoogeography of the area. Along with the report he presented a map of Dzhungaria (scale: 1 inch equals 40 *verst*), which was compiled by the Military-Topographical Department of the West Siberia Military District. The map was based on Pevtsov's geodetic work as well as topographer Skopin's route surveys. The map was a consolidated work, which included the latest materials on Russian surveys of the 1870s. Pevtsov's report included valuable ethnographic material on the locals of the region. His interest in ethnographic research continued to show in his subsequent expeditions to Inner Asia.

Whilst in Dzhungaria, Pevtsov amassed interesting geological, botanical and zoological collections, which he then went on to donate to the IRGO. Leading Russian experts were put in charge working through the accumulated material that Pevtsov had gathered. Academics F.B. Schmidt and K.I. Maksimovich took part as well as I.S. Poliakov, a keeper of the Zoological Museum of the Imperial Academy of Science.[13]

The scientific results of Pevtsov's trip were approved by the Council of the IRGO. According to P.P. Semenov, the Vice-Chairman of the Society, the journey had completed what scientists had long sought after – a thorough examination of the Lake Ulungur and the south-eastern extremity of Altai, as well as studies of the territory between Altai and Tian Shan and the mountain Bogdo Ula.[14] Pevtsov's successful expedition propelled him as a prominent and active representative of the IRGO. In 1879 he was awarded the small gold medal of Society. Nonetheless important was that the expedition had offered Pevtsov real valuable experience; he acquired organisational and research skills under field conditions, during his time in Dzhungaria as a commander of escort detachment of the commercial caravan.

Pevtsov and the West Siberia Branch of the IRGO

New opportunities for Pevtsov as an explorer of Inner Asia opened up as a result of the establishment of the *Zapadno-Sibirskii Otdel* (West Siberia Branch) of the Imperial Russian Geographical Society (*ZSO IRGO*), which began its work on May 10, 1877. The elected chairman was the Chief of Staff of the West Siberia Military District, Major-General I.F. Babkov, who was Pevtsov's immediate

---

13 See N.V. Slepkova, pp. 86–87.

14 P.P. Semenov, *Istoriia poluvekovoi deiatel'nosti Imperatorskogo Russkogo Geograficheskogo Obshchestva, 1845–1895*, Chast' II, Otdel IV, St. Petersburg: tipografiia V. Bezobrazova, 1896, p. 578.

FIGURE 2.3 Office of the West Siberia Branch of the IRGO and the Branch's Museum, Omsk, late 19th century. OSMLL, 4325.

superior in the district headquarters. Pevtsov himself was elected Governor of Affairs and Chairman of the Executive Committee (from 1879). Members of the branch included prominent researchers of Siberia and Inner Asia, such as N.G. Potanin, N.M. Iadrintsev, A.V. Andrianov and others.

Pevtsov was a capable and energetic officer and his new appointment was welcomed by members of the branch. Potanin believed that Pevtsov's position within the branch would bring great benefit to the area of geographical science.[15] During his time there Pevstov developed regulation drafts on branch activities (July 1877), edited scientific papers published in the *Zapiski* ZSO IRGO and carried out extensive correspondence. He also took the initiative to update the 50 *versts* per inch map of Western Siberia, which was created in 1848. As a result of his efforts, the branch library was formed, stocked and replenished with books and magazines. On 15 December, 1879, at the branch meeting, Pevtsov made news of his journey between the Russian Zaisan post to Kuku Khoto and back via Urga to the Russian town of Biisk. He donated the branch his collection of birds and minerals which he had accumulated during his latest expedition. In 1884, at the branch's request, Pevtsov processed astronomical

15 See Pis'mo N.G. Potanina k N.M. Iadrintsevu: 11/23 August 1879, Kobdo, in *Pis'ma Potanina*, T. 3, Irkutsk: izdatel'stvo Irkutskogo universiteta, 1989, p. 165.

observations of K.A. Lazarev, in order to determine the geographical location of places in Altai and on the upper Bukhtarma River and barometric determination of heights in the south-eastern part of the Semipalatinsk *oblast'*.[16]

On March 15, 1902, after Pevtsov's death, the ZSO *IRGO* dedicated a special meeting to his memory. Speaking at the meeting, the head of the Military-Topographical Department of the Siberia Military District, Major-General Iu. Schmidt, who knew Pevtsov on joint service, gave a brief outline of his life and work. As Schmidt noted, Pevtsov was well respected and left an impression even on those who did not know him well, recognised was his simplicity and unfailing responsiveness to the needs of others.[17] General Staff Lieutenant-Colonel K.P. Linda, a member of the branch, put together a biographical essay and brief description of Pevtsov's travels, which remains an important source in the study of the life and work of Pevtsov.[18]

Expedition into Western Mongolia and the Northern Chinese Provinces (1878–79)

By the end of the 1870s, tensions were high between Russo-Chinese relations, due to Russia's occupation of the Ili Valley. After regaining control of Western China, the Qing government raised the question of the return of the Ili Valley to China. In historiography this episode of Russo-Chinese relations was known as the "Ili crisis". China's attempts at regaining the valley, was met with great resistance from the Russian government. The Ministry of War and the Main Staff began to implement a series of measures aimed at preparing for an armed conflict with China in case final attempts to resolve the issue came to no avail. The Main Staff started working out specific plans for military action against China. The plan called for the protection of the Russian possessions in the Ili Valley, which envisaged the Chinese defeat at the Russian border having to 'use all efforts to create in Western China, a Kashgar and Dungans Muslim khanates'.[19] The atmosphere of impending conflict was strongly sensed in Russian

16 *Zapiski* ZSO *IRGO*, kniga VI, Omsk, 1884, pp. 1–6.

17 *Zapiski* ZSO *IRGO* za 1902, Omsk, 1903, p. 11.

18 K.P. Linda, M.V. Pevtsov i ego puteshestviia, in *Iubileinyi sbornik Zapadno-Sibirskogo otdela IRGO*, 1877–1902, Omsk, 1902.

19 RGVIA, F. 1396, Op. 2, D. 111, l. 104 ob.: Report, Military-strategical description of Dzhungaria and the western part of Kashgaria and a concept of action of Turkestan and West Siberia Military Districts in the event of war with China.

society. The biggest question for the Russian press at that time was – 'What is the likelihood of war between China and Russia?'[20]

Under such conditions, the Main Staff needed more precise geographic and military information on Western China. In early 1878 several reconnaissance missions were dispatched into Dzhungaria, Russian officers who took part included General Staff Colonel P.P. Matveev (1878), General Staff Colonel N.M. Przhevalskii (the 3rd Inner Asian journey, 1879–80), Lieutenant-Colonel N.S. Maevsky, Staff-Captain M.P. Tikhmenev and Lieutenant P.P. Kalitin (1880). In 1879, a medical officer A.E. Riegel, son of well-known Russian botanist E.L. Riegel, completed the trip from Kuldja to Turfan and delivered valuable information about the location and condition of Chinese troops.

Of great interest to the Russian Main Staff were the regions which connected Chinese inland provinces with Dzhungaria. In case of an armed conflict, these areas were expected to be used to bring reserves and weapon supplies for the Chinese army to the theatre of war in Dzhungaria and Kashgaria. In 1879, the Main Staff was presented with an opportunity to carry out a long-range reconnaissance of Western Mongolia and Gansu-Shanxi Province.

In early 1878, P.P. Semenov, Vice-Chairman of the IRGO, received a note from the well-known researcher of Mongolia G.N. Potanin, that Biisk merchants that autumn were determined to send a caravan of elk horns from Kobdo to Guihuacheng in Shanxi Province. The caravan was supposed to take the direct commercial path which ran close to the northern foot of the mountain range of Southern Altai, which had not yet been visited by Europeans. Potanin also reported that officer Pevtsov had volunteered to accompany the caravan.

The Pevtsov expedition in a way tackled the question of connecting Khangai with the Mongolian Altai. Potanin's proposal was supported by General Babkov, Chief of Staff of the West Siberia Military District. The Council of the IRGO requested that two military topographers and an escort of Cossacks should accompany Pevtsov along the way. The Governor-General of Western Siberia, N.G. Kaznakov obtained by petition the Supreme order to assign to Pevtsov two military topographers – Alexei Skopin (who had participated in the first Dzhungarian expedition with Pevtsov in 1876) and Ivan Chuklin. The expedition also allocated six Trans-Baikal Cossacks who were familiar with the Mongolian language.

The expedition's mission was to primarily carry out rough route surveys along the proposed route, as well as astronomical determinations of geographical coordinates, barometric readings of heights and descriptions of routes. Whilst additionally collating ethnographical, commercial and natural historical information on the places they visited.

---

20 *Golos*, 1880, № 198 and 200.

The expedition route cut through Inner Asia from the northwest to the southeast, having traversed the Mongol Altai to the Yinshan Mountains (Daqing Shan). Pevtsov's expedition departed on August 3, 1878 from the Altaiskaia *stanitsa*[21] and within three weeks had reached the town of Kobdo, from there they followed a south-eastern direction along the northern foot of Southern (Chinese) Altai. Between Kobdo and Kalgan Pevtsov travelled south of Ney Elias's route; Elias during his expedition into Western Mongolia between September and December of 1872, had passed along the old post road from Kalgan to Kobdo. To the south-east of Kobdo, on a stretch of about 280 *verst*, Pevtsov passed by the 'Valley of the lakes', located along the northern slopes of the Mongolian Altai, which was still yet to be explored by any European. As such Pevtsov was one of the first to cross the Altai region, where mountain ranges connected the Mongolian Altai and the Khangai Range and the Gobi Altai Mountains with Daqing Shan.

Passing the 'Valley of the lakes' the expedition crossed the Gobi desert at its most widest and dangerous stretch, eventually reaching the Chinese city of Guihuacheng (Kuku Khoto in Mongolian, contemporary Hohhot). Having parted with the merchants, the Pevtsov expedition progressed on to the nearby city of Kalgan or modern Zhangjiakou. Here the members of the expedition would spend almost two months recuperating and carrying out various scientific observations.

The return route home lay along the straight caravan route through Urga (modern Ulan-Bator) and the city of Uliastai, which was well known to travellers. On July 20, 1879, Pevtsov arrived at the settlement of Kosh Agach by the Russian border and thus ending his thirteen month journey.

In February 1881, Pevtsov presented a brief report at the IRGO general meeting summarising the main achievements of the expedition to Mongolia and China.[22] Understanding the autonomy of the Mongolian Altay and the absence of an orographic link with the Khangai ridge was the most significant result of the expedition. The expedition helped further refine the map of Inner Asia's third major mountain system – Khangai Mountains. Valuable information was obtained about the Central Gobi region, which had previously been crossed in a meridional direction by Prezhevalskii. Pevtsov on the other hand, made the journey in a different direction crossing from north-west to south-east. At the same time, Pevtsov made important discoveries in the area of geodesics, having established that many rivers rise from the Khangai ridge, whereas in the Mongolian and Gobi Altai there are almost no sources of rivers.

21 *Stanitsa* – large Cossack settlement.

22 *Otchet IRGO za 1880*, St. Petersburg: tipografiia V. Bezobrazova, 1880, p. 21.

He was also the first to study the geodesics of the basin of large lakes in the northwest of Mongolia, as well as the upper reaches of the Dzabkhan River.

The expedition held an important significance for cartography of Inner Asia. During his travel Pevtsov used a universal instrument by Pistor & Martins, three pocket chronometers by Viren, Dent and Brockbanks (the latter proved to be unreliable and was not used by Pevtsov), a barometer, thermometer, a large optical telescope of 100x, along with a telescopic support. With the aid of these instruments Pevtsov was able to determine the geographical location of 28 points and measured 44 heights. He would try to carry out the measurements in the same locations where they were previously conducted by his predecessors, such as in a private house in Kobdo (used by topographer Rafailov in expedition of Potanin, 1876–77), in Kuku Khoto in Splenger's residence (which housed a Catholic spiritual mission), in Kosh Agach where Captain S.T. Miroshnichenko made observations in 1869. Professor Major-General K.V. Sharngorst of the Nicholas General Staff Academy, calculated the field astronomical points which were collated throughout the expedition.[23] Evaluating Pevtsov's results, Sharngorst noted that with numerous readings Pevtsov had carried out a great service for the field of geography, especially with the limited means available at his disposal, pointing out that he would have done any professional astronomer proud.[24] However, Sharngorst also noted that some of Pevtsov's readings had errors, most probably due to observation difficulties, tool imperfections as well as Pevtsov's own errors. Pevtsov's astronomical observations in Urga were 'taken with caution' by Sharngorst, who considered them to 'carry less scientific weight'.[25] The Uliastai latitude was not determined by Pevtsov due to weather conditions.

During the travel in Mongolia and China, topographer Skopin produced route surveys (scale: 1 inch equals 5 *verst*) over the 4,070 *verst* in total, from Kobdo to the town of Kuku Khoto (2,200 *verst*), from Kuku Khoto to Kalgan (350 *verst*) and from Urga to Uliastai (1,500 *verst*). Astronomical points determined by Pevtsov and military topographer Rafailov (in Potanin's expedition of 1876–77), served as control points.[26] The survey was conducted from the town of Kobdo, as the route from the Russian border to Kobdo had previously

23 Otzyv K.V. Sharngorsta o zaslugakh polkovnika M.V. Pevtsova, in *Otchet IRGO za 1885 god*, St. Petersburg: tipografiia A.S. Suvorina, 1886, p. 14.

24 Ibid.

25 Ibid, p. 329.

26 Topograficheskie raboty v Omskom Voenno-topograficheskom otdele, in *Zapiski Voenno-topograficheskogo otdela Glavnogo shtaba*, Chast' XXXVIII, St. Petersburg: Voennaia tipografiia, 1883, p. 18.

been surveyed twice, by the General Staff Captain A.G. Prints (1863) and military topographer Matusovskii (1870).[27] In 1883, having obtained all data and material from the Pevtsov expedition, the Main Staff were presented with an opportunity to create a consolidated map of Mongolia, which included the works of Przhevalskii, Rafailov and Potanin. Along the way Pevtsov managed to put together a significantly large natural-science collection, including zoological (about 200 species of mammals, birds, fish and reptiles), mineralogical (around 100 rock specimens and minerals), botanical (approximately 180 species). Due to their late departure out of Kobdo (September 5, 1878) it was not entirely possible for Pevtsov to create a full-scale botanical collection on a route from Kobdo to Kalgan.[28]

In 1880 Pevtsov published the preliminary results of his expedition,[29] and two years later followed the full report of his travels.[30] The work contained new and valuable information on orography and geodesics, flora and fauna of Mongolia and Mongolian ethnography. The report annexes contained astronomical tables and barometric readings, lists of birds, all of which was put together by Pevtsov along his travel in Mongolia. 'The map of Mongolia' held a particular importance. It was compiled as a result of route surveys carried out by topographers and a variety of cartographic sources (a total of 20), as well as the Chinese geographical works *Menggu youmu ji* (*Notes on the Mongolian encampments*).

Details delivered by Pevtsov held not only an important significance for the geographical study of Inner Asia but also for the Russian military, for future war planning in the event of an armed conflict with China over the Ili region. On his return Pevtsov was awarded the Order of Saint Vladimir 4th Class, for successfully completing his assignment. This was the highest award bestowed upon Russian officers for completing remarkable reconnaissance missions in Asia. For his scientific work (and backed by General Sharngorst's review in 1886), the IRGO awarded Pevtsov the Count Litke Medal (the highest award bestowed by the Society). Military surveyors who took part in the expedition were also promoted in rank.

27 *Otshet ZSO IRGO za 1879 god*, Omsk: 1879, p. 12.

28 Pis'mo N.G. Potanina k N.M. Iadrintsevu, St. Petersburg, 15 November 1879, in *Pis'ma Potanina*, T. 3, Irkutsk: izdatel'stvo Irkutskogo universiteta, 1989, p. 148.

29 M.V. Pevtsov, Kratkii ocherk puteshestviia po Mongolii i Vnutrennemu Kitaiu v 1878 i 1879 gg., in *Izvestiia IRGO*, T. XVI, vyp. V. St. Petersburg: tipografiia A.S. Suvorina, 1890, pp. 435–457.

30 M.V. Pevtsov, Ocherki puteshestviia po Mongolii i severnym provintsiiam Vnutrennego Kitaia, in *Zapiski ZSO IRGO*, kniga V, Omsk: tipografiia okruzhnogo shtaba, 1883, pp. 1–354.

After the second expedition to Inner Asia, Pevtsov returned to Omsk, where he served as Staff Officer at the Headquarters of the West Siberia Military District. In 1881 he was honored rank of Colonel and in 1886 appointed as Senior Adjutant of the Headquarters of Omsk Military District. During this period he collaborated with the West Siberia branch of the IRGO and was primarily occupied with scientific research. Pevtsov set up a small observatory in the courtyard of his house, where he would be occupied daily with astronomical observations, honing his new method of determining latitudes. The results of his discovery were published in an article entitled *Determining the latitude by means of the height of two stars*.[31] In 1881, Pevtsov published a small print run of textbooks for military academies, entitled *The initial foundation of mathematical and physical geography*. Authority's in the field felt the textbook was of a very high scientific level. However, Pevtsov for some reason did not circulate the book for public sale and showed it to no one.[32]

In 1882 and 1883 Pevtsov as Government Commissioner participated in the boundary delimitation with China. In accordance with Articles 8 and 9 of the St. Petersburg Treaty (February 12, 1881), signed between Russia and China with regards to the Ili region, a change was agreed in the Russo-Chinese boundary. Previously the boundary was determined by the Protocol of Chuguchak (September 25, 1864) of the delimitation of Russo-Chinese state border from Shabin Dabaga to the Qunling mountain ridge. To work on boundary delimitation, Russia and China both agreed to appoint special border commissioners. Their main assignment was to agree on boundary lines and sign various protocols. The Russians appointed Chief of Staff of the Omsk Military District Lieutenant-General, Ivan Babkov (who held a similar position in 1862 during the delimitation with Western China) and Colonel Pevtsov as border commissioners.

On July 31, 1883, Babkov and Pevtsov together with the Chinese commissioners signed the Kabinsky protocol (named after the valley of the Kaba River where it had been signed), which fixated the direction of the Russo-Chinese border between Altai ridge and Saur (Sairin Ula) Mountains.[33] On August 23, 1883, Russian and Chinese commissioners signed off the description of the northern section of the Russo-Chinese border from the upper Ak-Kaba River to

31 M.V. Pevtsov, Ob opredelenii geograficheskoi shiroty po sootvetstvennym vysotam dvukh zvezd, in *Zapiski IRGO po obshchei geografii*, T. XVII, № 5, St. Petersburg: tipografiia A.S. Suvorina, 1887, pp. 1–115.

32 For more details see Linda, op. cit., p. 14.

33 See Protokol ob izmenenii russko-kitaiskoi gosudarstvennoi granitsy ot gory Mus-tau (gornye khrebty Saur) na iuge do istokov r. Ak-Kaba (Iuzhnyi Altai) na severe, in *Russko-kitaiskie dogovorno-pravovye akty. 1689–1916*, Moscow: Pamiatniki istoricheskoi mysli, 2004, pp. 147–150.

the mouth of the Alkabek River, which had been established by Protocol on July 31, 1883.[34] On September 1, Pevtsov signed the last Protocol, *The description of the southern section of the Russo-Chinese border from the mouth of the Alkabek River to Mount Mus-tau (mountain ranges of Saur)*.[35]

During the delimitation Pevtsov often met with Chinese commissioners and discussed not only issues of determining the direction of the borderline, but also various diplomatic issues involved. Negotiating with the Chinese commissioners was quite difficult and required a lot of patience and tact from Pevtsov. With regards to the atmosphere of the time, in the joint Russo-Chinese boundary commission camp in the Kaba Valley, the Russian media reported:

> With regard to the boundaries they are still ongoing, nothing solid has been set. 21st [May 1883], there was a minor accident involving the Chinese Commissioner [Sheng-tai], who found his porter and groom playing cards for money. For this, he decided to execute them both, but at the request of the Second Commissioner, he reduced the punishment for both and cut off their fingers on their left hand instead – one had four fingers chopped off, the other – two. Both were admitted to a Russian hospital, where they are being treated for their wounds.[36]

During delimitation work Pevtsov visited the bordering areas, in particular, northwest Dzhungaria. Here he produced surveys to supplement the cartographic materials of this unknown area. Pevtsov's work consisted of area surveying (scale: 1 inch equals 5 *verst*), which was ceded to Russia. The Omsk Military-Topographical Department unit took part in the topographic surveys, under the leadership of military surveyor Colonel S.T. Miroshnichenko. Pevtsov was tasked with visually surveying the largest possible area of Chinese territory to the east of the assumed new boundaries and defining astronomical points for plotting this survey. The Council of the IRGO and the Military-Topographical Department of the Main Staff supplied him with the necessary scientific tools. Pevtsov conducted a review of the near-border territory, as well as geodetic and topographical work for two seasons between June 7 and October 25, 1882 and from May 15 to October 25, 1883. During this period with the help of Stephan's bearing circle Pevtsov surveyed about 50 thousand square *verst*, he determined the geographical position of 8 points and took barometric

34 Ibid., pp. 151–153.

35 Ibid., pp. 154–156.

36 *Turkestanskii Sbornik*, T. 326, p. 149.

readings of several heights.[37] For works on delimitation he was awarded by the Imperial rent (*arenda*)[38] the annual sum of 600 rubles, as well as 2,250 rubles in silver.

Tibetan Expedition of 1889–1890

In January 1887 Pevtsov was transferred to the Main Staff, where he was made Staff Officer of the Asiatic Department of the Main Staff. This military body was responsible for the activities of Asian military districts and for the military planning in the East. It is believed that Pevtsov's transfer to St. Petersburg happened as the result of the support of the Assistant to the Chief of the Main Staff General M.F. Mirkovich.[39] However, it seems that the actual figure involved in this appointing was General Staff Colonel L.F. Kostenko, who occupied the important position of Senior Staff Officer of the Military-Scientific Committee of the Main Staff.

Colonel Kostenko was an active member of the IRGO and spent almost 20 years of service in Turkestan. He was well known for his military geographical works and expeditions to Turkestan. In 1885, Kostenko was posted to Omsk where he was temporarily assigned to the Commander-in-Chief of the Omsk Military District in order to prepare important military-geographical work on Dzhungaria. During this period he was introduced to Pevtsov, who provided great assistance in preparing scientific research. In the preface to his confidential work *Dzhungaria. A military statistical survey* Kostenko wrote about Pevtsov's assistance.

> Mr. Pevtsov's data completely satisfied the scientific requirements and holds an important significance as they yield valuable material for purely military purposes. Besides the research which had already been published, Mr. Pevtsov presented for my disposal handwritten notes, which are the results of reconnaissance carried out by the author after publishing his previous journeys. Extractions from his notes were prepared spe-

37 M.V. Pevtsov, Rezul'taty astronomicheskikh nabliudenii dlia opredeleniia geograficheskogo polozheniia mest i izmerenii barometrom vysot v severo-zapadnoi Dzhungarii, in *Izvestiia IRGO*, T. XXV, vyp. I, St. Petersburg: tipografiia A.S. Suvorina, 1889, p. 97.

38 *arenda* – in the Imperial Russian Army this was a monetary reward to any outstanding officer paid annually for a specific period of time, usually 12 years.

39 See Linda, op. cit., p. 17.

> cifically for myself, therefore, are in print for the first time, and I consider it my duty to express to Mr. Pevtsov my deepest gratitude.[40]

A few months after Pevtsov's transfer to the Main Staff, Kostenko was appointed as Head of the Asiatic Department of the Main Staff and thus Pevtsov's superior. Following his transfer to St. Petersburg Pevtsov was primarily involved in the affairs of the Asian military districts as well as Russian military planning in Asia. In the early period of his service in St. Petersburg, there is hardly any desire left in him to continue expeditionary activities in Inner Asia. However, in the Autumn of 1888 in the remote town of Karakol on the shores of Lake Issyk Kul events would occur which would yet again see Pevtsov return to Inner Asia.

On October 20, 1888, in preparation for the 5th Inner Asian expedition there was news of the sudden death of Przhevalskii. The Russian government had allocated enormous resources to this mammoth expedition, which aimed at exploring Tibet. Prior to his death, Przhevalskii's main goal in life was to reach Lhasa. Przhevalskii's death put the Council of the IRGO into a very difficult position. Large funds were spent on organising the expedition to Tibet, and its fiasco had the potential of inflicting serious damage on the international authority of Russia. The Council of the IRGO unanimously decided that Przhevalskii's expedition plans should continue, and was seen as 'the best monument, not made by human hands, erected over Przhevalskii's freshly dug grave'.[41] In view of the ensuing winter a decision was taken to push the expedition to spring of 1889.

Przhevalskii's death created somewhat of a 'personnel crisis' within the IRGO and thus an absence of researchers, capable of conducting large scale and remote expeditions into Inner Asia. P.P. Semenov believed that the leader of the new expedition should be an officer of the General Staff, with a scientific background, extensive travel experience and the ability to create route surveys and carry out astronomical work.[42] On the recommendation of the Assistant to the Chief of Main Staff Lieutenant-General Mirkovich, Pevtsov was appointed as head of the expedition, having had two previous expeditions into

40 L.F. Kostenko, Chzhungariia. Voenno-statisticheskii ocherk, in *Sbornik geograficheskikh, topograficheskikh i statisticheskikh materialov po Azii* (SMA), Vyp. XXVIII, St. Petersburg: Voennaia tipografiia, 1887, p. x.

41 RGVIA, F. 401, Op. 4, D. 53 (1888), ll. 2–2 ob.: A letter from the Vice-Chairman of the IRGO to the Chief of the Main Staff, 29 October 1888.

42 Ibid. l. 3.

Mongolia and China, and was 'completely prepared to carry out astronomical observations and determining geographical coordinates'.[43]

Pevtsov's appointment as head of the scientific expedition, by the Main Staff, proved to be very successful. Of all the Russian military researchers he was most prepared as a geoscientist and cartographer, he had experience travelling to remote areas and showed the necessary personal qualities – will, endurance, discipline and rational thinking. Pevtsov led the expedition at a time when geographical research into Inner Asia had reached a new epoch. The old approach of 'conquest' (as described by Przhevalskii) gave way to a more pragmatic and scientific approach. Despite Przhevalskii's accumulated important results, he viewed them as mere 'scientific reconnaissance' and indicated that for a more careful study of the countries he visited, a calmer more balanced researcher was needed. Pevtsov by his nature and mindset belonged to precisely such natures, he was one of the more useful and honest followers of Przhevalskii.

Pevtsov in comparison to Przhevalskii belonged to the new generation of researchers. He relied greatly on Przhevalskii's scientific heritage, but in its studies, style and methods, he went much further. Unlike Przhevalskii, Pevtsov did not pursue any political goals, in terms of expeditionary activity he did not set unrealistic tasks, and least of all did he desire to make a name for himself or acquire material benefits. As Kozlov rightly noted 'having completely surrendered to intellectual life, Pevtsov was never interested in the riches of the world, honour, prominence or success. He could have easily made a career if he were in any way ambitious'.[44] Unlike Przhevalskii Pevtsov was a man of a different cast. His scientific work was particularly different. He was prone to pre-organised, systematic and thorough scientific information, including geodesic, topographic, natural-scientific, historical and cultural. He was equally interested in the results of scientific work and the process itself, especially the technique of geographical research. During Pevtsov's time, Russian Inner Asian expeditions had drastically changed, with increasingly more expeditions becoming more civic and scientific, steadily losing the militaristic approach, which Przhevalskii had deemed necessary. Pevtsov moved away from the principle of military organised expeditions and began involving more civilian specialists (something Przhevalskii had advocated against).

---

43 Ibid., l. 9: Memorandum of the Main Staff. On resuming the expedition of the late Major-General Przhevalskii and some changes in staff of the expedition, 19 November 1888.

44 P.K. Kozlov, Pamiati Mikhaila Vasil'evicha Pevtsova, in *Izvestiia IRGO*, T. XXXVIII, Vyp. 6, St. Petersburg: tipografiia V. Bezobrazova, 1905, p. 619.

Pevtsov's research had a more pragmatic element to it, often associated with implementing specific military tasks – reconnaissance of areas involving potential military activities, most dangerous directions, strategically important communications which lead to China's inland territories, etc. Przhevlaskii's expeditions to Inner Asia, were undertaken in somewhat offensive style and encompassed huge geographical areas. At the same time, they were striking to the imagination since it was he who first broke down the general prejudice – the halo of inaccessibility surrounding the vast unmapped region of Inner Asia. Contemporaries often compared him to Yermak, the well-known Russian subjugator of Siberia. Hence Przhevalskii's obsession with reaching Lhasa at all costs becomes more understandable.

With Przhevalskii each journey would be its own project, in which other participants played the role of auxiliary personnel. For Pevtsov however these members were independent researchers to whom he would entrust critical tasks. If for Przhevalskii his expedition account included a comprehensive book under his own name, then for Pevtsov this meant including his closest associates as co-authors. Pevtsov never credited himself personally and provided the opportunity for other members of the expedition to prove themselves in the field of science. During the Tibetan expedition, each participant had carried out independent work and had the opportunity to publish their scientific work as an independent part of the general report of the expedition. Thus the tradition was laid, later to be continued by Roborovskii and Kozlov.

The proposed expedition to the northwestern outskirts of Tibet also included such figures as Lieutenant (*poruchik*) V.I. Roborovskii and Second Lieutenant (*podporuchik*) P.K. Kozlov, previously assigned to the expected 5th Inner Asian expedition of Przhevalskii. As advised by the Council of the IRGO, geologist K.I. Bogdanovich was included in the expedition. Przhevalskii's initial convoy consisted of 25 soldiers, however due to changes in the expedition programme this convoy was reduced to 12 men. This two year expedition was allocated a total sum of 46,000 rubles.

Whilst developing the new expedition program, the main issue was to identify key areas of Inner Asia, which could be further studied. As Przhevalskii's original plan to reach Tibet and the mountainous region of Kham was rejected, the Main Staff having consulted with the Council of the IRGO, developed a new area of research for the Pevtsov expedition. It was considered wise to focus their attention on surveying two regions: the northwestern part of Tibet and the Tian Shan oasis. The necessity to explore these vast lands was noted in the report of the Chief of the Main Staff addressed to the War Minister.

> In the opinion of the Imperial Russian Geographical Society, an area of high scientific interest as many important geographic discoveries made in Central Asia were carried out exclusively by Russian travellers. There-

> fore, regardless of the vast enrichment of data in the sphere of geographical science, the study of these countries becomes <... > a must for us from a national perspective, as English travellers who have already visited some areas of Inner Asia, do not have time to finish the research started by Russian travellers.[45]

Exploring the Tian Shan oasis was of interest to the Main Staff as this region was a 'stronghold' for Chinese troops in the event of a war with Russia. The expedition was set with the task of 'finding out whether there were enough local food sources for a large Chinese army, understanding how large the Chinese population was which would replenish army numbers as well as gathering information on the number of local Dungans and their relationship to the Chinese.[46]

On December 10, 1888 the programme for the so called Tibetan expedition (*Tibetskaia expeditsiia*) was set in motion and received approval from the Emperor Alexander III. In the months preceding the expedition, Pevtsov thoroughly studied available maps and literature relating to the projected route of the expedition. Well-known sinologist E.V. Bretschneider, at Pevtsov's request copied maps from the Chinese geographical atlas, published in 1863, of Eastern Turkestan, Dzhungaria and northwestern Tibet (scale: 1 inch equals 26 *verst*) along with a transliteration of all Chinese geographical names into Russian. Furthermore, Bretschneider extracted Chinese geographical writings from *Xiyu tuzhi* ('Description of the Western region') on the mountain systems of Eastern Turkestan, and made a list of works by European authors regarding areas of China and Tibet. Pevtsov acquainted himself with principal works by Russian and European authors on the subject of Eastern Turkestan including the works of N.Ia. Bichurin, Carl Ritter, Ch. Valikhanov, A.N. Kuropatkin, N.M. Przhevalskii, N.F. Petrovskii, Robert Shaw, T.D. Forsyth, H.W. Bellew, and others.

On April 20, 1889, the members of the expedition arrived into a small town of Karakol on the eastern outskirts of Lake Issyk Kul and prepared for the long journey ahead. On May 13 the expedition which included Pevtsov, his two assistants, 12 Cossacks, a preparator and two translators, left Karakol and moved in the direction of the Chinese border. The entire expedition load weighed more than 400 *poods* (including 15 *poods* of silver ingots) and was packed across 88 camels and 20 horses.[47] Having crossed the Bedel Pass Pevsov trekked into Eastern Turkestan. Here his route lay through the town of Maralbashi to

45 RGVIA, F. 401, Op. 4, D. 53 (1888), ll. 23–24: Report of the Main Staff, 2 December 1888.

46 Ibid., l. 24 ob.

47 RGVIA, F. 401, Op. 4, D. 53 (1888), l. 102 ob.: A report of the head of the Tibetan expedition to the head of Military-Scientific Committee of the Main Staff, 13 May 1889.

FIGURE 2.4 A view from the wall's watchtower towards the Yarkand oasis. COPY FROM M. PEVTSOV'S TRAVELOGUE.

the southeast to Yarkand and further along the main Khotan road through Khotan, Keriya and into Niya, where the expedition stopped over winter. Contrary to expectations, the Chinese authorities provided full assistance to the Russian travellers along the entire route of their expedition.[48]

Geologist Bogdanovich left Karakol for Eastern Turkestan a week before the main body of the expedition moved on. Having crossed the Russo-Chinese border, he reached Kashgar, where went on to Sarykol, in order to investigate the Chinese part of Eastern Pamirs. His main objective was to explore approaches to the mount of Mustagh Ata, an area on which Captain Grombchevskii had previously reported, as he travelled in the near vicinity of Mustagh Ata in the autumn of 1888 on his ways from Hunza. At the beginning of July 1889, having inspecting Mustagh Ata, Bogdanovich joined to the rest of Pevtsov expedition in the town of Yarkand.

In St. Petersburg the Pevtsov expedition was considered of great importance both for its scientific and military significance. The expedition explored the

48 Ibid., l. 113: A letter from the Russian Consul in Kashgar N.F. Petrovskii to the head of Military-Scientific Committee of the Main Staff, 23 June 1889.

FIGURE 2.5 A group of locals from the Niya oasis. COPY FROM M. PEVTSOV'S TRAVELOGUE.

Chinese Pamir, Raskam and Southern Kashgaria, an area which was of paramount interest to the Main Staff in view of the forthcoming Pamir crisis, the struggle between Russia and Great Britain for supremacy in the Pamir. Whilst on the expedition, Pevtsov was in permanent correspondence with Petrovskii, Russian Consul in Kashgar. Petrovskii forwarded Pevtsov's reports to the head of Military-Scientific Committee of the Main Staff, who informed the Chief of the Main Staff and the War Minister about the proceedings of the Tibetan expedition. The War Minister on a regular basis brought the extracts from the Pevtsov's reports to the attention of Alexander III.

Having arrived on October 17 into the Niya oasis, the Pevtsov expedition completed a short reconnaissance in the direction of the Tibetan Plateau in order to find a convenient crossing to further explore northwestern Tibet. After returning to Niya the expedition prepared for winter, having set up a separate house, which had furnaces and a bathhouse. During the stay in Niya the team continued their scientific research: under Pevtsov's guidance a reconnaissance party ventured out a peep into Kunlun (27 October – 17 November 1889). Roborovskii explored the paths leading into the Tibetan Plateau (18 February – 6 April 1890), whilst Pevtsov, Kozlov and Bogdanovich completed their trip into

FIGURE 2.6 Pevtsov's expedition camp on the southern slope of the Tian Shan. COPY FROM M. PEVTSOV'S *A JOURNEY ACROSS EASTERN TURKESTAN, KUENLUN, THE NORTHERN FRINGE OF THE TIBETAN PLATEAU AND DZHUNGARIA IN 1889–1890* (ST. PETERSBURG, 1895).

the heart of the Taklamakan desert (23–31 March 1890). In Niya, Pevtsov conducted regular astronomical, magnetic and meteorological observations; he drew route surveys and worked on an essay on the ethnography of Kashgaria.[49]

From May to early September 1890 the expedition explored the south-eastern part of the Kashgar Depression, the Kunlun and the adjacent southern areas of the Tibetan Plateau. They determined the general nature of the terrain, the geographical position of several points. Route surveys were carried out along all covered paths. Several heights including those of the Kunlun Mountains were barometrically and trigonometrically determined. During these works the expedition crossed the Kunlun mountain range five times, which made it possible for them to get acquainted with its geology and determine the approximate average height of the ridge.[50] From Lopnor, the expedition in full strength continued on *via* Korla and Karashar to Urumchi, along the river bed of the Konche Darya and the large freshwater Lake Bagrash Kul, which was yet

49 Pevtsov 1895, p. 141.

50 RGVIA, F. 401, Op. 4, D. 53 (1888), l. 161 ob.: A report of the head of the Tibetan expedition to the head of Military-Scientific Committee of the Main Staff, 4 September 1890.

still to be visited by any European. In November 1890, the expedition reached the town of Toksun, on the western outskirts of the Turfan oasis. Here the expedition made another important discovery; in the centre of the Asian continent they found an enormous crater, the Turfan Depression. Pevtsov had no idea that a few months prior to this another Russian expedition of Grum-Grzhimailo brothers had visited this area (the town of Lukchun) and came across a similar discovery. This discovery not only explained the climate pattern of this part of Eastern Turkestan but also enabled more accurate interpretations of weather observations across the Inner Asian highlands. The parallel data obtained from the Asian heartland was unknown to the world of science.

From Urumchi the expedition proceeded towards the Russian border along the well-known caravan road. It moved along the lower course of the Manas River, passing the large freshwater Lake Telli Nor and an extensive group of small lakes of Sary Kul, further past Lake Ayriknor and entered the Zaisan border post on January 5, 1891.[51]

The Pevtsov Tibetan expedition was a complete success and held great scientific and military value. During the expedition about 10,250 *verst* of route survey was taken, of which Pevtsov personally surveyed 5,000 *verst* (scale: 1 inch equals 15 *verst*) and the remaining 5,250 *verst* were surveyed by the other members of the expedition on a scale of 10 and 5 *verst* to the inch. The geographic locations (latitudes and longitudes) of 47 points were determined by Pevtsov (34) and Roborovskii (13), 10 magnetic observations were made, and about 350 heights were determined.

During the winter quarter in Niya, the expedition set up a fixed meteorological station where throughout the winter of 1889–90 Pevtsov observed air temperature, humidity, atmospheric pressure, direction and strength of wind. Bogdanovich, the geologist of the expedition, carried out geological surveys along the route of 2,000 *verst*. His collection of rocks and soil samples would later be transferred to the Museum of the Geological Committee.

The zoological collection of the expedition consisted of 60 species of mammals, 220 species of birds, 11 species of fish, 20 species of amphibians and reptiles and 200 species of insects. The collection was transferred to the Museum of the Imperial Academy of Sciences for further scientific work. The expedition's impressive herbarium contained up to 700 plant species was transferred to the Imperial Botanical Gardens.

The expedition collected valuable data on the general geography of the visited area, as well as ethnographic information on Kashgaria and Dzhungaria.

51 Ibid., ll. 176–176 ob.: A report of the head of the Tibetan expedition to the head of Military-Scientific Committee of the Main Staff, 3 April 1891.

During Pevtsov's stay at the winter quarters, he penned *An ethnographic sketch of Kashgaria*, which later became a separate chapter in the official report of the expedition. Lieutenant Roborovskii's photographic work along the expedition route was essential to the study of ethnography of the peoples of Inner Asia. He took photographic images of the different types of locals in Eastern Turkestan, which would later be used to illustrate Pevtsov's ethnographic essay published in the official report.

Collected military information over the course of the expedition consisted of confirmed dislocation and number of Chinese troops, the military-political condition of Kashgaria and Dzhungaria, the attitude of the local Muslim population towards the Chinese as well as the Chinese government's measures on colonisation of the region. In the Supreme report War Minister P.S. Vannovskii noted:

> The Tibetan expedition after two years of hard work had delivered to the War Ministry 9,600 *verst* of excellent surveys across a country hitherto unknown and assembled valuable materials on the geography of the adjacent region, which may become the theatre of war. This description represents a complete picture of the entire region, and provides a better understanding of terrain traversability as well as the wealth of the region; included is data on the population, their distribution and being, as well as information on Chinese troops, stationed in the region.[52]

The information gathered by the Tibetan expedition was of great importance, as it would be used to refine the cartography of Eastern Turkestan, Kunlun and northwestern Tibet. Maps were considerably refined, as a result of the geodetic work, including those of the Kunlun, the Tarim Basin (especially in the area between Kashgar and Khotan), the basin of the Tarim River and Lake Lop Nor. Bogdanovich was able to explore the Chinese Pamir, the ridges of Sarykol and Kashgar and the mount Mustagh Ata. Thus connecting his own surveying in Eastern Pamir and Western Kunlun with that of Grombchevskii, who explored these regions almost simultaneously with Pevtsov's expedition. The route surveys of both travellers helped to create the most up-to-date and precise map of the area directly adjacent to Hunza and Ladakh, the British possessions in the north of British India.

As Pevtsov's expedition route was closely adjacent to the northern frontier of British India, it was only natural that their actions were closely monitored by the Indian authorities. The progress of the Tibetan expedition appeared in British geographical publications, mainly in the form of translations from

52 Ibid., ll. 250–250 ob.: Report of the Main Staff, 5 July 1891.

the Russian press.[53] A famed Russian geographer M.I. Veniukov, who resided in France, also supplied the British geographers with information on activities of the Pevtsov expedition. He was well-informed as he was in direct correspondence with Roborovskii.[54] Pevtsov's exploration activities on the borders of British India, were relayed by the British Ambassador to St. Petersburg to the Prime Minister of the Marquess of Salisbury.[55]

Following the release of preliminary results of the Tibetan expedition Vannovskii lodged a request to Alexander III to award Pevtsov as well as the other members of the expedition. On the basis of this petition, Alexander III ordered to award Pevtsov, Roborovskii and titular counsellor Bogdanovich with higher ranks. In addition, a lifetime annual pension for each was granted, Pevtsov was awarded 500 rubles, Kozlov and Bogdanovich 100 rubles each. Kozlov was also awarded the Order of Saint Anna, 3rd Class.

On October 2, 1891 at the IRGO ceremonial meeting, P.P. Semenov welcomed Major-General Pevtsov with a speech. Pevtsov addressed the meeting with a brief account and a summary of the results of the Tibetan expedition.[56] On November 6, 1891, at the IRGO general meeting, Bogdanovich made a presentation on the geological structure of Kunlun and the geography of northwestern Tibet. He also presented the orographic characteristics of the Kunlun area, adjacent to Eastern Pamir and Karakoram. Bogdanovich noted the role of loess deposits and sands in the development of civilization in Southern Kashgaria, drawing attention to the presence of traces of ancient settlements in the desert. The study of ancient civilizations, lost in the sands of the Taklamakan desert, would only later be studied in depth by the likes of Sven Hedin and Aurel Stein. In his speech, Bogdanovich has also considered the issue of mining in Kashgaria, primarily for gold and jade.[57] In view of the major scientific results achieved by the team of the Tibetan expedition the Council of the IRGO made the decision to name the expedition in memory of the late Przhevalskii.

---

53 For the details see Progress of the Russian Expedition to Central Asia, in *Proceedings of the Royal Geographical Society* (*PRGS*), vol. XII (1890), pp. 19–36; On the Pevtsof Expedition and M. Bogdanovich's Surveys, in *Geographical Journal*, 1893, no 2, pp. 55–56).

54 See The Russian Expedition to Central Asia under Colonel Pevtsof, in *PRGS*, vol. XIII (1891), pp. 99–105.

55 *The National Archives* (*TNA*), Kew, FO539/54/106: Mr. Howard to the Marquess of Salisbury, St. Petersburg, October 19, 1891.

56 Pevtsov 1891, pp. 461–479.

57 K.I. Bogdanovich, Severo-zapadnyi Tibet, Kun'lun' i Kashgariia, in *Izvestiia IRGO*, T. XXVII, Vyp. VI, St. Petersburg: tipografiia A.S. Suvorina, 1891, pp. 461–504.

In November 1891 on the grounds of the review of the Assistant to the Chairman of the IRGO, Lieutenant-General I.I. Stebnitskii,[58] Pevtsov was awarded the Constantine Medal, the highest award bestowed by the Society. As noted in the IRGO report, Pevtsov was awarded the medal 'for scientific work, travels and supervising of the Tibetan expedition'.[59] Pevtsov's colleagues – Roborovskii, Kozlov and Bogdanovich were honoured the Przhevalskii silver medal 'for their work on the study of the nature of Inner Asia'. Pevtsov's merits in the field of geographical exploration of Inner Asia were also noted by foreign scientific societies. In April 1893 he was chosen by the Royal Geographical Society as Foreign Honorary Corresponding Member of the Society.[60]

On his return from the expedition Pevtsov was transferred at the disposal of the Chief of the Main Staff in order to process the scientific results of the Tibetan expedition. Roborovskii and Kozlov were assigned to assist him. Processing the results of the expedition presupposed a publication of several independent parts, which Pevtsov and his assistants still had to write. The entire work was called *Trudy Tibetskoi ekspeditsii 1889–1890 gg. pod komandovaniem M.V. Pevtsova* ('Transactions of the Tibetan expedition, 1889–1890, under the command of M.V. Pevtsov'). In 1892, Bogdanovich completed the second part of the official report, *Geologicheskie issledovaniia v Vostochnom Turkestane* ('The geological studies in Eastern Turkestan').[61] In 1895 Pevtsov published the first and most significant part of the report, *Puteshestvie po Vostochnomu Turkestanu, Kun'luniu, severnoi okraine Tibetskogo nagor'ia i Chzhungarii v 1889 i 1890 godakh* ('A Journey across Eastern Turkestan, Kunlun, the northern fringe of the Tibetan Plateau and Dzhungaria in 1889–1890').[62] In 1896 Roborovskii and Kozlov complied the final part of the report, *Ekskursii v storonu ot putei Tibetskoi ekspeditsii* ('Excursions away from the main route of the Tibetan expedition').[63] In January 1896, the War Minister Vannovskii presented Pevtsov's report as a gift to the Nicholas II, who was very pleased with the book and requested that Pevtsov be awarded a treasured gift for his work.[64]

58 Otzyv deistvitel'nogo chlena IRGO I.I. Stebnitskogo o puteshestviiakh i trudakh M.V. Pevtsova, in *Otchet IRGO za 1891 god*, St. Petersburg: tipografiia A.S. Suvorina, 1892, pp. 4–8.

59 *Otchet IRGO za 1891 god*, St. Petersburg: tipografiia A.S. Suvorina, 1892, p. 61.

60 See *The Geographical Journal, The Monthly Record*, Vol. 1, no 5, May, 1893, p. 452.

61 K.I. Bogdanovich, *Geologicheskie issledovaniia v Vostochnom Turkestane*, St. Petersburg: Imperatorskoe Russkoe Geograficheskoe Obshchestvo, 1892.

62 Pevtsov 1895.

63 V.I. Roborovskii, P.K. Kozlov, *Ekskursii v storonu ot putei Tibetskoi ekspeditsii*, St. Petersburg: Imperatorskoe Russkoe Geograficheskoe Obshchestvo, 1896.

64 RGVIA, F. 401, Op. 4, D. 53 (1888), ll. 301–301 ob.

On the basis of route surveys, astronomical and barometric determinations, made during the expedition, the Military-Topographical Department of the Main Staff compiled a *Map of Eastern Turkestan and the northern margin of the Tibetan Plateau* (scale: 1 inch equals 60 *verst*).

The zoological collection, assembled during the Tibetan expedition, was transferred to the Zoological Museum and was further worked on by the academician A.A. Strauch. Amongst the selected mammals, of particular interest were the skin and two skulls of wild yaks and a dipus sample, which had been previously described, but was not part of any museum collections. Pevtsov's gathering contained no newly discovered species, but, nevertheless, it had become a great addition to the already available samples from Inner Asia and provided a complete picture of the vertebrate fauna of Kashgaria and the northwestern Tibet. Pevtsov and his aides were honoured with gratitude for the donating the collection to the Imperial Academy.[65]

Pevtsov presented the results of his meteorological observations in Eastern Turkestan in a separate article, which was the first independent work on this subject.[66]

The Final Years

On completing the expedition report, Pevtsov continued his studies in astronomy, geodesy, and geography. Soon after his articles began to appear in a number of the scientific journals each of which touched on another important problem of the science of geography: including *O barometricheskom nivelirovanii* ('On barometric levelling'), *Spisok punktov Vnutrennei Azii, vysoty kotorykh opredeleny neodnokratno posredstvom barometrov razlichnymi puteshestvennikami* ('A list of landmarks in Inner Asia, whose heights were determined by barometric levelling'), *Instruktsiia dlia opredeleniia vysot posredstvom posledovatel'nogo barometricheskogo nivelirovaniia* ('The instructions for determining heights by barometric levelling'), *Sokrashchennyi sposob predvychislenii pokrytii nepodvizhnykh zvezd Lunoiu i solnechnykh zatmenii dlia dannykh mest* ('A simplified method of preliminary calculations of occultation and solar eclipses for a fixed location'), and others.[67] Pevtsov continued to develop new

65 See N.V. Slepkova, op. cit., p. 101.

66 Pevtsov 1894.

67 See M.V. Pevtsov, O barometricheskom nivelirovanii, in *Zapiski IRGO po obshchei geografii*, T. XXIX, № 2, St. Petersburg: 1895; Id., Spisok punktov Vnutrennei Azii, vysoty kotorykh opredeleny neodnokratno posredstvom barometrov razlichnymi puteshestvennikami, in

techniques and equipment for scientific observations, all of which were highly valued by specialists in the field.[68]

In November 1896, Major-General Pevtsov was promoted to the position of a general on special duties of the Main Staff. By the nature of his service, he was involved in the development of Russian military planning in Asia. He participated in drawing mobilisation plans for the Main Staff, worked on strengthening the combat capabilities of the Russian Asian military districts. Pevtsov took part in the development of operational plans of Russian troops in the Chinese campaign of 1900–01.[69] During this time he continued to be interested in the development of Chinese military forces in Eastern Turkestan and published an article on this topic in *Voennyi Sbornik* (Military Journal), an official publication of the War Ministry.[70]

The Tibetan expedition became Pevtsov's last adventure and would see him never to return to Inner Asia. When in 1893 the Council of the IRGO brought up the prospect of another Inner Asian expedition, Pevtsov declined to lead the expedition. As a result, Captain Roborovskii, an experienced yet relatively young junior officer was put in charge. The expedition of 1893–95 made history in the field of Russian geographical research of the area, as the first expedition headed by an officer who did not belong to the General Staff Corps. This signalled a weakening of interest of the Russian War Ministry in large scale and expensive studies of Inner Asia, or more precisely, in exploring areas which were not directly included in Russia's military priorities.

Pevtsov's refusal to lead the new expedition was in part to do with his deteriorating health. Previous burdens of camp life coupled with Inner Asia's rough climate started to take its toll on Pevtsov's health. From 1893 Pevtsov would regularly fall ill. In a letter to his colleague in Omsk, he noted 'my health recently has not been great, I often fall ill.' He had even considered relocating,

---

Izvestiia IRGO, T. XXXI, vyp. I, St. Peterburg: tipografiia A.S. Suvorina, 1895, pp. 44–56; Id., Instruktsiia dlia opredeleniia vysot posredstvom posledovatel'nogo barometricheskogo nivelirovaniia, St. Peterburg: 1896; Id., Sokrashchennyi sposob predvychislenii pokrytii nepodvizhnykh zvezd Lunoiu i solnechnykh zatmenii dlia dannykh mest, in *Zapiski IRGO po obshchei geografii*, T. XXXII, № 2, 1899).

68 See M.V. Pevtsov, Sifonno-reservuarnyi barometr dlia meteorologicheskikh stantsii i puteshestvii, in *Meteorologicheskii Vestnik*, 1895, № 12, pp. 421–428; Id., Pribor dlia napolneniia priamykh baromentricheskikh trubok rtut'iu kholodnym sposobom, in Ibid., pp. 429–431; Id., Ob ustranenii nekotorykh neudobstv barometra Parrota dlia puteshestvii, in *Meteorologicheskii Vestnik*, 1896, № 10, pp. 281–284.

69 NA RGO, F. 62, Op. 1, D. 28, ll. 2–4: A copy of a wire of the War Minister to the Chief of the Main Staff, 27 September 1900.

70 See M.V. Pevtsov, Vostochnyi Turkestan, in *Voennyi Sbornik*, 1895, № 4–6.

together with his wife and foster-child, from St. Petersburg, with its wet and cold climate somewhere warmer to the South. But alas urgent matters prevented him going through with his plan.

Along with his health issues Pevtsov experienced additional problems involving various lawsuits. For several years, he was being sued by a landowner in one *uezd* of the Novgorod *guberniia* as a result of an unsecured cheque that Pevtsov had issued as a guarantee for a land transaction. In March 1895 the court ordered Pevtsov to pay out roughly 3,000 rubles to the plaintiff, a huge sum at the time.[71]

In his last years, Pevtsov lived a secluded life and seldom left St. Petersburg. On weekends he would receive close friends at home, among those who visited were his companions from the Tibetan expedition, Roborovskii and Kozlov. They would fairly often appear at his small apartment on *Vasil'evskii Ostrov* in St. Petersburg and discuss current affairs and plans for future expeditions. In the meantime, Pevtsov's health gradually deteriorated. He often thought about his impending fate and worried about providing financial support to his loved ones, a wife and 18-year-old foster-child. P.K. Kozlov who frequented Pevtsov during his illness, recalls

> Pevtsov was often visited by relatives or friends, before whom his life slowly faded. 'Melancholy, melancholy, terrible melancholy, – muttered Mikhail Vasil'evich to me, – it is strange that nothing interests me anymore', premonition weighing on his heart ... A week before his death he was bedridden, on that he noted to his wife – "for now I have been laid in a bed yet soon I will be laid in a coffin ..!'[72]

His loving wife spared no expense in caring for her dear sick husband, nevertheless the disease prevailed. On February 25, 1902, Pevtsov passed away. The Main Staff took care of the funeral, as Pevtsov's family were in a rather desperate situation after his death. A few months later his wife passed away, unable to bear the grief from the loss of her beloved husband. Pevtsov was buried in the Smolensk cemetery in St. Petersburg, his grave has not survived to this day. Of all the outstanding Russian explorers of Inner Asia, the life and work of Pevtsov to this day remain the least studied, yet this does not clearly correspond to the scale of his scientific work and contribution to the field of geographical science. Pevtsov lived a short but rich life and in spite of his rank and title he

71 NA RGO, F. 62, Op. 1, D. 24.

72 Kozlov P.K., Pamiati Mikhaila Vasil'evicha Pevtsova, in *Izvestiia IRGO*, T. XXXVIII, Vyp. VI, St. Petersburg: 1902, p. 620.

remained a remarkably humble man, who had little interest in public attention and honour.

Thanks to M.V. Pevtsov's explorations, a vast region of Inner Asia opened up for geographical science, an area which Nikolai Przhevalskii's 4th expedition map was marked as "completely unknown". A distinctive feature of Pevtsov's travels were his use of scientific instruments and advanced geodetic methods, the most advanced at the time, which made it possible to determine dozens of new astronomical points, as well as the latitude and longitude of several locations. Pevtsov's travel routes in Inner Asia intersected with the routes of other travellers, this facilitated to build up complex geodetic networks for the purpose of cartography.

Pevtsov is credited with the exploration of the Dzhungaria and the Mongolian Altai regions, the Central Gobi, the margins of the Kunlun Ridge in the area from the Yurungkash River to the meridian of Lake Lop Nor where it joins the Tibetan Plateau. He managed to introduce significant changes to the map of the Kunlun. As a result of numerous determinations of heights, Pevtsov was in a position to make a general conclusion about the hypsometry of the Kunlun, and make judgments on the structure of the Tibetan Plateau. Pevtsov's material on the orography of the Western Kunlun from the meridian of Khotan to Kashgar, the Eastern Tian Shan, along with locating the limits of the Tarim Basin and the Lukchun (Turfan) Depression were completely new. Pevtsov managed to make significant changes to the mapping of water objects, especially the Tarim River system and the basin of Lake Lop Nor. Just as significantly important were Pevtsov's studies on the climatology of Inner Asia.

Throughout all of his Inner Asian expeditions Pevtsov covered a total route survey of 9,900 *versts*, determining the geographic position (latitude and longitude) of 68 points, measuring over 500 heights, and gathering a significant zoological collection which consisted of 78 mammals, 283 birds, 20 fish, 40 amphibians, as well as a vast herbarium including over 1,000 plants. Pevtsov was one of the first Russian explorers in Inner Asia to include ethnographic research into their expeditions.

CHAPTER 3

Vsevolod Ivanovich Roborovskii (1856–1910)

Alexandre I. Andreyev

> Yes, the time spent in travelling will be remembered by many of us as the happiest one on the life's trodden path.
>
> V. ROBOROVSKII

Despite Vsevolod Roborovskii's renown as a pioneer explorer of Inner Asia and one of the closest associates of Przhevalskii, his life has been scantily researched thus far. His biographers are few[1] and they focus mainly on the period of Roborovskii's active journeying with Przhevalskii, Pevtsov and Kozlov, between 1879 and 1895, giving little attention to his private life outside his travels. The same flaw applies also to the latest biography of Roborovskii, written by B.K. Vinogradov in 2012. While providing a broad historical panorama of the centuries-old exploration of Inner Asia by a host of Western and Russian travelers as a framework within which Roborovskii is conveniently placed, Vinogradov gives a rather sketchy account of Roborovskii's ancestry, his early life before he met Przhevalskii, and his dramatic final years, when his successful career as a scholar-traveler was interrupted by his fatal illness. Yet these shortcomings can perhaps be explained by the extreme paucity of source materials.

Early years

Vsevolod Roboroskii was born in St. Petersburg, on 26 April 1856, into a noble family of modest means. His father, Ivan Ivanovich Roborovskii (date of birth unknown; died around 1880), was a nobleman and there is no information whatever about his mother. All we know is that she died long before Vsevolod set out, in 1879, on his first journey, under Przhevalskii. The Roborovskii couple

 | DOI 10.1163/9789004376267_005

FIGURE 3.1
Vsevolod Ivanovich Roborovskii, 1880s.
NA RGO.

had three children, according to B.K. Vinogradov, – two older daughters, Maria and Elizaveta, and a son, Vsevolod.[1]

The origin of the Roborovskii family cannot be traced in the historical records. The name itself consists of two Russian words – *roba, rob'* (work, manual labour) and *rov* – pit, ravine, and could possibly refer, according to B.K. Vinogradov, to a person digging pits, a digger or navvy.[2] Many generations of Roborovskiis lived in the countryside, in the Vyshnivolotskii district of Tver' province, situated north of Moscow, and they possessed some land there, i.e. they were local squires (*pomeshchiki*). The Roborovskiis owned a small estate in the district, named Andreevskoe, bordering on the village Taraki. It consisted of the manor house, with ten house-serfs attached, and four peasants' houses nearby where 33 peasants lived. This estate, believes Vinogradov,

1 B.K. Vinogradov, *Isslerdovatel' Tsentral'noi Azii V.I. Roborovskii*, 61. This information is at variance with what Roborovskii himself related to Przhevalskii in his letter of 21 June 1887. There he stated that he had had two brothers (both dead by 1887) and that his younger sister, named Katia (Catherine), died on 19 June of the same year in the village Taraki. See NA RGO, F. 13, Op. 2, D. 219, l. 16 ob.

2 B.K. Vinogradov, Ibid., 139.

was presumably purchased by Ivan Roborovskii shortly after the abolition of serfdom in Russia in 1861.

Vsevolod's grandfarther, Ivan Mikhailovich Roborovskii (his dates are also unknown), belonged to the Orthodox clergy. He was educated in two institutions – Vladimir seminary and St. Petersburg Academy of medical surgery, having finally become a military doctor who served in the army and took part in the war with Napoleon in 1812. He had two sons, Nikolai and Ivan, Vsevolod's father. Nikolai was a clergyman in the church of the Archangel Mikhail in the village Ovsishche, close to Roborovskiis' family estate.[3] And this is basically all we known about Vsevolod Roborovskii's ancestors.

The boy grew up in St. Petersburg where he attended the 4-year Vilensky classical gymnasium in the Peterburgskaia Storona, the city's northern quarter. (He was in fact the only Petersburger in the cohort of scholar-travelers portrayed in this book.) The subjects he studied at school included the Christian religion (*Zakon Bozhii*), Russian language and two foreign languages (French and German), Latin and old Greek, mathematics, physics, history, geography, calligraphy and drawing. His favourite ones were geography and history, and he had much difficulty with mathematics and Latin.

Having an inquisitive mind, Vsevolod as a teenager had a habit of wandering in the environs of the city during holidays, observing carefully the animal life and collecting insects, lizards, fish and plants. This was his great hobby. And he continued his naturalistic pursuits in summer when his parents moved to their country estate, a few hundred miles south of St. Petersburg. Also, he was fond of reading and had a special liking for travel books which carried him away from the daily routine to some exotic distant lands.

This is practically all we know about Roborovskii schooling and his early years. At the age of 20, in late 1876, that is when Przhevalskii's account of his first expedition to Inner Asia had been already published, Vsevolod joined the 145 Novocherkassk infantry regiment of His Imperial Highness the Grand Duke Alexandre Alexandrovich (future emperor Alexander III) stationed in St. Petersburg. Yet three months later, in February 1877, his commanders sent him to the military (junker) college in Helsingforce (today's Helsinki). There, a year later, Vsevolod Roborovskii was given the rank of non-commissioned officer, and a few months later (in April 1878) that of ensign, upon which he returned to his regiment. All too soon, however, he found his service, with its regular drills, rather boring, which made him think of a socially more useful career. It was at this point that he met in St. Petersburg his college mate, Feodor Eklon.

3 Ibid., 59–61.

Eklon had just returned from Inner Asia, with Przhevalskii's Lopnor expedition (1876–1877), in which he participated as the assistant of the leader.

The many tales he heard from Eklon about his journey and adventures strongly impressed Vsevolod and inspired him with a desire to become a traveler himself. So he asked his mate at once to introduce him to Przhevalskii, hoping to take part in Przhevalskii's next expedition. Yet Roborovskii's first encounter with the famous traveler in August 1878 was rather discouraging. Przhevalskii told him from the very start about the many hardships and privations awaiting someone who wanted to journey in Asia. By doing so he actually put the young man to the test to know his worth. The testing had continued for about four months, when Vsevolod was occasionally calling on his 'examiner' in his Petersburg apartment. Przhevalskii, as a matter of fact, had many candidates willing to travel with him out of whom he had to choose just one person, someone he could fully trust and who perfectly fitted his strict requirements. And his choice ultimately fell on Roborovskii. Przhevalskii accepted him into the expedition team on 10 (22) December, a memorable day for Vsevolod. Przhevalskii's general impression of his new assistant was fairly positive – "a rather sensible man, a good draughtsman who knows how to do surveys, of good temper and perfect health".[4]

This verdict finally decided Roborovskii's fate. Przhevalskii then wrote a letter to his father who was residing at his country estate. (He would die some time during his son's journey).

Travelling with Przhevalskii

By the end of February, 1879, the entire party assembled in Zaisan, at the Russian-Chinese border, the starting point of Przhevalskii's Tibet-bound third Inner Asian expedition (also known as his First Tibet expedition).[5] Roborovskii joined it, together with Eklon, as the leader's second assistant. His main task in the field was to collect plants and make sketches en route, while Eklon was to prepare mammals and birds, being placed in charge of the zoological collection. Przhevalskii taught Volia, as he pet-named Vsevolod, his special method of herborization. This envisaged the usage of the large format drying paper so that plants could be preserved 'in full length', with their stalks unbent, as well

4 Dubrovin 1890, 269, 270.

5 For more details on this expedition see Przhevalskii's biographical essay in this volume, section "Towards the Forbidden Land (1879–1880)".

as the special manner of their drying up in the open.[6] This was a rather elaborate method, yet Roborovskii easily mastered it. His trophies as a plant collector would amount to 12,000 plants of 1,500 species, some of which were new. He also compiled a list of geographical locations of the 406 collected plants, according to the botanist V.L. Komarov.[7]

During his botanical excursions Roborovskii displayed much enthusiasm, fervor and in some cases even courage to fulfill his task. As he would later confess:

> My fascination with botany reached a point when I, quite often and risking my life, would climb the mountains to get a little flower which seemed almost unreachable. However, if I had not seen it before and believed that it was a new species, I would exert all my efforts – the stones were slipping from under my feet and I was about to fall into the precipice at any moment. Still I managed to get hold of the flower that had attracted me.[8]

At the same time Roborovskii sketched everything that caught his (and Przhevalskii's) eye – animals, plants (primarily the new and rare species), landscapes and indigenous people, for want of photography. As Przhevalskii noted, "The types of men and women, visiting us, were sketched by V.I. Roborovskii, who did it on the sly, by making use of good opportunities".[9] Many of these pencil sketches, which numbered over 200 sheets altogether,[10] would be reproduced in Przhevalskii's travel account of his journey, *Iz Zaisana cherez Khami v Tibet i na verkhovia Zheltoi Reki* (From Zaisan across Hami toward Tibet and the upper reaches of the Yellow River, 1883), where they served as perfect illustrations of his exciting narrative.

One more type of work Volia was charged with was collecting ethnographic material by questioning the natives about their customs, mode of life, etc. with the help of an interpreter, and he showed much diligence and patience while doing this work.

---

6 Przhevalskii 1888, 61, 62. See also M.N. Kozhevnikova, *The Botanical Collections*, in Andreyev (ed.) 2013, 234, 235.

7 This was entitled "The List of plants, collected in 1879 in Dzungaria, East Tian Shan, the Hami Desert, the Nanshan Mountains, and Tsaidam", see Komarov 1908, 39.

8 Roborovskii 1892, 222.

9 Przhevalskii 1948, 217.

10 In a letter to Przhevalskii of 1 November, 1880, from Kiakhta, Roborovskii informed the expedition leader that he had made 240 drawings during the journey. All these drawings he sent to Przhevalskii as he did not regard himself their owner. See NA RGO, F. 75, Op. 1, D. 219, l. 6 (Roborovskii's letter to Przhevalskii of 1 November 1880 from Kiakhta).

FIGURE 3.2 Roborovskii's drawings of the rare species of animals, from N. Przhevalskii's *Iz Zaisana cherez Khami v Tibet* (St. Petersburg, 1883). Tibetan bear (bear-pike, Ursus lagomyiarius n. sp.).

FIGURE 3.3 Wild yak (Poëphagus mutus n. sp.).

FIGURE 3.4 Wild horse (Equus ferus Przewalskii).

FIGURE 3.5
Orongo antelope (Pantholops Hodgsoni).

FIGURE 3.6
Dzeren (Mongolian dzeren, Antelope gutturosa).

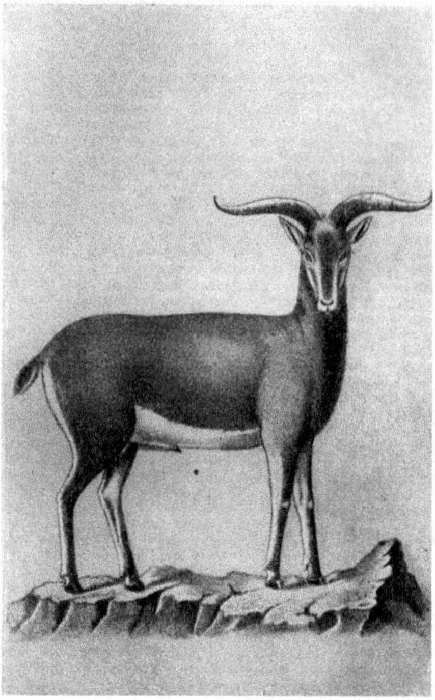

FIGURE 3.7
Kuku-yaman ('Blue ram', of the goat-antelope subfamily, Pseudois Nahoor).

With Przhevalskii's First Tibet Expedition Roborovskii trekked through the entire Inner Asia, with its vast steppelands, deserts, and high mountains – Dzungaria, the Great Gobi, East Tian Shan, Tsaidam, and finally, Tibet, covering the distance of 7660 km. Having ascended the Tibetan plateau, he reached the upper Yangtze River, however the party was unable to make its way further to Lhasa, being halted by a Tibetan frontier cordon outside Nagchu. Still the travelers succeeded in conducting a series of pioneer explorations on the Tibetan plateau, and Roborovskii's contribution to these was of no little importance.

This first journey in Inner Asia was a serious trial for Roborovskii, both physical and emotional, yet he stood the test. This made a good start of his career as a military scholar-traveler, largely owing to Przhevalskii's special type of field schooling. More than once he risked his life as happened, for example, when the expedition camp was attacked by a group of militant *egrai* (Tibetan) nomads. Seeing that one of the attackers was going to strike the interpreter, Abdul Yusupov, with a spear, Roborovskii immediately rushed forward and snatched the weapon out of his hands, having thus rescued his companion, an incident vividly recounted in Przhevalskii's travelogue.[11]

11 Przhevalskii 1948, 199, 200.

FIGURE 3.8 The Nomohun-gol gorge in the Burhan Budda range, the boundary of Northern Tibet, a drawing by Roborovskii.

Upon his return to St. Petersburg in early 1881, Roborovskii was awarded – together with the expedition leader and his first assistant, Eklon, – the Small Gold Medal by the Imperial Russian Geographical Society, the organizer of Przhevalskii's expedition, in appreciation of his excellent service; and a life pension of 200 rubles annually. He was also promoted to the rank of second lieutenant and, with the permission of the War Ministry of 17 February, 1881, left at the disposal of his military commander, the colonel of the General Staff Przhevalskii.

•••

Roborovskii spent the summer of 1881 in Sloboda, Przhevalskii's new estate on Lake Sapsho, in his native Smolensk Province. Przhevalskii had purchased it shortly before and tried to turn into a refuge from "civilization", as the place was located amidst "wild nature" so much admired by the traveler. They passed their time shooting and fishing, or strolling along the shore of the lake, a most scenic spot.

Having finished his account of the recent expedition (which would make a thick volume published in May 1883), Przhevalskii began to make plans for his next journey, the Second Tibet Expedition. On 9 February he submitted his project to the Council of the Imperial Russian Geographical Society which readily approved it. The object of his fourth journey was a more thorough examination of the Tibetan plateau, with Lhasa as his final destination, and Eastern Turkestan (mainly Lake Lopnor and the Tarim basin). Przhevalskii's two assistants this time would be Roborovskii and Petr Kozlov, a 20-year old volunteer of the 2nd Sophia infantry regiment who replaced F. Eklon. (Eklon, who was originally taken on the expedition staff, changed his mind at the last minute, having decided to get married).

On this journey, Roborovskii acted as Przhevalskii's first assistant. His principal tasks were plant collecting as well as taking pictures en route with the help of the photographic camera specially designed for Przhevalskii by V.I. Sreznevskii.[12] This was quite a new job for a traveler, yet Roborovskii handled it fairly well. His photos, like his pencil sketches, depicted various types of landscapes, towns, villages, the images of local dwellers and samples of vegetation of the terrain the expedition passed through. Apart from these tasks, he also assisted Przhevalskii and Kozlov in shooting birds and collected insects as well as flower seeds.

This second journey was another gruelling experience for Roborivskii. While travelling in Tibet, the party again was attacked twice by groups of roaming Tangut and Golok brigands, and Roborovskii had to take part in fighting, which was quite fierce. On another occasion he nearly drowned when taking a flock of sheep across a mountain river Dji-Chu. Yet he also had some happy moments, contemplating the magnificent mountain scenery, as happened, for example, when he and Przhevalskii ascended a holy mountain in the vicinity of the Orin Nor Lake, in northern Tibet, where annual sacrifices were made by the locals. "We have spent about an hour on the summit of the sacrificial mountain, enjoying the panoramic views which opened before us and trying to imprint on our memory their minutest details", Przhevalskii recorded in his diary.[13]

The passage through the North Tatung Mountains, which led to Lake Kokonor, and on to Tsaidam and Tibet, was the most pleasant part of the entire journey. This was a land of great beauty, with its fine forests, pure mountain streams, gorgeous Alpine meadows, surrounded by formidable cliffs, and the rapid and winding Tatung-gol River down below, but also a land of plentiful fauna and flora. Recalling the time spent in the Tatung Mountains, Kozlov would say: "In no other place and never before we have been so utterly happy, so pure in heart, so receptive to everything beautiful, as in these charming

12 On this see Andreyev (ed.) 2013, 146–148.

13 Przhevalskii 1948, 85.

places ... amidst the living virgin nature".[14] Exploring the place, Roborovskii once happened upon a little cave of a Buddhist hermit who was very unhappy when a stranger disturbed his blissful solitude.

The travelers spent the winter of 1885 – some 50 days – by the shores of Lake Lopnor, which was a large reedy bog. They found the scenery rather dull, making a sharp contrast to the beautiful mountains of Tatung in Gansu which had so enchanted them. There was very little wildlife there, and the atmosphere was constantly filled with brownish dust. When there were sandstorms, the whole place would sink into hazy gloom. As Przhevalskii would note, Lopnor, which attracted him so much in his vision, disappointed him when he got there in person.

However, despite the unfavourable weather conditions, Przhevalskii carried on his usual field work. Accompanied by his assistants, he regularly went hunting, watched the seasonal migration of birds in early spring, and interviewed the locals, who were quite friendly as they remembered him well from his first visit to Lopnor in 1877. Also, Roborovskii busied himself with his new job of a photographer. Thus, he once sailed down the Tarim in several boats, together with Przhevalskii, and the Lopnor ruler, Kunchikan Beg, with a retinue, to a village Uitun to take pictures of the local dwellers, Kara-kurchins.

At the final stage of the expedition the party passed through the oases of Eastern Turkestan – Charkhalyk, Niya, Keriya, Chira, Khotan, and others, heading for the Russian frontier. The return journey was not entirely smooth. In Khotan, an ancient oasis town on the Silk Road, Przhevalskii had a quarrel with two Chinese soldiers who came to his camp to inspect the expedition luggage. In protest, he sent a group of ten Cossacks under Roborovskii, fully armed, with bayonets attached to their rifles; they were to 'take a walk' through the town intended as a military demonstration to scare the Chinese authorities. (And it did scare them!) And later, when the party was already on the way home, a mishap occurred – Roborovskii fell down, when saddling a horse; he badly hurt himself and had to be carried in a bullock-cart.

Przhevalskii's second Tibet expedition ended at Karakol, on 7 November, 1885. It was another big success, largely owing to his faithful and enthusiastic assistants, Roborovskii and Kozlov. Roborovskii's main contribution was a huge plant collection of 799 specimens from Mongolia, Gansu and Northern Tibet, 543 of which were gathered in 1884 and 257 in 1885.[15] The collection was donated by Przhevalskii to the Botanical Gardens in St. Petersburg where it went into the hands of one of the leading Russian botanists, Academician

14 Kozlov 1929, 45.

15 See M.N. Kozhevnikova, *Botanical collections*, in: Andreyev (ed.) 2013, 244.

K.I. Maksimovich, for expertly study and description. Some of the plants in it were completely new species, such as *Androsace tapete* Max., a plant remarkable for its shoot, which Maksimovich called the "dwarf sapling of the Tibetan highlands". The most unique part of the herbarium was a collection of the small Tangut plants. Also, Roborovskii delivered to St. Petersburg dozens of photographs, taken throughout the entire expedition route, which would be reproduced later in Przhevalskii's travel account *From Kiakhta to the upper reaches of the Yellow River...* (1888).

These two Tibet expeditions under Przhevalskii finally shaped Roborovskii as a scholar-traveler, one of his 'school', – a land surveyor and naturalist. He fully mastered Przhevalskii's method of 'route reconnaissance' – his system of semi-military expedition teams aimed for complex – all-embracing – en route field explorations. The latter envisaged topographic (route) surveys, various instrumental observations and collecting specimens of flora, fauna, minerals and ethnographic artefacts; as well as obtaining varied oral information from the locals, a procedure being basically a method of intelligence gathering.

Upon his return to Russia, Roborovskii was promoted to the rank of lieutenant and left at Przhevalskii's disposal for the subsequent work on the collected materials. On 28 February, 1886, he was awarded the order of St. Vladimir of the 4th degree and an additional pension of 200 rubles annually. Also, on 2 April, he joined the IRGO as its full member.

After the expedition Roborovskii spent his time in St. Petersburg, trying to fill up the gaps in his education. He regularly attended the Botanical Gardens and the Zoological Museum, where he had an opportunity to expand his knowledge of Inner Asian flora and fauna by consulting the best experts in natural sciences. By doing so he actually followed the example of Przhevalskii himself who is known to have taken lessons in field geology from the eminent geologist I.V. Mushketov, prior to his forth expedition.

Encouraged by Przhevalskii, Roborovskii also made an attempt, in August 1886, to enter the Nicholas Military Academy (NAGSh). Despite his long and assiduous preparations for the entry examinations, he flunked one of them (that in fortification), or, to be more exact, he simply evaded the exam because of his poor knowledge of the subject. Despite the failure, Roborovskii was permitted by the head of the Academy, General M.I. Dragomirov, thanks to Przhevalskii's intercession on his behalf, to attend classes in the senior course of the Academy as an unregistered student and undergo special training in land-surveying, which he did in 1886–1887.[16] Trying to console Vsevolod, who was

16 This fact unknown to Roborovskii's biographers is evidenced by Przhevalskii's correspondence with Roborovskii; see, for example, Roborovskii's letters to Przhevalskii of 26

still utterly upset with his failure to join the Academy which meant he would not receive the diploma of this highly privileged school, Przhevalskii would admonish him paternally: "Having many abilities and a strong character, one can achieve much without wearing the Academic badge. There are many examples to this".[17] Roborovskii maintained close relations with Przhevalskii, or Psheva for short, after the expedition and occasionally visited him at his new Sloboda estate, together with Kozlov and some other expedition members. He was his dear friend and a trusted lieutenant, one of Przhevalskii's intimate circle, his "expedition family". So Przhevalskii did not hesitate when he invited Roborovskii, in the spring of 1888, to take part in his next – fifth – expedition. They travelled together then from Sloboda to St. Petersburg where he assisted the leader in procuring the necessary equipment and then on to Karakol, on Lake Issyk Kul, the starting point of the journey. However, Przhevalskii's plans were frustrated by his sudden death at Karakol, which was a severe blow to Roborovskii and other members of the expedition staff.[18]

Przhevalskii's death was bitterly mourned by the whole of Russia, and Roborovskii promptly responded to the tragic event with a short essay, "The last hours of N.M. Przhevalskii", published in the Transactions of the Imperial Russian Geographical Society later in the year.[19] In this he related in detail the story of Przhevalskii's demise – they two went chasing after pheasants outside Karakol, and Przhevalskii, being flushed with the hunt, drank some unboiled water from the river. This caused his fatal disease diagnosed as typhoid fever by the doctors. Several years later (1892) Roborovskii would publish his recollections of Przhevalskii, from the time of their first meeting in St. Petersburg through their travels in Inner Asia and Przhevalskii's death.[20]

Przevalskii regarded Roborovskii and Kozlov as his direct followers who would continue the exploration of Inner Asia along the lines of the program he evolved in his last travelogue (1888).[21] An interesting detail: Lying on his deathbed, Przhevalskii gave his two best sporting guns – the Lancaster carbine to Roborovskii and the Purdey rifle to Kozlov. These farewell gifts of his they would treasure as sacred relics through the rest of their lives. He also asked

August 1886 and 8 June 1887 (NA RGO, F. 13, Op. 2, D. 219, Ll. 14, 14 ob., 31, 31 ob.), Przhevalskii's letters to Roborovskii of 7 September 1886, 31 May and 5 July, 1887 (NA RGO, F. 13, Op. 1, D. 139, ll. 3 ob., 47, 48, 88) and by Przhevalkii's published letters to Teleshov, Przhevalskii 1940, 613, 618 (letters of 24 September 1886 and 9 March 1887).

17 NA RGO, F. 13, Op. 1, D. 139, l. 47 (Przhevalskii's letter to Roborovskii of 31 May 1887).

18 On details of Przhevalskii's death see his biographical sketch in this volume.

19 Roborovskii 1888, 277–280.

20 Roborovskii, 1892.

21 See Przhevalskii 1888 (Chapter 1: How to travel in Inner Asia).

Roborovskii to take his last photo in the coffin, after his death, with his favourite Lancaster placed next to him, which he did.

In the Tibet Expedition Led by M. Pevtsov, 1889–1901

After Przhevalskii's death Roborovskii temporarily remained in charge of the expedition, until December 1888. He refused, however, to replace Przhevalskii as its leader, so the Imperial Russian Geographical Society, jointly with the War Ministry, had to look for someone else, and their choice fell on the 45-year old Colonel Mikhail Vasilievich Pevtsov, an adjutant of the Omsk Military District. An experienced scholar-traveler and excellent land-surveyor, he had already made his name through his two journeys in Dzungaria (1876) and Mongolia (1878–79).[22]

Pevtsov reorganized Przhevalskii's 5th Expedition – he considerably reduced the scope of explorations, having made his main object the Kunlun Range, with the adjoining southern strip of the Tibetan highlands, as the General Staff commissioned him to locate the convenient routes to the Tibetan plateau.[23] Pevtsov likewise gave up on Przhevalskii's overly ambitious plan for reaching 'the forbidden' Lhasa. He also cut down the expedition escorts from 25 to 12 people and added one more assistant to his team, a professional geologist K.I. Bogdanovich.

An important fact to note: concurrently with Pevtsov, the IRGO sent two more geographical expeditions to Inner Asia in 1889 for a period of about a year and half each – that of brothers G.E. and M.E. Grum-Grzhimailo to explore Eastern and Central Tian Shan, and another one under B.L. Grombchevskii, whose routes were to cover the western Kunlun, between the meridians of Kashgar and Yarkand. Incidentally, Pevtsov would meet Grombchevskii and his party in the oasis Niya (in Chinese Turkestan), in February 1890, which allowed him to link up his route surveys with those of Grombchevskii. All this clearly evidenced that Russia increased its expeditionary activity in the region in the post-Pzhevalskian period.

The Tibet expedition under M.V. Pevtsov started in Karakol, on 13 May, 1889, and terminated in Zaisanskii Post on 3 January, 1891. As on his two previous journeys under Przhevalskii, Roborovskii was again collecting plants, yet this time he also made some instrumental observations and conducted

22 For more details about M.V. Pevtsov and his career as a scholar-traveler, see his biographical essay by M.K. Baskhanov in this volume.

23 Vinogradov, 81, 82.

geographical investigations, much on his own. Pevtsov was explicitly unlike Przhevalskii – he did not have his heroic streak in the character, and his leadership was not as tough and absolute as that of his predecessor. He used to discuss things with his assistants and make his suggestions to them, rather than give strict orders. Thus, Roborovskii had largely a free hand in his activities in the field, with the leader's due approval, having made five independent surveys or "excursions" from the expedition camp, all in 1890.

The second of these leading over the Saryk-tuz pass (5270 m high) in the Kunlun into the Tibetan plateau turned out to be the hardest trip he had ever made. He was accompanied by only one Cossack, Bessonov, the most robust of the expedition escort, as none of the locals agreed to join him. They said that the plateau was a cold wasteland where death and suffocation, *is-bar,* reigned supreme. And indeed, the territory Roborovskii traversed, about 100 versts (60 miles) southward, was an absolutely barren land, with no animal life, except for a few starved-out *orongo* antelopes. "It was the first time that I have found myself in such a horrible wilderness, and my desire to get acquainted with it led me perhaps a bit too far than it should", he would later write in his travel account.[24] The horses of Roborovskii and Bessonov perished during this trip, so they had to walk on foot for about 40 km on the rocky ground to return to their camp.

Roborovskii's contribution to Pevtsov's expedition was quite substantial. Apart from his herbarium of the flora of Kunlun amounting to 7,000 samples (nearly 700 species) and his rather valuable notes on the geographical spreading of these plants, he surveyed about 2,500 versts (1,650 miles) of terrain during his independent excursions, determined the geographical coordinates of 16 points and measured 40 elevations.[25] His surveys alone were of great benefit to cartography. For example, Roborovskii surveyed the entire course of the river Cherchen (Cherchen-darya), which starts in the middle part of the Kunlun and was originally discovered by Przhevalskii. He followed Cherchen downstream up to the point where it flowed into Lake Karaburankul (or Karaburan, according to Przhevalskii) and upstream, to its sources situated on the slopes of the Przhevalskii Range. In the upper reaches of Cherchen he discovered a clayey elevation named Belkhash, stretching from the Altyntag to the Moscow (Achikkol) Range. Also of importance was his complete survey, jointly with

24 Roborovskii 1896, 40–41.

25 Pevtsov 1892, 31–32. B.V. Yusov gives somewhat different figures in his biographical sketch of Roborovskii, saying that he determined the coordinates of 13 points and measured 34 elevations (Yusov 1949, 13).

Kozlov, of Lake Bagrashkol located in the spurs of the eastern Tian Shan, which until then was shown on the maps on the basis of oral information.

As recognition of his achievements, the IRGO awarded Roborovskii in 1891, along with K.I. Bogdanovich and P.K. Kozlov, a silver Przhevalskii medal. (This was a new award instituted by the Society in the same year in commemoration of Przhevalskii).

Leading His Own Expedition, 1893–1895

After the end of the expedition Roborovskii spent much time in the Botanical Gardens, sorting out his large herbarium to be expertly analyzed by botanist Maksimovich. There he befriended a gardener and purveyor of medicinal plants, Aleksandr Lapin, his wife, Khristiana, of Baltic-German extraction, and their daughter, Lydia. (The latter was a young widow, having lost her husband, Osipov by name). Vsevolod fell in love with Lydia and she returned his feelings. They got engaged in early 1893, yet their marriage had to be postponed as Roborovskii was preparing for a new journey into Inner Asia, this time under his own leadership.

The IRGO made plans, in the spring of the same year, to send two expeditions to the region, led accordingly by Potanin and Roborovskii and both having Sichuan as their destination and main object. Sichuan was a south-western province of the Celestial Empire in the basin of the Yangtze River, known for its rich wildlife; its exploration, like that of southern Tibet with its capital city of Lhasa, was one of Przhevalskii's priorities and his unrealized dream, which, quite naturally, became a strong attraction for Roborovskii and Kozlov.

Roborovskii's main assistants on the two-and-half year journey were Kozlov and Veniamin Ladygin (a former teacher in a Dungan school at Karakunuz, being a fluent speaker of Chinese, Manchu and Turkic); they were accompanied by an eight-man military escort and two guides. The party was well equipped – apart from their usual military outfit, they carried a large set of instruments for various field observations provided by the Main Staff and the IRGO, as well Przhevalskii's brothers, Vladimir and Evgenii, plus some commodities which had never been used by travelers before, such as special rubber bags for keeping drinking water (called *makintoshi*), manufactured by the Macintosh Company.[26]

26 Every rubber bag contained 5 buckets (*vedro* in Russian) of water, i.e. 13,5 gallons (1 *vedro* makes 2,7 gallons).

The expedition started in Przhevalsk (Karakol) on 15 June, 1893. Prior to its departure, Roborovskii and his team visited Przhevalskii's grave and served a mass for him to pay homage to the great traveler as well as to raise their spirits. The expedition caravan included 25 pack and 10 reserve camels, 15 horses, 5 rams, a goat and three dogs. From Karakol the travelers proceeded to the Big Yulduz plateau in the central Tian Shan, little known to European geographers (earlier visited by Przhevalskii); descended into the Turfan plain,[27] where Roborovskii set up a meteorological station at Lukchun, following the example of Pevtsov, who had organized a similar station in Niya; thence the party moved along the northern fringe of the Hami desert toward the northern Nanshan and further on to the Amne Machin Range, making the eastern extremity of the Kunlun. The Amne Machin mountains are considered sacred mountains by Buddhists. They are known for their many temples, shrines and refugees for pilgrims. It was from here that the travelers planned to cross into Sichuan (The entire route of Roborovskii's journey is shown on the map of Kozlov's expeditions reproduced in the next chapter).

The explorations began in the Tian Shan and were continued in the Turfan plain. In the vicinity of the village Lukchun, Roborovskii rented a fanza from a local district official, Beshir-Ahun, and turned it into a store-house. There he accommodated his incipient collections and supplies for the return journey. On the flat roof of the fanza, he built a wooden weather station which would be run for two years (between 1 November 1893 and 18 October 1895) by a Cossack under-officer Nikolai Shestakov, whom Vsevolod specially trained for the job. He also marked the roof with the cardinal points to help his assistant determine the direction of winds. Apart from making meteorological and astronomical observations at the station, Shestakov was to collect various ethnographical materials (for which purpose he was supplied with sufficient monetary resources), seeds of local plants, and skins and skeletons of wild camels, horses and other animals.

As soon as the weather station was ready, Roborovskii conducted a serious of geodetic observations, the most important of which were the leveling of the Lukchun depression, made together with Kozlov and Shestakov, according to Pevsov's instructions, and the determination of the altitude of Bogdo-Ula, the highest mountain in the Tian Shan. Before his departure from Lukchun, Roborovski left with Shestakov a horse named Gegen and his favourite pointer Yashka, so that he would not feel lonely.

After a short period of respite at Toksun, one of the settlements at the bottom of the Turfan depression, the expedition split: Kozlov trekked southward

27 The Turfan plain or depression situated near the oasis of Turfan (130 m below the sea level) was discovered by G.E. Grumm-Grzhimailo in the course of his 1889–1890 expedition.

to the Kyzyl-Synyr oasis, the only one in the Kashgar desert, by a road which had not been trodden by Western travelers since the time of Marco Polo; and thence to the Konchedarya river and Lake Lopnor, first visited by Przhevalskii in 1877. In the meantime (in November 1893), the main party under Roborovskii headed eastward across the Hami desert toward the oasis Sa-chow where it was eventually to catch up with Kozlov's party.

On 10 January, 1894, Roborovskii arrived at Shazhou, also known as Dunhuang, which he first visited with Przhevalskii on his third journey (in 1879–1880). The oasis town used to be the gateway to China on the Northern Silk Road, and it would become a great attraction to western travelers, in the early 20th century, after the discovery nearby of the many caves of the Thousand Buddhas (Qian-fo-dong). The expedition spent about three months in Shazhou, at the northern foot of the majestic Nanshan, on the bank of the Danhe river, flowing down from the Humboldt Range. The travelers investigated the fauna of the place, watched the springtime passage of migratory birds, and they greatly enjoyed "the awakening of nature". They also visited the site of the sacred Qian-fo-dong caves and another one, that of "the caves with the burkhans", 27 versts north of the town, on the bank of the Danhe river.[28]

Roborovskii's general impression of Dunhuang was strongly negative: "I have not seen anywhere a dirtier and a more disgusting place as Dunhuang. All its narrow streets and lanes are filled with all kind of filth... Everywhere by the walls of houses there are heaps of rubbish, in which pigs and dogs are digging".[29]

The peculiar character of Roborovskii's expedition was that it was not just following its well-charted track, but was constantly making additional 'reconnaissance excursions', off the main route, in various directions, covering hundreds of miles. From the very start, the expedition actually had two leaders, since Kozlov did a large amount of field work entirely on his own. He made 12 excursions altogether, having surveyed 8,000 versts (about 7, 000 miles), which comprised about one half of the entire surveying work (the other half was done by Roborovskii). In his diary Kozlov made a special point to say that he and Roborovskii were both "properly prepared (for the journey)" and therefore "simultaneously led two expeditions, one may say".[30] (The Roborovskii expedition would later come to be known as the 'Expedition of Przhevalskii's Companions' – *Ekspeditsiia Sputnikov Przheval'skogo* – in geographical literature)

28 Roborovskii 1949, 164–185.

29 Ibid., 182.

30 Kravklis 2006, 151.

The party left Shazhou on 11 May, 1894, heading for the western Nan Shan to continue the work started earlier by a host of Russian travelers.[31] This was a large mountainous country twice as big as the Caucasus or the Alps, consisting of narrow ranges stretching in the north-west direction. The system of field explorations here was as follows: moving up the river Sharogoldzin (upper Danhe), the travelers made stops every 100 versts (over 60 miles) for some 20–30 days. These stops were made "at the best spots", those rich in fodder for caravan animals and having water springs, which served as convenient starting points for "fleeting reconnaissance trips" to either side, made by Roborovskii and Kozlov. When undertaking such explorations, Roborovskii would leave his second assistant, Ladygin, at the main base, where he was replenishing the botanical and entomological collections and making systematic meteorological observations at the temporarily organized weather stations, while his extra assistant, Kurilovich, was shooting birds.[32]

This system allowed Roborovskii and Kozlov to conduct a most thorough survey of the Nanshan, which lasted for four months. Roborovsky surveyed the terrain lying to the south-west of the main route, and Kozlov that to the north-east by means of their extensive sidelong excursions. The travelers passed through all the valleys separating the principal ranges of the Tian Shan and also crossed some of these (the Humboldt, Ritter and other ranges) by transversal routes. These surveys allowed Roborovskii and Kozlov to gain a good understanding of the complicated orographic system of the western Nanshan, to determine the borders of the region, and the sizes and the peculiar features of these ranges.

Having left the Nanshan, the expedition crossed the Tsaidam plains and struck south-east, towards the much-coveted Sichuan. Ahead of them lay the formidable snow-clad Amne-Machin Range, venerated by Buddhists as 'sacred mountains', the major barrier on their way. Having ascended the range, the party bivouacked close to the Manlun pass, beyond which spread out the valley of Huang He, inhabited by the militant Goloks. Here a most severe trial awaited Roborovskii – on 28 January (9 February), 1895, he had a stroke at night. As a result the right part of his body was paralyzed.

The expedition had to be interrupted and to return to Russia. As Roborovskii would later comment on the event in his travel account, "With much sorrow, having to abandon my desire to see Sichuan, the land of promise for our

---

31 These pioneer explorers were Sosnovskii and Matusovskii (1875), Przhevalskii (1872, 1879–80, 1883), Potanin (1886), Grumm-Grzhimailo (1890) and Klementz (1892).

32 Ibid., 190.

expedition, I was contemplating the situation in the dead of the nights, and decided to turn back...."[33]

The sudden attack of apoplexy (i.e., brain haemorrhage) which temporarily disabled Roborovskii was apparently an outcome of his long-time exposure to high altitude and climatic rigours. He was not feeling well for the greater part of the journey, suffering mainly from headaches and fits of dizziness as well as skin and other disorders. And in March 1894, when he was making an excursion to the Anembar Ula, his party was caught in a violent sandstorm which nearly killed him and his companions. So his health had been seriously shattered long before he ascended the Amne-Machin.

On 5 February, 1895, the travelers set out from their camp. About a week later they were unexpectedly attacked by a large group of Goloks and Tanguts. Roborovskii, despite his condition, took part in the skirmish – seated on the ground, he somehow managed to fire his Berdan rifle, fixed upon a box in front of him.

Being unable to sit on horseback, Roborovskii had to walk, or rather hobble along, supported by one of the Cossacks. He could not do any writing as his right hand was paralyzed. Yet, luckily, his ailment began to recede in a short while, and he gradually began to recover during the two month long bivouac in Kurlyk at Tsaidam on their way home. Roborovskii resumed his regular observations and other field work en route – he would collect plants and insects, take daily measurements of the temperature with the help of a thermometer, and make notes in his travel diary. In Kurlyk the party celebrated Easter on 2 April, following the Orthodox traditions, "with the red-colored eggs" for breakfast. (These were stolen from goose nests in the nearby bogs, which somewhat embarrassed Roborovskii, and painted red, the way they do in Russia.)

On their return journey the travelers passed through northern Tsaidam, Hami, Shazhou, and moved further on towards Lukchun, where Shestakov had been eagerly waiting for them at the meteorological station. Here, at Lukchun, Roborovskii spent about a month, between 24 September and 18 October. During this period he visited Turfan again to express thanks to the head of the district for his taking care of Shestakov, but, more importantly, he from Turfan he made a tour of many ancient sites in the area.[34] Thus he surveyed the ruins of Old Turfan, Idygot-Shari (between Turfan and Lukchun), Assa-Shari,

33 Roborovskii 1949, 309.

34 See V.I. Roborovskii, *Kratkaia Zametka o drevnostiakh v Liukchunskoi kotlovine* (1897), NA RGO, F. 75, Op. 1, D. 9, ll. 9–10 ob.; also ll. 5, 5 ob., 6, 12 (List of items sent by V.I. Roborovskii to the IRGO); published in Izvestiia IRGO, XXXIII, 1897, 435–437; see Roborovskii 1897.

the Moslem monastery Tuyuk-Mazar (between Lukchun and Idygot-Shari) and other places. While passing through these, Roborovskii succeeded in collecting a number of antique items, whatever he could find or purchase from the locals, such as coins, statuettes (*burhans*), samples of pottery, various decorations, as well as fragments of manuscripts in the Old Uighur and Turkic languages. (This was not a fully safe trip though, as Roborovskii was once bitten at night by a scorpion.) On the other hand, Shestakov too had obtained, during his two-year stay at Lukchun, various ethnographic and archeological artefacts, including ancient manuscripts, clay *burhans*, etc. These finds, especially the Uighur scrolls, when delivered to St. Petersburg, made quite a sensation among Orientalists and prompted the Academy of Sciences subsequently to send an expedition to the Turfan region under archeologist and ethnographer D.A. Klements.

From Lukchun Roborovskii's expedition headed for Russia in two separate parties led by Kozlov and Roborovskii; these arrived at Zaisan (in today's Kazakhstan), on 7 and 21 November, 1895, respectively. (Roborovskii proceeded through Urumchi, Manas, and thence trekked northward by a new path running across Dzungaria, while Kozlov, accompanied by Shestakov and a guide, followed another route, lying to the east of the one chosen by Roborovskii).

After the expedition was disbanded, on 7 December, Roborovskii and Kozlov left Zaisan, and they finally reached St. Petersburg on January 2, 1896. There the travelers were given a warm welcome by the public and especially the scholarly community. They were also received in audience by the new Emperor, Nicholas II, a great admirer of Przhevalskii, at the Winter Palace (on 13 January). Having listened to the fascinating story of their travels, the tsar gave an order to promote the expedition leader to his next military rank, that of captain. (The heir of the throne, Nikolai Aleksandrovich, who became the emperor of Russia in 1894, showed much interest in Roborovskii's expedition. He personally read all letters and telegrams that its leader dispatched to St. Petersburg – to the Main Staff and the Geographical Society. Moreover, when he learnt about the incident with the Tanguts and how bravely a handful of Russians fought against a 200-strong group of their assailants, he issued an order to award the expedition's lower ranks the cross of St. George, the highest decoration for soldiers and under-officers in Russia.[35])

On 7 February, the Vice-Chairman of the IRGO, P.P. Semenov, speaking at the general meeting of the Society, highly appraised the work done by Roborovskii and Kozlov in the expedition "as the followers of Przhevalskii", he said, they "have justified his behests in a deserving manner". And a year later (1897) the IRGO would give Roborovskii their highest award, the Konstantin Gold

35 Roborovskii's letter to Kozlov of 19 November [1895], NA RGO, F. 18, Op. 3, D. 589, l. 15.

Medal, as a token of appreciation of his outstanding scholarly achievements. In his letter of recommendation for the nomination, the IRGO's secretary, A.V. Grigoriev, stressed that Roborovskii's journey was "one of the most remarkable and most fruitful expeditions of recent times".[36] Having returned to Russia, Roborovskii also learnt that he had been elected, on 9 March 1895, while he was still on the Asian track, an honorary member of the Royal Dutch Geographical Society,[37] an event that signified recognition of his scholarly merits by Western geographers.

Overall, Roborovskii's expedition made several substantial contributions in the following fields:

Cartography: The expedition traversed 16,000 km in little known or completely unknown territories of Inner Asia. This exceeded the total length of the routes of Przhevalskii's four expeditions, or any other single expedition mounted by the IRGO. Roborovskii's and Kozlov's topographic surveys were based on 30 astronomical points determined in the field; as a result about 222,000 sq. km were put on the map by cartographers.

The expedition's greatest merit perhaps was its extensive surveying of the western part of the Nanshan, until then a large 'white spot' on the maps of Inner Asia. In the opinion of Soviet geographer N.M. Shchukina, "this finally provided an insight into the complicated orographic system of Western Nanshan, to determine the outlines of all its major ranges, their sizes and peculiar features".[38] Much more than that – the travelers located the sources of many rivers flowing from the Nanshan and determined the sizes of lakes (those in the Nanshan and Eastern Tsaidam). They also surveyed a number of ranges lying to the south of the Turfan depression and the depression itself was crossed by a network of routes in the course of which its sizes and altitudes were determined.

Meteorology: The meteorological observations conducted at the weather station in Lukchun over a period of two years and at those established in other places were of great value for the study of the climate of the region. General A.A. Tillo, one of the leading Russian geodesists and cartographers of the time, specially commended the work of the Lukchun station set up in the heart of the continent, its elevation being close to ocean level: "For the first time the

36 Grigoriev 1897, 15; also quoted in: Kravklis, 2006, 169.

37 The date comes from the Society's journal: Tijdschrift van het KNAG (*Koninklijk Nederlands Aardrijkskundig Genootshap*), 2nd series, Part XII, Leiden, 1895, 308–309 (information of the executive secretary of KNAG, Eelko Postma). Russian biographers of Roborovskii usually give a later date (1897).

38 Shchukina 1955, 145.

barometric measurements were conducted three times a day, and it is only now that we have come to know about the daily changes of atmospheric pressure right in the center of Asia".[39] In the opinion of Soviet geographer B.V. Yusov, the observations at Lukchun provided "absolutely new data for the climatology of Inner Asia. Thanks to these observations, it became possible to draw the isobars on the map of Inner Asia (i.e., lines connecting places of equal pressure) with an accuracy of 1–2 or 3 mm, instead of 10–20 mm, according to previous data. This was a great advance in geography".[40]

Zoology: The expedition's zoological collection included 250 skins and 30 skeletons of rare species of animals (mammals); 1200 samples of birds, 450 reptiles and fishes, and about 30,000 insects.

Botany: The botanical collection numbered 1300 species (25,000 samples) of plants and about 300 seeds of the cultural and wild plants.

Geology: The expedition collected 350 samples of rocks and minerals.

On top of that, the expedition journals and Roborovskii's en route reports dispatched to the IRGO contained a wealth of information on the wildlife in the explored regions; and the various ethnographic materials and notes, as well as archeological finds, including a bagful of manuscripts in the Uighur and Turkic languages, were likewise of no little value.

Final Years

After the completion of the expedition, Roborovskii stayed in St. Petersburg for some time. He was visiting doctors and sorting out his huge collections which arrived at the IRGO in March 1896. He did not feel strong enough though to get down to the exhausting work of writing up his journey as was a must for an expedition leader, so his military superiors granted him a four-month leave to restore his health. On the recommendation of doctors, Roborovskii went, in early May, to a popular medical spa town in Germany, Bad Wildbad, situated in Schwarzwald. This was his first trip to Western Europe, and he used the opportunity to visit his main attractions there, the German zoological and botanical gardens in the cities he passed through. He was completely ravished with the famous *Tiergarten* in Berlin, the largest in Europe; there he saw many rare species of animals and plants, such as pheasants and the *Lophophora* cacti, known

39 A.A. Tillo, *Glavneishie rezul'taty meteorologicheskikh nabliudenii na stantsii ustroennoi IRGO v Pritian'shan'skoi vpadine v gorode Liukchune*, in: Roborovskii 1899, 9; quoted in Andreyev (ed.) 2013, 137.

40 Quoted from Andreyev (ed.) 2013, 137.

as the Peyote to the American Indians, which had so strongly attracted Przhevalskii to Sichuan at the end of his life.

Having returned to St. Petersburg, Roborovskii got married to his fiancé, Lydia Osipova, in the fall of 1896. Together they then made a trip to the Black Sea coast of the Crimea and the Northern Caucasus, visiting the most scenic places there – the famous spa resorts, mineral water springs and nature reserves, such as Oreanda, Yalta, Livadia, Batumi, Kislovodsk and others. As a result of the trip, the medical treatment and the comfort of family life, Roborovskii's condition improved so that he was able to take up the pen. The writing progressed rather slowly as he was often interrupted by recurring fits. In the meantime, a daughter was born to the couple, named Vera (on 27 May, 1898). Her god-farther at the christening was Mikhail Pevtsov, one of Roborovskii's close friends, since the time of the Tibet expedition. His other friend, Petr Kozlov, resided in Moscow, together with his wife, Nadezhda Kamynina, whom he married in the same year. The three would occasionally see each other, being intimately united for the rest of their lives by their exploratory missions and, above all, by the dear memory of the great Przhevalskii.

When finished, Roborovskii's travel account was published in the three-volume edition of the 'Works" of his expedition, as its first volume (1900).[41] The edition was dedicated by Roborovskii to the memory of the late emperor Alexander III, "under whose patronage a long series of scientific explorations of the unknown lands of Asia was carried out". The second volume was written by P. Kozlov and described his separate reconnaissance excursions (*Otchot pomoshnika nachal'nika ekspeditsii – P.K. Kozlova*); the third one presented 'The Scientific results of the Expedition' (*Nauchnie rezul'taty Ekspeditsii V.I. Robotovskogo*), edited by A.A. Tillo. This third volume gathered miscellaneous materials, such as the meteorological observations at the Lukchun station and in other places (presented by A.A. Tillo); a list of the measured absolute heights; the astronomical observations (A.R. Bonsdorf), the description of the levelling of the Lukchun depression by V.I. Roborovskii; the magnetic measurements (A. Rodd), and, finally, the lists of the collected samples of mammals (determined by E.A. Bikhner), birds (by V.L. Bianki), reptiles, amphibian and fishes (by A.M. Nikol'skii), and the lepidoptera (by S.I. Al'feraki); as well as the lists of

41 The full title of the edition is "The Works of the Expedition of the Imperial Russian Geographical Society to Central Asia, conducted, in the years 1893–1895, under the leadership of V.I. Roborovskii" (*Trudy Imperatorskogo Russkogo Geograficheskogo Obshchestva po Tsentral'noi Azii, sovershennoi v 1893–1895 gg. pod nachal'stvom V.I. Roborovskogo*), St. Petersburg, 1900–1901.

the samples of Tibetan and Mongolian flora delivered to the Botanical Gardens (by K. Maksimovich), and of the collected rocks (by V.A. Obruchev).

Not included in the volume were the materials which were not yet fully described, namely the collection of clams and that of the archeological finds at the Lukchun depression. Appended to Roborovskii's travel account were seven maps, printed at the D.M. Rudnev cartographic factory in St. Petersburg, and 21 illustrations made from his photographs.

An employee of the factory, D.K. Ivanov, remembered Roborovskii as a tall, well-proportioned and good-looking man, who possessed much innate strength and fortitude, despite his illness. His face, with a little light brown beard, was unhealthy, yet it was manly and handsome. He was a good storyteller and was fond of jokes, and the stories he told about his travels and adventures captured people's attention and earned him the nickname of 'John Chrysostom'. "He was always surrounded by groups of listeners, spell-bound by his words; he inspired them with his energy, his passion for travelling, and his great love of nature", in the words of his biographer B.V. Yusov.[42]

Roborovskii was very social, soft-hearted, and unusually modest, a distinctive feature of his character noticed by many. He was a good mixer and had many friends among villagers. The old residents of Taraki recalled how Roborovskii had brewed the barley beer on the bank of the river Volchanka, according to some old recipes, assisted by peasants.[43] He usually came to Taraki in the summer as his house there was only a summer cottage where nobody lived in the wintertime. When he was strong enough, Roborovskii made plans for building himself a new house, somewhere on a hillock in the coniferous forest, with a stream flowing nearby, after Przhevalskii's example.

In 1901 Roborovskii was promoted to the rank of sub-colonel, but two years later (on 11 June, 1903) he was dismissed from the military service, with the rank of colonel, due to his ailing condition. His illness recurred in the 1900s and it rapidly progressed and worsened, turning him eventually into a poor cripple. Roborovskii's once mighty organism was slowly failing – first his legs became numb and then he lost the ability to speak.

He spent his final years on his family estate, the village Taraki, where he was taken care by his loving wife, his relatives and doctors. He lived long enough to learn about Pevtsov's death in 1902 and Kozlov's two brilliant expeditions in 1899–1901 and 1907–1909. Roborovskii died of a new stroke on 23 July 1910. He was buried at the Ovsishche village cemetery, close to the chapel of Michael the

42 Yusov 1949, 26.

43 Vinogradov 2018, 120.

FIGURE 3.9
Roborovskii's tombstoneat the Ovsishche cemetery, Udomlia district.
COURTESY B.K. VINOGRADOV.

Archangel. The red granite tombstone had a laconic and modest inscription: 'Colonel Vsevolod Ivanovich Roborovskii, born 26 April 1856 – died 23 July 1910'.

Roborovskii's death was mourned by the scholarly community and the general public in Russia. The IRGO published his short biography, written by Kozlov, in its *Transactions*. In the biography Kozlov especially emphasized Roborovskii's closeness to Przhevalskii – as one of his pupils and followers, he "fulfilled the behest of the teacher and drew closer to Przhevalskii's glorious wreath".[44] (Kozlov's last contact with Roborovskii was presumably in early 1907 when he was preparing for his Mongolia-Sichuan expedition; it was then that Roborovskii presented him the Lancaster carbine, the one he had received from the dying Przhevalskii[45].)

The popular illustrated magazine *Niva* also carried an obituary in which it portrayed Roborivskii as an "explorer of unknown lands", a field surveyor who

44 Kozlov 1910, 357–367.

45 This famous rifle is kept presently at the Russian Geographical Society in St. Petersburg.

made numerous meteorological and astronomical observations, and a naturalist who assembled valuable collections of rare species of flora and fauna. "He did much to expand our knowledge of the nature and life of Central Asia".[46]

Very little is known of the later life of Roborovskii's widow, Lydia. We know that she had kept in touch with Kozlov until 1914 at least, when he went to the front after the beginning of the Great War. Her final years she spent in Leningrad, together with her daughter Vera. Both of them died in the besieged city during World War II. Their apartment was then pillaged, which resulted in the loss of some of the Roborovskii family memorabilia. According to Vinogradov, there are no living direct descendants of Roborovskii in Russia today.

•••

Vsevolod Roborovskii is remembered nowadays as one of the heroic Russian pioneer explorers of Inner Asia and a close associate and follower of Nikolai Przhevalskii. His most significant contributions were in geography, cartography, meteorology, botany and zoology. His name was given to a number of species of Inner Asian fauna and flora discovered by him.[47] The best known today in the feathered kingdom are perhaps Roborovskii's hamster (*Phodopus roborowskii*) and Roborovskii's rosefinch, a little bird in the genus *Carpodacus* of the finch family (originally named *Leucosticte roborowskii przhewalskii, 1887*); and in the vegetable realm one can mention such species as *Adientum Roborowslii* (maidenhair fern), *Cacalia Roborowskii* (tall smooth herb) and *Lagochilus Roborowskii* (double hare plant). There is also the Roborovskii Glacier in the mountain juncture Tabyn-Bogdo-Ola in the Mongolian Altai, close to the Russian-Mongolian border.

46 *Niva: Illustrirovannyi zhurnal literatury, politiki i sovremennoi zhizni*, 23 July 1910, p. 698

47 B.K. Vinogradov lists 28 species of plants and those of 10 animals, named after Roborovskii (Vinogradov 2018, 211, 212).

CHAPTER 4

Petr Kuzmich Kozlov (1863–1935)

Alexandre I. Andreyev and Tatiana I. Yusupova

> How many times have I been happy standing face-to-face with the magnificent nature of Inner Asia.
>
> P. KOZLOV

Petr (Pyotr) Kuz'mich Kozlov (1863–1935) was another Russian pioneer explorer of Inner Asia.* Like V. Roborovskii, he was a close assistant and friend of the great Przhevalskii who started the brilliant era of the Russian exploration in the region. According to British historian and archeologist Evart Barger, both Przhevalskii and Kozlov rank among the most outstanding Western travelers of the 'period of *Asienforscher*', a 'gold age of discovery', along with other great explorers from Western Europe and America, such as Sven Hedin, Paul Pelliot, Edouard Chavannes, Ferdinand von Richthofen, Aurel Stein, Albert Grünwedel, Albert von Le Coq, Raphael Pumpelly and Ellsworth Huntington.[1]

Early Life and Education

In his short autobiographical sketch written in the late 1920s Petr Kozlov stated that his whole life was spent "under the badge of exploration of nature and man of Inner Asia". He had just one desire – to wander around in the wide expanses of deserts and mountains of the great Asiatic continent (Kozlov 1927).

Kozlov was born on 3(15) October 1863 into a poor, semi-literate family in Dukhovshchina, a little town in Smolensk Province (west of Moscow). His

* The chapter is a slightly revised version of P. Kozlov's biographical sketch published by the same authors, see *Andreyev A.I., Yusupova T.I.*, Pyotr Kuz'mich Kozlov, 1863–1935, *Geographers Biobibliographical Studies*, Vol. 34 (Ed. by H. Lorimer and Ch.W. Withers), London; Oxford, New York; New Delhi; Sydney: Bloomsbury Academic, 2015, 127–164.

1 Barger 1944, 1.

 | DOI 10.1163/9789004376267_006

FIGURE 4.1
Petr Kuz'mich Kozlov, ar. 1912.
NA RGO, F. 18, OP. 7, D. 1.

father Kuz'ma Egorovich belonged to a lower middle class; he was a herdsman who drove cattle from Ukraine to central Russia. The boy occasionally accompanied his farther on these trips and that was how he took to the life of a wayfarer, seeing scenes and constantly moving from one place to another. His mother Praskovia Nikitichna was, in his own words, an "embodiment of meek temper and kindness". When the boy was nine, his father sent him to the local parish school which was turned, in 1875, into a six-year town school, with a new teaching staff which included some talented pedagogues of new formation. One of them was Vasilii Porfirievich Vakhterov, a Darwinist and the founder of the theory of evolutionary pedagogics. He strongly influenced young Petr, having aroused in him keen interest in geography and natural sciences and acquainted him with Przhevalskii's books *Puteshestvie v Ussuriiskom krae, 1867–1869* (1870) and *Mongolia i strana Tangutov* (1875), describing his journeys in the Far East and Inner Asia.

Petr left the school in 1878 and immediately started to make preparations for entering the Pedagogical institute at Vilno (today's Vilnius, Lithuania). However, his parents could not sponsor his further education, so he had to take the job of a clerk at the distillery in the village Sloboda in the same Smolensk Province. Incidentally, it was there, at the shore of the scenic Lake Sapsho, that

Przhevalskii had settled, after the completion of his third expedition to Inner Asia (1879 – 1880), having purchased an estate there. Kozlov met the traveler by chance in Sloboda and this meeting settled his destiny. During their friendly chat Przhevalskii sensed a kindred soul, the 'nomadic spirit' in the youth and invited him to take part in his next expedition. Moreover, he offered to prepare him for the journey. As Kozlov would later recall in his memoirs, Przhevalskii became his second father, who educated him and taught many things needed by a field-surveyor; he shaped him as an explorer. Under Przhevalskii's tutelage, Kozlov learnt how to conduct field survey of many kinds, such as route sketching, taking instrumental measurements, shooting animals, collecting plants and minerals.

In January 1883 Kozlov passed the examinations, as an external pupil, at the Smolensk real school, and shortly after joined the 2nd Sofiiskii infantry regiment in Moscow. Knowing that Przhevalsky recruited his expedition staff from military men alone, he had to undergo a three months' training as a volunteer of the regiment.

In his later years Kozlov would continue his education trying to acquire knowledge of the things which might be useful to him in his travels. He associated with the leading scientists in St. Petersburg, then the capital of Russia, and often visited the Imperial Russian Geographical Society (IRGO), the Pulkovo Astronomical Observatory, the Zoological Museum and the Botanical Gardens. "The years of my settled life at home, he wrote in one of his autobiographical essays, I spent in perfecting my knowledge of the natural sciences, ethnography and astronomy".[2]

With Przhevalskii in His Last Expeditions, 1883–1885, 1888

Przhevalskii's fourth journey to Inner Asia, also known as his second Tibet expedition (1883–85), had as its principal objective the exploration of Northern Tibetan plateau. His final destination was Lhasa, the residence of the Dalai Lamas of Tibet, which he failed to reach on his previous journey (1879), being halted by a Tibetan frontier cordon, some 165 miles north of the holy city. Kozlov joined the party as the leader's junior assistant, along with sub-lieutenants Vsevolod Roborovskii and Fedor Eklon. Other members, forming the escort, were grenadier-soldiers and Cossacks from Transbaikalia.

The expedition caravan trekked from Urga (today's Ulaanbaatar) south across the Gobi Desert towards the Nanshan Mountains and Eastern Tibet (Kham). There Kozlov was spell-bound by the majestic grandeur of the place,

2 Kozlov 1927, 4.

FIGURE 4.2
Petr Kozlov, with the Order of St. George Cross, after his first expedition under Przhevalskii, ar. 1885. NA RGO, F. 18, OP. 7, D. 1.

especially by the swift and winding Tatung River flanked by the rocks, of which he had read before in Przhevalskii's books. He immediately fell in love with Tatung and would later express his admiration in the words: "At Tatung my soul woke up consciously for the first time – I learnt to apprehend the beauties of Alpine nature".[3] From Nanshan the party travelled to Lake Kokonor, thence to the Tsaidam (Qaidam) marshland and finally ascended the Tibetan Highlands, having crossed the Burhan Budda range. There the expedition explored the upper reaches of Asia's two greatest rivers, Huang He and Yangtze. However the travelers soon found themselves in trouble when their base camp was attacked by a group of militant Ngolok nomads. Kozlov had to engage in fighting, for the first time in his life. Later, when the party camped out at Lake Orin Nor (Hnora mTtsho, Lake Russian, Ngoring Hu), he fought against another band of Ngoloks. Przhevalskii admired his courage and fearlessness and personally awarded him the Order of St George Cross 4th class for bravery.

After these two clashes Przhrvalskii had to give up on his original plans and moved no further into the unhospitable no-man's land. He returned to Tsaidam and proceeded to Lake Lopnor to observe the springtime migration of

3 Kozlov 1947, 107.

birds. From Lopnor his expedition went to the Takla Makan Desert and visited several oases on its outer rim (Cherchen, Niya, Keriya and Chira), on the final leg of their journey.

The scientific results of Przhevalskii's 4th expedition were many. The travelers explored the northwestern fringe of Tibet, the Kunlun and Lake Lopnor, mapped the source of the Huang He river and the upper Yangtze, discovered a series of hitherto unknown mountain ranges. They also found a pathway leading from Tsaidam to Takla Makan and crossed the 'burning' desert. Kozlov's own contribution was mainly meteorological and barometric measurements and he collected zoological specimens. His field work under Przhevalskii, who taught him how to discern birds by their voices and plumage, was especially important for him as a novice naturalist and zoologist.

After the expedition Kozlov would record: "Inner Asia has become my life's goal. Nature in general and that of Inner Asia in particular captivated me completely".[4] Upon his return from the journey, Kozlov entered the infantry military college in St. Petersburg, his tuition being paid by Przhevaskii. Having graduated with the rank of sub-lieutenant in 1887, Kozlov was attached to the 1st Ekaterinoslav grenadier of the life-guards regiment at Moscow and, a year later, was again placed under Major-General Przhevalskii to take part in his new expedition.

The journey was again Tibet-bound with Lhasa as its destination. This was not accidental, since Tibet and Chinese Turkestan (especially Kashgaria) had become, in the 1880s, the new playgrounds of the Great Game, the Anglo-Russian geopolitical rivalry in Asia, a 'Cold War' for the Victorian Age. As a Russian patriot and a senior officer attached to the General Staff, Przhevalskii anxiously watched developments in Inner Asia and gathered political and military-statistical intelligence in addition to his scientific work.

Przhevalskii's fifth expedition, however, was a fiasco. At its start he caught typhoid fever and died suddenly on 19 October 1888 at Karakol, close to Lake Issyk-Kul. Przhevalskii was buried at the lake, according to his will, and a year later the town was renamed into Przhevalsk in commemoration of Przhevalskii's great merits. "The death of my teacher and friend, Nikolai Mikhailovich Przhevalskii", Kozlov would say later, "was a great shock to me. This pain though did not kill my will to live, but furthered my spiritual growth, as I understood at once that I remain, from now on, all by myself and must hold sacred the precepts of my teacher".[5]

4 Kozlov 1913, 91.

5 Kozlov 1927.

Kozlov's new mentors and patrons, after Pzhevalskii's death, were Petr Petrovich Semenov (Tian-Shanskii), the explorer of Tian Shan and the Vice-Chairman of the Imperial Russian Geographical Society; Aleksandr Vasilievich Grigoriev (1848–1909), the Society's secretary, Mikhail Vasilievich Pevtsov, an explorer, and zoologists Valentin L'vovich Bianki (1957–1920), and Evgenii Aleksandrovich Bikhner (1861–1913), who described and studied his collections.

After Przhevalskii: Traveling with Pevtsov and Roborovskii

In 1889 the IRGO placed Mikhail Vasilievich Pevtsov, Colonel of the General Staff and a member of the Society, in charge of Przhevalskii's 'orphaned' expedition. Pevtsov had previously led two expeditions to Dzungaria, Mongolia and Eastern China. He was also known as the author of "The Basic Foundations of Mathematical and Physical Geography" [*Nachal'nye osnovania matematicheskoi i fizicheskoi geografii*] (St. Petersburg, 1881) and for his own method of definition of geographical latitude by the appropriate heights of two constellations. Pevtsov had three assistants on the journey: the junior officers Roborovskii and Kozlov, and a geologist Karl Bogdanovich (1864–1947).

The new leader radically altered Przhevalskii's expedition program. Instead of exploring the southeastern corner of Tibet (Kham) and heading to Lhasa he focused on Tibet's northwestern fringe and aimed to cross in several directions the hitherto unexplored area between the Russian border and the Kunlun range. On this trip Kozlov had an opportunity to work individually, being put in charge of the zoological collection by Pevtsov, while Roborovskii was collecting plants and Bogdanovich minerals. Quite often they were send out on separate 'lateral excursions', off the main caravan route, to undertake topographic survey and collect specimens of fauna, flora, rock, etc. Kozlov made four such excursions.

The Tibet expedition under Pevtsov lasted from May 1889 to the end of 1890. Despite its reduced program, it was fairly productive of scientific results. Kozlov's personal contribution was his zoological collection and the mapping of the 270 kilometers long Konchedarya riverbed along with the northern shore of Lake Bagrashkol. He also collected ethnographic material, including legends he had heard from the local settlers at Bagrashkol, the Kalmyks. Kozlov's report of his journey consisted of two separate narratives: *Vverkh po reke Kon'chedarie* (Upstream the river Konchedarya) and *Po beregu ozera Bagrashkol* (Along the bank of Lake Bagrashkol); both were included in the third volume of the

Works of the Tibet Expedition.[6] These were the traveler's first published works. This journey shaped Kozlov as an explorer – as field surveyor, geographer, and naturalist. For his achievements he was awarded in 1891 the Przhevalskii silver medal recently instituted by the IRGO, and in October of the same year he was elected a full member of the Society upon the recommendation of Semenov and Grigoriev.

Kozlov's next journey to Inner Asia took place in 1893–1895 when he joined the expedition under Vsevolod Roborovskii (1856–1910). This journey came to be known as 'The Expedition of Przhevalskii's Companions' in the records of the Geographical Society as Kozlov in fact shared its leadership with Roborovskii. Its program envisaged the exploration of the eastern part of the Kunlun range, the Lukchun depression south of Turfan, and the north-eastern Tibet, especially the basin of the Yangtze River, visited previously by Przhevalskii.

On this journey Kozlov had a chance to work independently, making 14 separate excursions about 1,000 kilometres in total. He altogether surveyed about 8,000 versts (5,280 miles), nearly half of the entire expedition sketching work. The most remarkable was his two and half month trek from Lukchun (where the travelers established a meteorological station) southward, across Lopnor, and further to Shadzou. On this trip alone Kozlov surveyed 1,750 versts (1,155 miles), which, in the opinion of contemporary geographers, was an expedition in itself.[7] Much attention, as before, was given by Kozlov to collecting zoological specimens. The results of his own work were published in 1899 in a volume of the Works of V.I. Roborovskii's expedition, entitled *Otchet pomoshnika nachalnika ekspeditsii P.K. Kozlova* (The report of the assistant of the expedition leader P.K. Kozlov).[8]

The expedition worked in very unfavorable conditions. The hurricane winds at the Tsaidam Basin knocked down the travelers and pack animals (horses and yaks), and they all suffered in the high altitude of the Tibetan Highlands. When the caravan reached the source of Huang He, Roborovskii collapsed with a stroke and Kozlov had to take command and bring the expedition to an end, which he did in November 1895. Overall the Roborovskii – Kozlov expedition was quite successful. The zoological collection alone consisted of 250 skins of mammals, 1,200 birds, 450 reptiles and fishes, and about 30,000 insects. What made it particularly valuable were some rare specimens of wild camel (*Camelus bactrianus ferus*), wild horse (*Equus Przewalskii*) and snow-cock. The

6 Kozlov 1896.

7 *Otchot IRGO za 1902 god*, (1903), Part I, IRGO, St. Petersburg, 10–12.

8 Kozlov 1899.

latter was obtained by Koslov in Nanshan and subsequently named *Tetraogallus himalayensis Koslowi* by zoologist V.L. Bianki.

Another joyous occasion was Kozlov's marriage. On 29 April 1896 he married Nadezhda Stepanovna Kamynina. She was a Muscovite, a close friend of Elena Przhevalskii, Nikolai Przhevalskii's niece, who lived in Moscow. Nadezhda Kamynina (1870–1942) came from a noble family of modest means, was well-educated, played piano, and, importantly to Kozlov, was sympathetic to his semi-nomadic way of living. The couple settled in Moscow in Nadezhda's apartment. Two children were born to the Kozlovs, Vladimir (in 1897) and Olga (in 1903).

Leading the Tibet Expedition, 1899–1901

In 1898 Kozlov came up with his own project for the further exploration of Inner Asia. This was largely based upon Przhevalskii's program roughly sketched in his account of the fourth expedition in a separate chapter entitled 'How to travel in Inner Asia'.[9] According to this program, the territories which needed to be first explored, were the northern, northeastern and southern fringes of the Tibetan plateau, with its Alpine provinces of Amdo and Kham, the three most "tasty morsels" for a "genuine explorer", in Przhevalskii's words. Lhasa also figured prominently on this agenda, yet it still remained 'the forbidden city', inaccessible to Western travelers. Other areas included the southeastern tract of Mongolia, with the Alashan Mountains, and the mountainous northern Mongolia, bordering on Russian Siberia.

Accordingly, Kozlov's priorities on this journey were three: Eastern Tibet (Kham), Mongolian Altai and the Central Gobi. He had two assistants (travelling companions), Aleksandr Nikolaevich Kaznakov (1871–1933), entrusted with the task of collecting insects and molluscs, and Veniamin Feodorovich Ladygin (1860–1923), in charge of plants and butterflies. At the same time, the latter, owing to his excellent knowledge of the Chinese and Mongolian languages, as well as of the Turkish dialects spoken in Eastern Turkestan, was to collect various ethnographic and historical materials, and to act as interpreter in all dealings with the local Chinese authorities. Both Kaznakov and Ladygin were also to make 'lateral excursions' on their own, as Kozlov had under Pevtsov and Roborovskii. As for the expedition leader, he took upon himself the important tasks of keeping the geographical, meteorological, and natural history diaries, surveying of the expedition main route, and taking the astronomical measurements. The escort of the expedition consisted of sixteen soldiers

9 Przhevalskii 1888, 1–66.

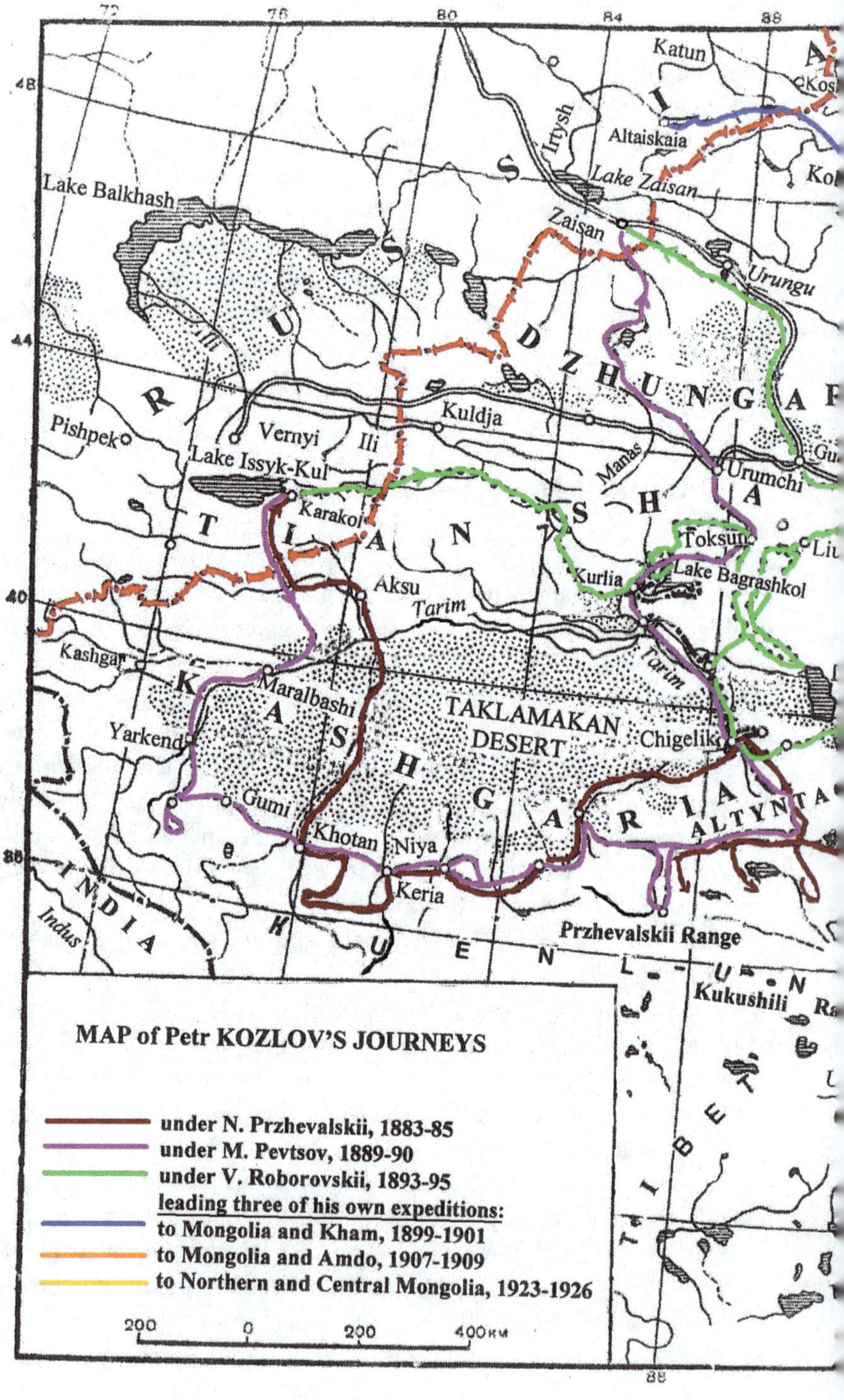

MAP 4.1 The expedition routes of P. Kozlov, 1899–1926, including the route of V. Roborovskii P. Kozlov's expedition (1893–1895). Compiled by A. Andreyev and T. Yusupova.

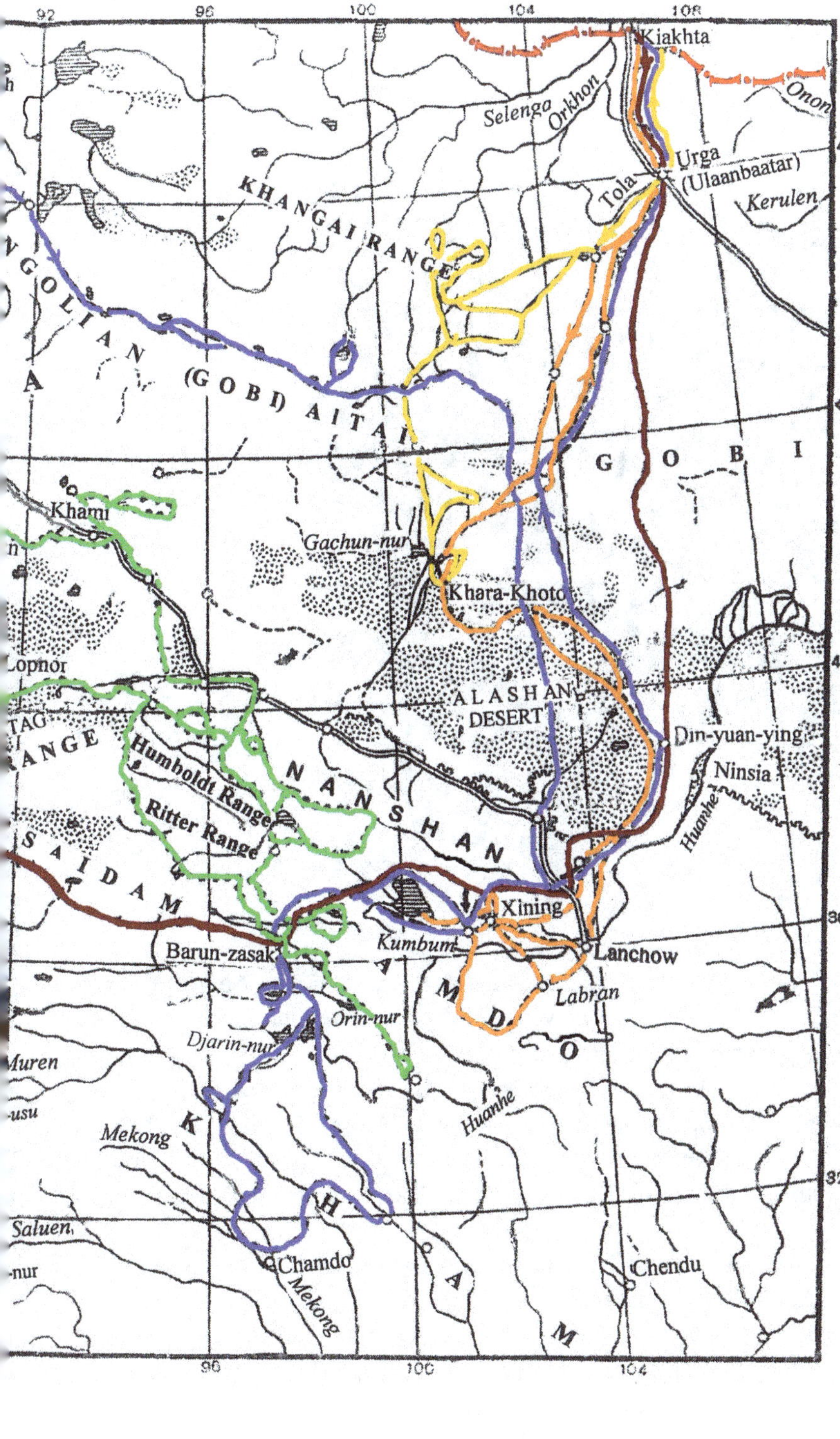
Kiakhta
Selenga
Orkhon
Onon
Urga
(Ulaanbaatar)
Tola
Kerulen
KHANGAI RANGE
NGOLIAN (GOBI) AITAI
GOBI
Khami
Gachun-nur
Khara-Khoto
Lopnor
ALASHAN DESERT
Din-yuan-ying
Ninsia
Huanhe
TAG
RANGE
Humboldt Range
Ritter Range
NANSHAN
SAIDAM
Xining
Kumbum
Lanchow
Barun-zasak
AMDO
Labran
Orin-nur
Djarin-nur
Muren
Huanhe
Mekong
KHAM
Saluen
Chamdo
Mekong
Chendu
92
96
100
104
108
48
44
40
36
32

FIGURE 4.3
P. Kozlov in the Barchu gorge of the Nanchin Ridge (Eastern Tibet), in the course of his Mongolia and Kham expedition, 1900.
NA RGO, F. 18, OP. 7, D. 434.

from different parts of Russia, placed under the command of Sergeant Gavriil Ivanov (?–1922), an experienced veteran who had taken part in three previous journeys of Przhevalskii, Pevtsov and Roborovskii.

The expedition began in the Altaiskaya village in July 1899, passed through Mongolian Altai, Central Gobi, the province of Gansu, Tsaidam, and Eastern Tibet. The return journey was made along the already well-known routes through the Alashan desert and via Urga to Kiakhta on the Russian-Mongolian border.

The results of Kozlov's Tibet expedition were impressive. The party thoroughly explored the Mongolian Altai, the Central Gobi and the portion of Eastern Tibet, known as Kham. The Altai was explored all along its northern and southern foot, and crossed several times, while the Gobi was traversed along four different routes. In eastern part of Tsaidam, at the northern foot of Tibet, a depôt of the collections and the provisions was set up, and the camels

FIGURE 4.4 A group of Tibetans from the Lhado district, 1900. NA RGO, F. 18, OP. 7, D. 789.

were left there as the journey across Tibet was only possible with oxen. A meteorological station was established there, upon the recommendation of one of the leading Russian geodesists and cartographers General Aleksei Andreevich Tillo (1839–1900). Four men, under Sergeant Ivanov, were left at the station, and the conduct of meteorological observations was entrusted to Egor Muraviov, who had received the necessary preliminary training. The Tsaidam meteorological station worked for 15 months without interruption, the records of the instruments being taken thrice a day, and once every three months every hour for twenty-four hours in succession. This was the first time that such work was done in Inner Asia, and the observations of the Tsaidam station provided a solid basis for calculating the altitudes in Tibet.[10]

The expedition lasted for two and half years. It brought back to St. Petersburg in 1901 about 8,000 miles of route survey; the positions of 40 localities determined astronomically; geographical, historical, ethnographical and commercial information about the regions visited; more than 400 photographs; meteorological observations; and rich natural history collections: about 1,200 geological specimens; nearly 1,400 species of plants (over 30,000 specimens);

10 Tachalov 1907, 1–3.

300 skins of mammals, ten skeletons, 1,500 birds, 500 fishes and reptiles, and 30,000 insects. All these collections were delivered to St. Petersburg where they were arranged and placed in the hands of specialists and different scientific bodies.[11]

The only setback of the journey was Kozlov's failure to reach Lhasa. He was stopped, on the very border of Lhasa district, by Tibetans who implored him not to proceed further towards the holy city. Later, two officials from Lhasa arrived in the expedition camp, allegedly from the Dalai Lama, to ascertain whether the travelers were Russian or British, informing Kozlov that all Western visitors were unwelcome to Lhasa. Nevertheless, Kozlov was able to explore thoroughly the territory of Kham where he made important geographical discoveries. He mapped the uncharted upper reaches of the Huang He and Yangtze rivers as well as a number of ranges to which he gave names – those of the Russian Geographical Society (between Yangtze and Mekong), of the Water Parting (between Huang He and Yangtze), of Tibet's first explorers (Pundit A-k, Dutrei de Rhins, Woodwill Rockhill), and several others. (These place names were subsequently discarded and replaced by indigenous toponyms on most maps.)

Of no little importance were also his pioneer limnological explorations. Kozlov carried a small canvas boat, dredges, thermometers and other hydrological instruments for this work. Kaznakov and Ladygin made soundings on the two Kobdo lakes, upper and lower, in Mongolian Altai, as well as on the Tibetan plateau (Alyk-nor).

The many achievements of Kozlov's Tibet expedition were highly praised by the Imperial Russian Geographical Society which presented the traveler with the Konstantin Gold Medal (1902), their highest award. The results of the expedition were published in several volumes, in 1905–1908. These included Kozlov's account of the journey (vol. I), accounts of his assistants Kaznakov and Ladygin (vol. II), and contributions by specialists who assessed the astronomic and meteorological observations (N.A. Tachalov, A.A. Kaminskii, vol. III) and described the ornithological and entomological collections (V.L. Bianki, Theodor Becker et al., vol. V, VII), and that of the Tibetan algae (K.S. Merezhkovskii, vol. VIII). Volumes IV and VI dealing with the specimens of mammals and reptiles were not published.

Kozlov's expedition to Mongolia and Kham aroused considerable interest among Western geographers. Thus the *Petermanns Geographische Mitteilungen* carried a series of his 'bivouac reports' reprinted from the Russian *Izvestia*

11 Kozlov 1902, 597–598.

IRGO,[12] and the *Geographical Journal* of the Royal Geographical Society of London published, in 1902, Kozlov's account of his travels which he had sent to this body.[13] The scientific results of the expedition were also presented at the 8th International Geographical Congress in Washington in 1904.[14]

Meeting with the Dalai Lama

While organizing his Tibet expedition, Kozlov made the acquaintance of the Thirteenth Dalai Lama's emissary, a Buryat lama Agwan Dorzhiev (1853–1938). He came to St. Petersburg in 1898 to establish contacts with the Russian government and the Tsar, Nicholas II, seeking their support to counter British encroachments on Tibet. Dorzhiev was an influential figure in Lhasa, the chief political advisor of Tibet's ruler, and it was largely owing to him that the latter adopted a pro-Russian orientation in the late 1890s. Having learnt about Kozlov's plans, Dorzhiev expressed his willingness to help the traveler obtain permission to visit Lhasa; however, he was unable to do so, as he was away from Tibet at the time of Kozlov's journey.

Dorzhiev's shuttle diplomacy on his several visits to Russia was fairly successful, bringing about a Russo-Tibetan rapprochement, albeit short-lived. What Dorzhiev and the Dalai Lama primarily sought was a protectorate treaty with the Russian government or a formal agreement that would safeguard Tibet against any potential aggressor.[15] Dorzhiev's activities, however, seriously alarmed the viceroy of India, Lord Curzon, who, in 1903, sent a British mission to Lhasa under Francis Younghusband (1863–1942), ostensibly to 'frustrate' Russian intrigue in Tibet. As a result, the Dalai Lama, together with Dorzhiev and some of his ministers, fled to Urga in Outer Mongolia, close to the Russian border.

As soon as Kozlov learnt from Russian newspapers about the British mission he immediately submitted a memorandum to the General Staff in which he proposed to send a similar Russian mission ('expeditionary force') to counter the British Indian authorities. The mission, consisting of Buryat Cossacks, was

---

12 Kozlov 1901, 90–94, 160–163, 182–187, 205–208, 237–239.

13 For Kozlov's correspondence with the Royal Geographical Society of London, see: NA RGO, F. 18, Op. 3, D. 160, ll. 1–23 (13 letters, 1902–1931); RGS/IBG Archives, London, Kozloff's Correspondence Blocks, 1881–1910, 1911–1920, 1921–1930.

14 See: "The scientific results of the Russian expedition to Kham, by Capt. Kozlov", in *Report of the Eight International Geographical Congress*, 1904 (1905), Washington, 737–740.

15 Andreyev 2003, 167–168; Andreyev 2003a.

to incline the Tibetan government, following the British example, to open their country to Russian Buddhist pilgrims and commercial goods, by resorting to either diplomatic negotiations, or, if need be, to 'appropriate' military action. In February 1904, however, the Russo-Japanese war broke out; Kozlov's proposal was shelved by the Russian military until a more opportune moment.

A year later, in spring 1905, despite the continued hostilities in the Far East, the Foreign Ministry jointly with the General Staff sent Captain Kozlov, as their spokesman, to the Dalai Lama, 'under the banner of the IRGO'. He was commissioned to meet with Tibet's pontiff and discuss with him the possibility of Russian assistance to Tibet. By that time Kozlov had come up with a rather dashing project of his own – that of helping the Dalai Lama return to Lhasa by providing him a Russian cossack escort to be led by Kozlov himself. The Foreign Ministry, however, turned down his project, while he was still in Urga, out of fear that such a mission might seriously damage Anglo-Russian relations.

Nonetheless the two-month stay in Urga, from May to July, allowed Kozlov to enter into friendly relations with the Dalai Lama who welcomed the idea of the Russian escort and invited Kozlov to come to Lhasa with a scientific expedition. The Lama was impressed by Kozlov's recently published travelogue *Mongolia and Kham* and he particularly admired the map of Tibet he had never seen before. "Now Lhasa is open to you", he frankly told the Russian traveler, when parting. In his travel diary Kozlov recorded enthusiastically: "The Dalai Lama, whom I see quite often, treats me amiably and friendly, and I am positively happy that one of my long-cherished dreams, to see the ruler of Lhasa and Tibet, has been realized. Yet it pains me and makes me feel bitter to know that the fulfillment of my second best dream, that of visiting Lhasa, is being opposed by our own diplomats".[16]

In subsequent years, when the Dalai Lama moved from Urga to inland China, Kozlov, together with Dorzhiev, continued to press the authorities to send a Russian expedition with scientific and diplomatic goals to Lhasa. Basically this was his revived project for a Russian escort to bring back the Dalai Lama to his homeland.

As a pupil and ardent admirer of Przhevalskii, Kozlov inherited much of his mentor's 'imperial vision' of Asia and his views of Russia's 'civilizing mission' on the continent. He certainly shared Przhevalskii's obsession with Tibet, the high prize of the Great Game, trying hard to enter Lhasa at the head of a Russian scientific expedition. In his 1905 Urga diaries Kozlov spoke harshly about Russia's expansion in the Far East which led to the dramatic clash with Japan. As a Great Gamer, he opted for a more advantageous 'object-region', the

16 Zhitomirskii 1989, 121; Kozlov 2004, 98–127.

neighbouring Mongolia and Eastern (Chinese) Turkestan, and advocated Russia's 'peaceful conquer' of these countries by means of commerce and culture. He particularly emphasized the importance of an alliance with the Dalai Lama that would help Russia "assert her rights" against those enjoyed by Britain in that part of Asia. In the same diaries he spelled out his political credo: "We (i.e. Russians – A.A., T.Y.) must commit ourselves to Mongolia and East[ern] Turkestan and walk hand in hand with the Dalai Lama".[17] Yet, when Kozlov was writing these words, the Great Game was already on the wane and, in 1907, it effectively ended (as it seemed then) after Russia and Britain signed an agreement that carved out their spheres of influence in Central Asia (Persia, Afghanistan and Tibet). Thereafter Kozlov adopted a more conciliatory stance towards the British and even talked of a need to cooperate with London on the Tibetan issue.

He was a politically minded person, a pragmatist ready to serve the interests of his country in the best way he could, yet he can hardly be seen as a proponent of 'conquistador imperialism' the way Przhevalskii was.[18] At the same time Kozlov had a strong romantic streak in his character as demonstrated by his emotive descriptions of the beauties of 'wild' nature unspoiled by man, especially that of Eastern Tibet, that one finds in his travel diaries and books. Yet, above all, he was a military man, which certainly colored his personality and views. In 1906 he was assigned on detached service at the Main Directorate of the General Staff (GUGSh) and put at the disposal of the Chief of the General Staff Lieutenant-General F.F. Palitsyn (1851–1923).

The Mongolia-Sichuan Expedition (1907–1909) and the Discovery of Khara-Khoto

In the spring of 1907 Kozlov submitted his new expedition project to the IRGO. His primary objectives this time were Central and Southern Mongolia, those parts not yet visited by Europeans; the Kokonor area with the high Alpine Lake Kokonor, and the northwestern corner of Sichuan known for its rich fauna and flora. Lhasa was not on his agenda, because the Dalai Lama was still away from his country and because British and Russian expeditions to Tibet were prohibited for a period of three years by the 1907 Anglo-Russian agreement. Kozlov's program was approved by the Society and the Tsar himself.

The expedition also had some additional undeclared tasks, namely the collection of economic information, the military reconnaissance of the route

17 Kozlov 2004, 116.

18 Schimmelpenninck van der Oye 2001, 41.

between Kiakhta and Tsaidam, that is, between Russia and Tibet, but, first and foremost, the search for the ruins of the 'dead city' of Khara-Khoto ('Black City') in southern Gobi, believed to be the capital of the Tangut kingdom Xi Xia (Western Xia). This was a major medieval Tangut settlement known to Europeans since the days of Marco Polo. All searches hitherto for Khara-Khoto by Russian travelers in the late nineteenth century had proven unsuccessful. Moreover, the Torgout Mongols who inhabited the lower Edzin-gol where the ruins lay hidden by the sands strongly denied their existence.

Surprisingly, when Kozlov's expedition had been already mounted, he received a letter (in mid-May 1907) with sensational information from a young Buryat trader Tsokto Badmazhapov (1879–1937), who had participated previously in his Tibet journey. Badmazhapov wrote to Kozlov that while passing with his caravan through the Edzin-gol area he had stumbled during a sandstorm, apparently by sheer luck, over the ruins of a medieval town called Khara-Khoto by local dwellers. He also forwarded a detailed account of his passage across the desert along with a description of Khara-Khoto and several photographs (of rather poor quality), the first ever taken of the legendary 'dead city', to the Russian Geographical Society.[19] This stunning news immediately gave a strong additional impetus to Kozlov's journey. In his travel diary Kozlov would confessed that he was thinking of Khara-Khoto all the way from St. Petersburg to Mongolia.

The expedition staff of the Mongolia-Sichuan expedition included, apart from Kozlov, his three assistants – geologist Aleksandr Aleksandrovich Chernov (1877–1963), topographer of the General Staff Captain Petr Yakovlevich Napalkov (1874–1940), and collector of plants and insects Sergei Silvestrovich Chetyrkin (1876–1938) and a ten-man escort headed by Gavriil Ivanov. At the start of the journey the party traversed Mongolia across the Gobi, from north to south-east, heading towards the oasis Dyn-yuan-in (Denyuanging) where Badmazhapov owned a house.

On his way Kozlov stopped at the *hoshun* (encampment) of the Mongol prince Torgout-beile which was a short distance from Khara-Khoto. Kozlov was eager to discover the mysterious ruins and he succeeded in securing cooperation of the prince who provided him with pack animals, diggers and a guide to take his party right to Khara-Khoto. Kozlov's dream came true on 19 March (New Style: 1 April) 1908 when, together with several companions (Chernov,

19 See: Ts. Badmazhapov, *Tridzatipiatidnevnaia poezdka ot rezidentsii kniazia Alasha Vana do stavki Torgoud-Beile (1907)*, NA RGO, R 97, Op. 1, D. 30 (Quoted in: A.I. Andreyev, *O tom, kak byl dvazhdy otkryt mertvyi gorod Khara-Khoto,* in: Andreyev 1997, 61–91). See also: Yusupova 2008, 112–129.

FIGURE 4.5 P. Kozlov with the members of the Mongolia and Sichuan expedition, 1908, Dyn-yuan-in (China). Seated, left to right: P. Napalkov, A. Chernov, P. Kozlov, Ts. Badmazhapov. NA RGO, F. 18, OP. 7, D. 146.

Napalkov, Ivanov and Arya Madaev), he reached the remains of the 'dead city'. From afar they saw its cob walls with the odd-looking Buddhist memorials, *suburgans,* mounted on them, silhouetted against the dazzling blue sky. This is how Kozlov described his first encounter with Khara-Khoto:

At last we have seen the City itself, spread out on a low terrace made of the rough gritstone. Above the northwestern corner of the fortress rose the main spire-like *suburgan*, one in a row of smaller size neighboring structures, also mounted on the wall, as well as those outside the fortress. As we approached the city we came across a great number of pieces of broken pottery; the sight of the fortress-city became obscured by the high-rising sand mounds. Finally we came upon the terrace and lo, here Khara-Khoto revealed to us all its outer charms.[20]

Kozlov stayed with his party in Khara-Khoto for a few days in order to conduct a preliminary excavation of the city. His early findings included fragments of porcelain, various household items and articles of trade, the copper *choh*-coins, Chinese paper money (the earliest known), and numerous religious

20 Kozlov 1948, 77.

FIGURE 4.6
Ruins of Khara-Khoto, a view from northwest, March 1908. NA RGO, F. 18, OP. 7, D. 920.

objects. It turned out that the wall *suburgans* were filled at their base with a large number of small clay figures of various Buddhist divinities, so called *tsatsa*.

During this first trip to Khara-Khoto Kozlov determined its geographic coordinates (41°45'40" N. and 101°5'15"E.), as well as its absolute altitude (810 m), while Napalkov drew the plan of the ruins.[21] After the excavation Kozlov immediately sent a letter to the IRGO, in which he gave a description of the locality, briefly sketched its historical past and enclosed a list of his finds and a few photographs.[22]

From Khara-Khoto the expedition moved to Dyn-yuan-in where the party spent about two and half months exploring the Alashan range and oasis. Kozlov also established a meteorological station with a depôt in Dyn-yuan-in. Then the travellers split into three groups. Kozlov and Chetyrkin with the main caravan headed in the south-east direction, over ridges and gorges of Nanshan, towards Sinin (Xining), the residence of the Manchu governor (*tsin-tsai*), who ruled the nomads of Kokonor and northeastern Tibet (Amdo). Napalkov with three others travelled along the right bank of Huang He via Lanzhou to Sinin. This party undertook topographic survey, conducted ethnographic studies and collected insects. A third party under Chernov conducted geological surveys along the route leading across the Gobi to Lanzhou and further to Lake Kokonor, where it was to join Kozlov's caravan, in August 1908.

21 Ovchinnikova 1964, 123.

22 *Izvestiia IRGO*, 1908, Novosti (News), 453–458.

FIGURE4.7
P. Kozlov in his study in Dyn-yuan-in (China), 1908.
NA RGO, F. 18, OP. 7, D. 740.

FIGURE 4.8 Meteorological booth in Dyn-yuan-in. Kozlov's Mongolia and Sichuan expedition, 1908. Standing: Ts. Badmazhapov (on the left) and meteorological observer M. Davydenkov (on the right). NA RGO, F. 18, OP. 7, D. 750.

Exploration of Lake Kokonor (Ch. *Qīnghai Hú*, Tib. *mTsho sugon po*, Mong. *Heh nuur*) was the expedition's second major object. Kozlov took several soundings of the 'Blue Sea-Lake', the largest in Asia, surveyed its shores and sent Chernov and Chetyrkin across the lake, on a folding cork-and-tarpaulin boat, to explore the island Kuisu. Kozlov was completely charmed by the lake: "Calm or agitated, Kokonor is always exceedingly beautiful. I sat for hours on its shore, or walked at a distance up and down from our encampment, never tired of gazing at the boundless expanse of water, nor wearied of the monotonous splash of its breakers, which reminded me of the southern shore of the Crimea".[23]

Kozlov's third task, the exploration of Sichuan, however, was not fulfilled. He had to abandon his original plan for the sake of Khara-Khoto, where he intended to return to make further excavations. Instead, he chose to explore more thoroughly the Amdo highlands, bordering on Sichuan. The area was originally visited by Przhevalskii, yet its central part, inhabited by militant nomads, remained a blank spot. A trip to Amdo also provided Kozlov with an opportunity of visiting the famous Buddhist monasteries of Labrang and Kumbum, in early 1909. As a traveller in Inner Asia, he would occasionally visit Mongolian and Tibetan monasteries en route, and that was how he began gradually to take interest in Buddhism, its traditions, rituals and monuments. This interest only grew with time, especially after he had met the Dalai Lama.

In general Kozlov was a good mixer, easily making friends with people wherever he travelled, especially with men of influence – the Buddhist abbots and incarnate lamas, such as Choibsen Khutuktu, as well as representatives of local Qing authorities, to whom he always gave presents by underscoring his affiliation with the Russian Geographical Society. This simple tactic proved effective as it allowed him to get access to places inaccessible to other Russian or Western travellers and collect some interesting religious and ethnographic artefacts.

While visiting Kumbum (in February 1909), Kozlov had a chance to meet the Dalai Lama for the second time. The latter treated him as an old friend and invited him to Lhasa again: "I hope you will visit me in Lhasa ... I'll wait for you in Lhasa, and then you will make not just one but several excursions in the environs, along the radii from Tibet's capital, where you will find some unexplored corners, in terms of nature and population".[24] Thus Kozlov became the only Western traveller to whom the Dalai Lama opened his 'forbidden city'.

23 Kozlov 1909, 400.

24 Kozlov 1923, 499.

The rest of Kozlov's journey was spent in Khara-Khoto; he returned there at the request of the Society's secretary Grigoriev to excavate it more thoroughly, in May and June 1909. The work was carried out by two teams of diggers whose cooperation Kozlov has duly secured: a Mongolian and a Russian one. The Mongolians under Madaev focused on the ruins of houses (*fanzas*) while the Russian team guided by Kozlov himself did all the general excavation both inside and outside the city fortress. The most valuable findings were made in one of the *suburgans* (stupas) dubbed the 'Great Suburgan' located outside the western city wall. The structure, after it was totally dismantled from top to bottom, yielded a whole library of books, scrolls and manuscripts numbering about 6,000 pieces altogether (in Chinese, Mongolian, Tibetan and Tangut languages),[25] three hundred Buddhist icons or *tangkas*, that is, paintings on canvas, silk and paper.[26] In the same *suburgan* were also found the metal and wooden statues, the so-called *burkhans*, printing blocks, models of *suburgans* as well as a beautiful piece of tapestry, a sample of traditional Tangut weaving.

Unfortunately, some of the Buddhist statues were too big and could not be transported safely to Russia. So before leaving Khara-Khoto Kozlov buried these earthen collossi in the sand by one of the walls of the 'dead city', hoping he would return to pick them up in the future. (However, when he returned, in 1926, he was unable to locate these treasures.)

With this final excavation of Khara-Khoto, the expedition ended. Its results were most productive and numerous but also sensational, as far as Khara-Khoto was concerned. These included a traverse survey of more than 6,700 miles of route, the heights of inhabited points, of passes, valleys, and places where the expedition stayed, determined barometrically. There were also meteorological readings taken both in the main caravan and on lateral excursions and, besides these, data provided by the meteorological station in Dyn-yuan-in during the year-long period. The island in Kokonor, Kuisu, was explored and the depth and composition of the bottom and of the water of this Alpine basin were ascertained.

Also, abundant collections were acquired – geological, botanical, zoological, as well as a very full collection of ethnographic objects, comprising Mongolian and Tangut garments, and many religious articles, including some hundreds of painted representations of the deities of the Buddhist pantheon, and a collection of Tibetan religious books, Chinese works of art and Chinese antiquities in the form of vases of original design, images, perfuming pans, all unearthed in

25 Men'shikov 1984; Kychanov 2008.

26 Samosiuk 2006.

FIGURE 4.9
P. Kozlov at the exhibition in the IRGO, February 1910. NA RGO, F. 18, OP. 7, D. 79.

FIGURE 4.10 The ethnographic items at the exhibition in the IRGO, February 1910. NA RGO, F. 18, OP. 7, D. 79.

Khara-Khoto. Numerous photographs were also taken (personally by Kozlov) of views and types in the places the expedition passed through.

In the autumn of 1909, the findings were delivered to St. Petersburg to the newly built premises of the Imperial Russian Geographical Society. There, in February 1910, they were displayed to the public, along with other naturalist and ethnographic collections gathered by Kozlov. After the closure of the exhibition the bulk of Khara-Khoto's collection was transferred to the ethnographic department of the Russian Museum in St. Petersburg, and, in 1934, to the Hermitage Museum, while its remaining part which included mainly the xylograph-books and manuscripts went to the Asiatic Museum of the Russian Academy of Sciences (today's Institute of Oriental Manuscripts). The samples of Buddhist painting from Khara-Khoto were then subjected to a most thorough examination by the eminent Russian Orientalist Sergei Feodorovich Oldenburg (1863–1934), which resulted in the publication of a volume devoted to Khara-Khoto's iconography.[27]

Of paramount importance was the book and manuscript collection. The deciphering of the Tangut written language undertaken in the 1920s by the Soviet Sinologist Nikolai Aleksandrovich Nevskii (1892–1937) laid the foundations of a new Oriental discipline, Tangutica. Overall, the Khara-Khoto findings provided an extraordinarily valuable historical and cultural resource which has considerably expanded our knowledge of the history and culture of the peoples of Inner Asia.

Kozlov's Later Activities

The great success of the Mongolia and Sichuan expedition won Kozlov the acclaim of the scientific world and the Russian public. He was welcomed enthusiastically as a heroic explorer the way Przhevalskii had been in his own days. In September 1910, Kozlov went to England, at the invitation of the Royal Geographical Society of London, where he gave a talk on his expedition illustrated by lantern-slides. The Society awarded him their annual Founder's Medal in 1911 for "explorations in the Gobi desert, Northern Tibet and Mongolia". In his speech at the presentation ceremony on 22 May 1911, the President, Major Leonard Darwin, hailed Kozlov as the explorer who had brought back from his journeys "a harvest of results, geographical, geological, and archeological". "In fact there can be no doubt", continued Darwin, "this is the name well worthy of

27 Oldenburg 1914; Samosiuk 2006.

an award from any geographical society in the world".[28] Kozlov, however, could not attend the event, and his medal was accepted by the Military Attaché of the Russian Embassy, General Ermolov. In 1911, Kozlov also received the gold medal from the Italian Geographical Society, and the Hungarian GS made him an honorary member. Two years later, on the initiative of Prince Roland Napoleon Bonaparte, the traveler was awarded the Petr Chikhachev Prize (Le prix Tchihatchef) of 3,000 francs by the Academy of Sciences of France for his studies and publications on Inner Asia.[29]

Kozlov's own account of his Mongolia-Sichuan expedition, however, was published after considerable delay, in 1923, under the title *Mongolia i Amdo i mertvyi gorod Khara-Khoto* (Mongolia and Amdo and the dead city of Khara-Khoto).[30] It would be his most successful book, an absorbing story of his discovery of the lost Tangut city buried in the Gobi sands. In the meantime, the Royal Geographical Society of London brought out his narrative in their Geographical Journal, in 1909–1910, as did the Italians.[31] The travelogue would also be translated in several other European and Asian languages in subsequent years: German (1925 and 1955), Czech (1929), Slovak (1951), Japanese and Chinese (2000s).

The period after Kozlov's return from his expedition was a turbulent one in Asian history. The Xinhai revolution in China (1911) overthrew the Qing dynasty, and the country became a republic. The event significantly affected the Manchu vassal territories of Mongolia and Tibet with both declaring their independence. Kozlov hailed these developments in a short essay, by emphasizing Russia's close commercial and political ties with neighboring Mongolia; as for Tibet, he pointed out that the country had "turned its gaze in the hour of need to Russia only" and that the Geographical Society believed in the sincerity of Tibet's invitation made to its members to visit Lhasa "for scientific research".[32]

The same years also saw changes in Kozlov's life. In the summer of 1910, while staying with his wife and children at Berck Plage in Normandy, he met an 18-year old Elizaveta Pushkareva (1892–1975), daughter of a physician, Vladimir I. Pushkarev from St. Petersburg, and fell in love with her. He would later say that he sensed in Elizaveta Przhevalskii's 'nomadic spirit' he valued so

28 Meetings of the Royal Geographical Society, Session 1910–1911, (1911), Anniversary Meeting, May 22, 1911, *Geographical Journal*, XXXVIII (1), July 1911.

29 Yusupova 2010b.

30 Kozlov 1923.

31 Kozlov 1909–1910, 1911.

32 Kozlov 1913b.

much, while Elizaveta, on her part, was strongly impressed by the strikingly virile personality of the world-renowned traveler. As a result Kozlov divorced his wife Nadezhda and, in 1912, married the young woman who reminded him so much of his lamented master. He moved to Petersburg, the couple settling in Smolny Prospect 6, which would become their permanent residence.

In the summer of 1913 the couple made a trip to Ascania-Nova, a nature reserve in the Kherson steppes in Taurida Province (today a part of Ukraine). Ascania-Nova was founded in 1828 as a private estate by Prince Ferdinand von Anhalt-Köthen and later (1875) was turned into an acclimatizing zoological garden by Ascania's new owner, Friedrich von Falz-Fein (1863–1920). It was inhabited by a great variety of animals from all over the world (zebras, antelopes, ostriches, yaks, Siberian stags, etc.). Elizaveta would later call it 'an earthly paradise'. In 1899, largely owing to Kozlov's mediation, Falz-Fein obtained from Mongolia (Dzungaria) some wild horses (stallions and mares) of a rare species *Equus ferus Przewalskii,* and they bred well at Ascania, the only place they did so outside their natural habitat. The couple would visit Ascania again in 1914. Kozlov wrote several essays about the nature reserve in popular Russian magazines. In one, he raised the vital question of nature preservation, by expressing his wish that the government should issue a law that would make Ascania-Nova a state-protected territory.[33]

Also, in 1913, the Kozlovs traveled to Lake Issyk-kul in connection with the 25th anniversary of Przhevalskii's death celebrated by the Imperial Russian Geographical Society. Kozlov was commissioned by the Society to lay a silver wreath on the pedestal of the imposing monument erected to the traveler on the shore of Lake Issyk-Kul, facing the majestic Tian Shan Range. The result of this trip was some biographical essays devoted by Kozlov to his great teacher.

While staying at Issyk-kul, Kozlov began to teach his wife how to shoot a rifle and collect and prepare birds as he was already contemplating his next journey to Inner Asia and wanted her to accompany him. The expedition was due to start in July 1914, its main objectives being a further study of the ruins of the 'dead city' and a more thorough exploration of the upper basins of the Yangtze, Mekong and Salweuen rivers on the Tibetan Highlands. The IRGO approved his program and allocated the necessary funds. But Kozlov's plans were overtaken by the outbreak of World War I in August 1914. Kozlov, as a senior army officer, was sent to the war theatre, but did not see action. He was posted as the military commandant of Tyrnov (Tarnov, now in Poland) and, in 1915, of Yassy (in Rumania) and Ternopol (in today's Ukraine), while his wife worked as a nurse on a hospital train. In 1915 Kozlov was recalled from the front and sent to

33 Kozlov 1914, 35.

FIGURE 4.11 Petr Kozlov with his wife Elizaveta as a nurse, shortly after the outbreak of World War I, 1914. COPIED FROM *GEOGRAPHERS BIOBIBLIOGRAPHICAL STUDIES, BLOOMSBURY ACADEMIC* (VOL. 34, 2015).

northern Mongolia as head of the expedition for the procurement of meat cattle for the acting army. This allowed Elizaveta to make several trips across Eastern Siberia and Mongolia in the company of her husband, and thus acquaint herself with these regions. As a student of ornithology she had an opportunity to observe and describe the migration of birds and make her first ornithological collection.

Kozlov's military career reached its peak at the end of 1916. On 16 December he was promoted to the rank of major-general,[34] which placed him on an equal footing with his great mentors, Przhevalskii and Pevtsov, who had held the same rank. In early 1917 he returned with his wife to Petrograd (renamed St. Petersburg in 1914). They witnessed the downfall of the monarchic rule in Russia after the abdication of Nicholas II which was followed by the Bolshevik October Revolution later that year and the civil strife between 1918 and 1922. To the Kozlovs, these dramatic developments, which brought about devastation and chaos to the country, were a severe trial. Yet they survived.

34 RGVIA, F. 407, Op. 1, D. 102, l. 140. The handwritten list of major-generals promoted to this rank between 22 May 1916 and 9 February 1917.

Kozlov easily found common ground with the new rulers of Russia, though he never publicly declared his allegiance to their proletarian ideology. Soon enough he acquired an influential patron amidst the Bolshevik elite, Nikolai Petrovich Gorbunov (1892–1938), the secretary of Vladimir Lenin, who was Elizaveta's childhood buddy and best man at the wedding. Kozlov's activities were also patronized by the Academy of Sciences, whose leaders began actively to collaborate with the Bolshevik government, the only means to survive for the scholarly community. Thus, at the end of 1917, Kozlov was sent by the Academy of Sciences to Ascania-Nova in the capacity of the government commissioner (*commissar*) to look after the protection of the nature reserve, with V. Lenin's approval of his commission forwarded to him by Gorbunov. He then moved with his wife to Ascania where they stayed for about two years. This period was a dark hour for Russia. The country was engulfed in the civil war and the territory of Ascania-Nova more than once turned into a real battle ground. So Kozlov made every effort to protect this wildlife sanctuary from destruction. He appealed personally to Lenin as well as to the Soviet government of Ukraine pleading the country's new rulers to take Ascania under the state protection. Upon his return to Petrograd, Kozlov published an article *Sovremennoe polozhenie zooparka Askania-Nova* (The current situation in the Ascania-Nova zoological garden)[35] to attract the public attention to the deplorable condition of the nature reserve, and his wife penned a whole book entitled *Askaniia-Nova, zoopark v iuzhno-russkikh stepiakh* (Ascania-Nova, a zoological garden in the South Russian steppes).[36]

In 1920 Kozlov went to Siberia to inspect the local branches of the Geographical Society, on the commission of the People's Commissariat of Education or *Narkompros*. In the same year, his new book *Tibet and the Dalai Lama* was published to be followed three years later by his magnificent travelogue *Mongolia and Amdo and the dead city of Khara-Khoto* (1923), in the writing of which he was largely assisted by his wife.

As life in Russia was gradually returning to normal, Kozlov began to make plans for a new expedition to Inner Asia.

The Last Expedition to Mongolia, 1923-1926

In mid–1922, he submitted a proposal for a new expedition to Tibet to the new Chairman of the Russian Geographical Society, Yulii Mikhailovich Shokalskii (1856–1940). In fact, this was a revival of his 1914 project aborted by the world

35 Kozlov 1919.

36 Kozlova 1923.

war. In his proposal of 19 August 1922, Kozlov underscored his three main objectives: further excavation of Khara-Khoto; establishing a meteorological station with a depôt at Tsaidam, at the encampment of his old friend Dzun-Tsasak; and more thorough exploration of the basins of the Yangtze, Mekong and Salween. He said nothing about a projected trip to Lhasa, though Lhasa was on his mind as his final destination.

The Society approved Kozlov's project and the Soviet government allocated the lavish sum of 100,000 gold roubles for the journey and additional 4,000 for presents to the Dalai Lama, which allowed the traveler to equip his party well. This generous level of support for a scientific expedition was well justified. Soviet Russia, from its early days, pursued an active Eastern policy, with special attention to Tibet, which was seen as a springboard for the ideological penetration of British India; this was to usher in a new round of the Great Game. The Bolshevik leaders, well aware of Kozlov's friendly relations with the Dalai Lama and the Lama's long-standing pro-Russian feelings, attached political importance to his new journey. They needed Kozlov and hoped that his visit to Lhasa would further that incipient Soviet-Tibetan dialogue which had begun in 1922 in the course of a secret mission to Lhasa of one of their diplomatic emissaries, Vasilii Khomutnikov (1891–1945).[37]

The expedition staff, apart from Kozlov, included his five senior assistants: Sergei Anatol'evich Glagolev (1893–?; geographer), Nikolai Vasilievich Pavlov (1893–1971, botanist), Elizaveta Kozlova (ornithologist), Sergei Aleksandrovich Kondratiev (1896–1970; naturalist and student of musical folklore), and Elena Petrovna Gorbunova (1888–1973), physician (Gorbunov's sister); four junior ones; the caravan leader Pantelei Teleshov, who had participated previously in Przhevalskii's last two expeditions; five men of escort and three interpreters.

The party arrived in Urga in September 1923 and started to make preparations for the journey to Tibet, yet these were soon halted upon the demand of the authorities in Moscow. The reason was a denunciation of the expedition leader and some of his companions of anti-revolutionary sentiments spelled out in a letter to the OGPU (precursor of the KGB) by one Martynov. According to him, Kozlov, being a former "tsarist general", might spend the funds allocated to him on anti-Soviet agitation, while working 'in the areas within the British sphere of influence'. Consequently, the Central Committee of the Bolshevik Party established a special commission which purged Kozlov's expedition staff, attached a political commissar to his party (Daniil Mikhailovich Ubugunov (1897–1938), and eventually rescinded their journey to Tibet on a formal

37 Andreyev 2003.

pretext of the 'unfavorable political situation' in the region and the difficulty of obtaining visas from the Chinese authorities.[38]

As a result Kozlov remained in Urga with his 'reorganized' party and had considerably to curtail his original plans. In the spring of 1924, having learnt about some old burial mounds at the Sudzukte gorge in the Noin-Ula (Mong. *Noyon Uul*) Mountains, north of the Mongolian capital, he started the excavation there, which eventually led to some sensational findings later that year.

Thus, quite unexpectedly, Kozlov got bogged down for many months in Mongolia. "Every cloud has a silver lining", he wrote to the Geographical Society's secretary, Academician Vladimir Leonidovich Komarov (1869–1945) in Petrograd, "and perhaps this forced halt, the delay of the expedition, will bring the RGO and science in general much more use than the one we counted upon".[39] The travelers discovered a whole ancient necropolis – about two hundred mounds at Sudzukte and the neighbouring gorges and started to unearth some of them with the help of Chinese diggers. The work was begun at the end of March when the soil was still frozen so the diggers had to build fires to thaw out the icy crust. Since the burial chambers lay beneath the level of the subsoil waters, the latter had to be pumped out by manual pumps. All excavation work at this early stage was directed by one of Kozlov's senior assistants, Kondratiev, who was not a professional archeologist. Therefore, as soon as the news of Kozlov's discovery reached Moscow, the Soviet Russian Academy dispatched two archeologists to Mongolia, Sergei Aleksandrovich Teploukhov (1881–1934) and Grigorii Iosifovich Borovka (1895–1941), from the Hermitage and the Russian Museum in Leningrad accordingly, along with two more specialists, mineralogist Vladimir l'ich Kryzhanovskii (1881–1947) and the soil scientist Boris Borisovich Polynov (1877–1952). And in early 1925 Kozlov returned briefly to the USSR to report of the Noin-Ula excavations to his sponsors at the Academy of Sciences and the Geographical Society in Leningrad. There he presented his discoveries to academic audience and the general public and was hailed as a Soviet hero.

The objects unearthed from the mounds were fragments of decorative silk and woolen tissues, carpets, wooden items, some fine Chinese lacquer cups, dating from the first century A.D. belonging to the Hun (*xiongnu*) high nobility. The finds were numerous and mostly in good condition, despite the fact that they had been buried for nearly 2,000 years. These finds caused a sensation in scholarly circles worldwide, particularly the carpets with the masterly embroidered images in the so-called 'animal style'. The Noin-Ula archeological

38 Andreyev, Yusupova 2001, 51–74.

39 Zhitomirskii 1989, 158.

excavation in 1924–25 added to Kozlov's fame, as the artifacts he unearthed shed light on the history of the peoples of Inner Asia, their artistic tradition, as well as the history of ancient handicraft.[40] The excavation of the Hun 'royal' mounds was continued in subsequent years: in 1926–27 by Simukov, who had stayed in Mongolia after the expedition, being taken on the staff of the Mongolian Scientific Committee, then in the mid-1950s and in 1961–64 by the Mongolian archeologist Ts. Dorjsuren (1923–1994), and four decades later, in 2006, by a team of Russian archeologists led by Natalia Polos'mak from the Institute of Archeology and Ethnography (Siberian Branch, Russian Academy of Sciences).[41]

In addition to the Noin-Ula excavations, Kozlov's expedition conducted other geographical, paleontological and exploratory work in Mongolia. It eventually split into two groups, one headed by Glagolev which trekked south in the direction of the Mongolian Altai, and then westwards to Khara-Khoto. Two other members of the group were Andrei Dmitrievich Simukov (1902–1942) and the youthful Nikolai Przhevalskii (1908–?), grand-nephew of the famous traveler, another newcomer. The other party, headed by Kozlov and which included his wife and several junior assistants, travelled southwest down the Tola River valley which they surveyed en route. At the end of August the travelers left the river valley and moved onto the upper Ongin-gol in the Khangai Range. There Kozlov pitched a camp, close to the Sain-noion-kure (monastery), and wintered for five months, until mid-March 1926.

This period was most productive for Kozlov's party. In the Olun-sume gorge on the Ongin-gol they surveyed the ruins of a monastery and undertook preliminary excavations. Then they visited the tombs of the thirteen generations of the Khans, the owners of the Sain-noyon khoshun. The party also made many lateral excursions from their base camp. During one of these trips they came across a large waterfall in the right tributary (Ulaan River) of the Orkhon river, the only one in Mongolia, which Kozlov named Expedition Waterfall, its original name known to local dwellers being Ulaan-Tsutgalan. At the final leg of the expedition, in the spring and summer of 1926, Kozlov, together with his assistant Vasiliy Gusev and a Mongolian interpreter, made a paleontological excavation at the Holt River which yielded some very interesting material relating to the Hipparion fauna of the Tertiary age. The collection they gathered there included the bones of rhinoceroses, giraffes, goats and various rodents, as well as of the extinct tridactyl (three-toed) horse, hipparion, the most typical representative of the ancient fauna. At the same time Elizaveta Kozlova

40 Yusupova 2010a, 2011.

41 Yusupova 2010.

stayed at Lake Orok-nor watching the spring migration of birds and surveying the shores of the lake, while Glagolev undertook further excavations at Khara-Khoto.

The expedition ended in September 1926 in Ulan-Bator, where all its parties, upon the completion of their special fieldwork, assembled. Its scientific findings were many and important. Apart from the most valuable archeological finds at Noin-Ula and elsewhere, the travelers traversed some 3,500 km with the route sketching. They explored several lakes whose soundings were taken and samples of their fauna and flora collected. Meteorological measurements were also taken at the points where the travelers stayed for any time – in the Noin-Ula and Khangai Mountains, in the Gobi, and the Edzin-gol river. Plant and animal collections were gathered: 750 specimens of plants, sixty species of mammals, 350 species of birds, numerous reptiles and over 2,000 insects.

Kozlov's Mongolian journey in 1923–26 was the first Soviet expedition to Mongolia (Mongolian People's Republic, since 1924). Quite importantly, Kozlov's close working contacts with the Mongolian Scientific Committee and members of the Mongolian government laid the foundation for later Soviet-Mongolian scientific cooperation. Already in 1925, the Soviet government (*Sovnarkom*) established a special Mongolian Commission to promote "a systematic and all-round scientific study of Mongolia".[42] The Commission, two years later, would be placed under the authority of the Academy of Sciences and would send several expeditions to Mongolia in subsequent years. Two of these, in 1929 and 1930, under Elizaveta Kozlova, explored the Alpine zone of Khangai, Eastern Kentei and the source of the Kerulen River.

Because of his good relations with the new Mongolian leaders, Kozlov was able to take a number of his most valuable finds out of the country although he deemed it expedient to hand over some of the archeological items and collected samples of flora and fauna to the newly established Mongolian national museum in Ulan-Bator. Throughout his expedition he cooperated closely with the Mongolian Scientific Committee and was elected an honorary member of the institution (1924).

The Mongolian expedition, being the center of public attention, helped Kozlov restore his contacts with Western friends and colleagues, which had been disrupted by the First World War and the Russian revolution. In November 1923, while in Urga, he renewed acquaintance with the famous Swedish explorer Sven Hedin, then returning home from a lecture tour in America via China. Kozlov and Hedin knew each other quite well for many years as the latter was often a visitor to St. Petersburg where he presented news of his

42 Yusupova 2006.

FIGURE 4.12 Petr Kozlov and Sven Hedin in Urga (Mongolia), 21 November 1923. NA RGO, F. 18, OP. 7, D. 1263.

journeys to the Russian Geographical Society. Hedin maintained warm relations with several Russian explorers and scientists, recognizing their fundamental contribution to exploration of Inner Asia.[43] The work of Kozlov became especially important to Hedin and his archeologist Folke Bergman during the Sino-Swedish expedition (1927–35) which conducted further excavations in the Edzin-gol area and Khara-Khoto.

Kozlov's other friend was the German explorer Wilhelm Filchner, who arranged with a publishing house *Neufeld & Henius* in Berlin, in 1925, for the publication of Kozlov's two latest works in a single volume, *Mongolie, Amdo und die tote stadt Chara-Choto,* prefaced by S. Hedin and himself.[44] Both Hedin and Filchner spoke highly of Kozlov – the former praised him as "a genuine explorer, very accurate and fully reliable", and the latter pointing out his remarkable abilities as an organizer who "knows the prerequisites of success of his

---

43 For the Hedin – Kozlov correspondence, see A.I. Andreyev, *Russkie pis'ma iz arkhiva Svena Gedina v Stokgol'me* [Russian letters from the Sven Hedin Collection in Stockholm], *Aryavarta Journal,* St. Petersburg, 1997, 47–48; Kozlov's letters to Hedin (1903–1923) are in the Sven Hedin Collection at the National Archives, Stockholm.

44 NA RGO, F. 18, Op.3, D. 724, Filchner to Kozlov, 23 December 1924, Berlin.

FIGURE 4.13
Elizaveta Kozlova outside a yurt in Urga (Mongolia), 21 November 1923.
NA RGO, F. 18, OP. 7, D. 1269.

expeditions – harmony among expedition members and timely elimination of obstacles".[45]

While in Mongolia, Kozlov made friends with the American paleontologist Roy Chapman Andrews, the leader of several fossil-hunting expeditions to Mongolia and China in 1922–30. Andrews visited the excavation site at Noin-Ula, on Kozlov's invitation, in 1924, and he was honored by the opportunity to meet "this great explorer of the old school" whom he had long admired.[46] The excavations also provided an occasion for Kozlov to get in touch with the eminent French Orientalist Paul Pelliot. In early 1926 he sent Pelliot an imprint of an ancient Chinese seal found in one of the mounds at Noin-Ula, asking him to translate the inscription.[47]

45 Kozlov 1925, VII, VIII, XI; Yusupova 2014.

46 Andrews 1932, 235.

47 Kozlov 2003, 703, 712, 814.

FIGURE 4.14 P. Kozlov presenting the results of his Mongolo-Tibetan expedition at the Academy of Sciences, Leningrad, 11 March 1925. NA RGO, F. 18, OP. 7, D. 1257.

As a geographer, Kozlov particularly valued his contacts with the Royal Geographical Society of London. These were re-established in 1923, despite the fact that Great Britain was regarded by the Soviets as a 'stronghold of imperialism'. On 3 July, shortly before he set out from Petrograd, Kozlov sent a letter to the Secretary of that Society, Sir Arthur Robert Hinks, in which he briefly outlined the program of his explorations in Mongolia and Tibet. "As a naturalist, I am infinitely attracted by the South-Eastern Tibet, allured by its rich vegetation and animal life", Kozlov wrote (in English). He promised to inform the London body of the progress of his work and to send the Society copies of his reports to the Russian Geographical Society and the Academy of Sciences, as he had done for his two previous journeys.[48]

48 RGS/IBG Archives, London, P.K. Kozloff: Correspondence Block, 1921–1930, Kozlov to Hinks, 3 July 1923, Petrograd.

Final Years

Kozlov's expedition to Mongolia was his last. That journey seriously affected his health. Back in Leningrad, Kozlov failed to produce the promised detailed account of it and published only a short report.[49] Nonetheless, despite his age and failing health, he made plans for another, seventh, trip to Inner Asia. On 15 November 1927 Kozlov presented his project to the Academy of Sciences and to the Russian Geographical Society in which he spelled out the principal task of the mission as an extensive exploration of the Yangtze (Mur-usu) river-head, "the last blank spot on the maps of Asia". The project was approved by both institutions, but the Soviet Foreign Ministry (*Narkomindel*) and the OGPU strongly objected to it, for political reasons.[50] This, however, did not discourage Kozlov. A year later he would turn for help to Gorbunov asking him to organize the expedition to the "sacred land" in order to fulfil Przhevalski's "last behest". This time Kozlov wanted to travel to Tibet by air, on a dirigible, the way Amundsen did during his 1926 polar expedition, in the company of Gorbunov (who explored the Pamir with the Soviet-German expedition in the same year) and Tibet's representative in the USSR Agvan Dorzhiev. It was for this reason that he, together with his wife and two other members of the Mongolian expedition, undertook a test flight on a Junkers aircraft over Ulan-Bator, in September 1926, after which he noted in his diary: "I wish I could fly like this towards the far-off South-West, to Tibet and Lhasa! And this is not just a dream, thousand times no".[51]

In 1928 Kozlov was elected a full member of the Academy of Sciences of Ukraine. And in 1933, when he was already 70, the Russian Academy requested him to lead an expedition to Lake Issyk-Kul and the Nanshan Mountains, yet his failing condition prevented him from accepting the offer.

Kozlov's final years were spent between Leningrad and Strechno, a village near Novgorod, where, following Przhevalskii's example, he built a hut to become his refuge. There he would make long promenades in the woods, hunt and write essays, being occasionally visited by his wife and children. His contemporaries remembered him as a man of great courage, determination and ambition, a keen lover of nature and a patriot of his country. All his life he worshipped Przhevalskii, being a staunch follower of his system and methods of field work as well as his life's principles and precepts.

---

49 Kozlov 1928.

50 Yusupova 2003, 52–56.

51 Kozlov 2003, 945.

Kozlov died on 26 September 1935 of cardiac asthma in a sanatorium at Petrodvoretz. He was buried at the Smolenskoe Lutheran cemetery in Leningrad. Russian Orientalist Nikolai Nikolaevich Poppe (1897–1991) noted in his obituary that Kozlov's travels constitute "unique phenomena in the history of geographical science" and that his personality is equally unique. His Khara-Khoto and Noin-Ula collections alone provide the most valuable material that sheds light on the early history of Inner Asia. In general, P. Kozlov's journeys and their results "present a monument which, in the words of a [Roman] poet, is 'aere perennius' (more lasting than bronze); these journeys have become part of the history of science and will never be buried in oblivion".[52]

After Kozlov's death, his wife handed the most valuable items remaining in their apartment to various museums in Leningrad (the State Hermitage, the Museum of Anthropology and Ethnography), Smolensk and Kiakhta. The traveler's archive, including his travel diaries and voluminous correspondence, she donated to the Russian Geographical Society, the main sponsor of his expeditions. Elizaveta Kozlova lived for forty years more and died in 1975. Having made a career as an ornithologist, under the tutelage of her husband and the eminent zoologist Petr Petrovich Sushkin (1868–1928), she worked for several decades (since 1931), until her death, at the Zoological Museum of the Russian Academy. Her most important publications were *Avifauna Tibetskogo nagoria, ee rodstvennye sviazi i istoria* [The avifauna of the Tibetan Highlands, its genetic connections and history] (Trudy Zoologicheskogo institute AN SSSR, IX, 964–1028) and a monograph *Ptitsy zonalnykh stepei i pustyn' Tsentralnoi Azii* (Birds of the zonal steppes and deserts of Central Asia, Leningrad, 1975).

The Kozlovs' apartment, by a decree of the Leningrad municipal authorities of 5 December 1988, was turned into a memorial museum which, shortly after, was affiliated with the Institute for the History of Science and Technology (St. Petersburg Branch, Russian Academy of Sciences). Formally inaugurated in 2002, the Kozlov Museum tells the story of the Russian exploration in Inner Asia; it holds Kozlov's book, photographic and documentary collections, as well as some valuable memorabilia.

Kozlov as a Field Surveyor, Geographer and Naturalist

Kozlov was a versatile explorer, who made a considerable and lasting contribution to many branches of science, including geography, cartography, zoology, botany, geology, ethnography, and finally archeology and Oriental studies. This

52 Poppe 1936, 748.

versatility can be explained by the fact that the exploration of 'unknown' Inner Asia was, at its earliest stages, largely descriptive and unspecialized, as travelers observed, recorded and collected whatever came their way. It was regarded by Przhevalskii and his sponsors at the Russian Geographical Society and the General Staff as the 'preliminary scientific reconnaissance of uncharted territories'. The primary task of this reconnaissance was to erase the numerous blank spots from the map of Asia. Kozlov's interests lay mainly in geography and the study of wildlife although, in later years, he also devoted much attention to ethnographic studies seeking to produce an overall picture of the Asian heartland as the unity of nature and man.[53] He was not properly speaking a cartographer, despite his valuable route sketches, neither was he an archeologist, his finds at Khara-Khoto and Noin-Ula being largely serendipitous. Yet his fame of the 'discoverer of Khara-Khoto' and the first to excavate the site as well as the Hunn burial mounds was well-deserved by someone who spent many years in the field and was always searching for the unknown.[54]

Shaped as he was by Przhevalskii, Kozlov fully adopted his mentor's methods of field work, the system of expedition staffing and equipping, as well as Przhevalskii's program of what he called the 'fleeting route reconnaissance'. The latter included: a) the survey by eye and sketching of the entire caravan route (sketch survey); b) astronomical determination of the latitude and longitude of points; c) barometrical determination of altitude; d) meteorological measurements; e) undertaking natural collections (zoological, botanical, mineralogical, ethnographic).[55] Kozlov was the first traveler who navigated the Central Asian lakes. Also, he sought to gather as much additional information (political, economic, military, environmental etc.) as possible, by means of questioning local inhabitants – Mongols, Kalmyks, Tibetans, and people from passing caravans. This strategy worked well, allowing him and his assistants to obtain much new data, such as, for example, on the administrative and political system in Tibet, the country's southeastern province (Kham), the 'forbidden' Lhasa, the Dalai Lama and so on, which he included in the account of his Mongolia and Kham expedition (1899–1901).

In the course of his travels Kozlov considerably modified and improved Przhevalskii's well-tested system of 'route reconnaissance'. Firstly, he sent out assistants on lateral excursions from the main caravan, thus broadening the area of exploration and increasing the scientific 'harvest' of his expeditions. He thus combined Przhevalskii's original 'linear' route reconnaissance with the

53 Andreyev 2013, 20, 155.

54 Yusupova 2010a, 2011.

55 Dubrovin 1890, 584.

'areal' surveying of terrain, the method employed by Pevtsov, Roborovskii and himself. As a result of this latter 'system of journeying', he wrote in *Mongolia and Kham*, "one gets to know better the physical phenomena of nature, as well as inhabitants of a given country".[56] Secondly, Kozlov introduced the practice of 'stationary observations' in the areas which interested him the most. For this purpose he established meteorological stations, with depôs attached, in the Turfan depression, Tsaidam, Alashan, Kham, and the Nanshan Mountains. The systematic meteorological observations and hypsometrical measurements at those stations were usually carried out for one or two years. As a result Kozlov collected a great amount of important data needed for the study of climate in Inner Asia, while his astronomical determination of coordinates at these bases was of great value in topographical surveying.[57]

In the course of the discussion of the preliminary results of his Mongolian expedition at the Academy of Sciences in Leningrad, in 1925, Kozlov agreed with his critics concerning his method of extensive 'route reconnaissance' – "the method of journeying from Przhevalskii to Kozlov", in the words of Sergei F. Oldenburg – and admitted, too, that the great epoch of exploration of Inner Asia by this method was over. Henceforth the region, Mongolia in particular, would be explored by parties, specializing in one type of fieldwork and acting from a single permanent base.

Kozlov was well aware of the tendency towards increasing specialization, generated by rapid scientific and technical progress, which required a new system of fieldwork by specialists. His Mongolia and Kham expedition (1899–1901) anyway pioneered limnological and hydrological field investigations in his use of specialist equipment. Kozlov also initiated archeological and paleontological excavations, which were not on the formally-approved expedition agendas. His use of photography was an important addition to visual observations and was particularly helpful in ethnographic research.

This transition from superficial and hasty ('fleeting') route reconnaissance to more sophisticated 'stationary' work had been foreseen by Przhevalskii in his 1888 travelogue, in the chapter "How to travel in Central Asia". There, in a section entitled 'The tasks of future explorers', he presciently outlined two types of forthcoming explorations in Central Asia – "the scientific reconnaissance of the remaining uncharted areas" and the "detailed study of more accessible countries, already reconnoitered by means of fleeting journeys". He especially recommended the use of "special stations in the most advantageous points", where several scientific parties could engage in their individual work.

56 Kozlov V. 1947, 44; Ovchinnikova 1964, 17.

57 Ovchinnikova 1964, 181, 182.

Those branches of science which should be given more attention by explorers, according to Przhevalskii, were geology and anthropology, as well as meteorology and entomology.[58] Kozlov, as Przhevalskii's true heir, focused precisely on these disciplines in subsequent years.

Kozlov's major contributions as a land surveyor was his route sketches, which were very detailed and accurate and covered about 40,000 miles as well as his many astronomic and hypsometric measurements needed for mapmaking.[59] The final result of this major topographic endeavour begun by Kozlov and other Russian land surveyors were the new physical maps of Inner Asia based on the latest and scientifically reliable orographic and hydrographic data produced by cartographers.[60]

Kozlov was a restless soul and had a natural predilection for traveling – what he called a 'nomadic spirit' – and which constantly put him on the move towards some distant lands in Asia. His major accomplishments as a geographer were his many discoveries made on the Tibetan plateau, in Amdo and Kham, Mongolia (northern, central and southern), and in Eastern Turkestan (Xinjiang). These included the mountain ranges he discovered, named and sketched (in the Mongol and Gobi Altai and the Kunlun), the lakes he surveyed and sounded (Kokonor, Edzin-gol, Bagrashkol etc.) and the upper beds of the Huang He, Yangtze and Mekong rivers. Finally, he obtained a great deal of new data on the relief of Central Gobi and defined the borders of the extreme west of the Gobi desert known as Dzungarian Gobi.

He also helped solve the enduring Lopnor problem – that of the alternating shape and location of the 'Wandering Lake' in Western China. He visited Lopnor several times and in 1897 he discussed the issue with Sven Hedin, then visiting St. Petersburg, which revived the old Richthofen–Przhevalskii controversy. Kozlov accounted for Lopnor's changing location by the periodic and alternating migration of Tarim and Konchedarya, the rivers feeding the lake. In 1898, he published an article devoted to the hydrography of Lopnor[61] in which he juxtaposed Richthofen's and Hedin's arguments, Przhevalskii's firsthand evidence and his own observations. Kozlov dwelt upon the subject again in his last publication, a short paper *Kochuiustchie reki Tsentral'noi Azii* {The Wandering rivers of Central Asia).[62] Its writing was occasioned by Colonel

58 Przhevalskii 1888, 64, 65.

59 Ovchnnikova 1964, 183.

60 Shchukina 1955, 196–225.

61 Kozlov 1898, 60–116.

62 Kozlov 1935, 599–601.

R.C.F. Schomberg's visit to the lake in 1928, by which time Konchendarya had migrated back to its old dry bed.[63]

Finally, Kozlov was a first-rate naturalist, a keen observer of natural phenomena as evidence his lively and detailed characterization of Asian landscapes and an excellent collector of fauna and flora. His first collections were made under Przhevalskii and Pevtsov, and his first publication dealt with the survey and faunal description of the upper Konche-darya, the tributary of Tarim, and of the northern shore of Lake Bagrashkol, and was included in the Works (*Trudy*) of the Tibet Expedition under Pevtsov (1895). In the expedition under Roborovskii Kozlov was in charge of the zoological collection. In three subsequent expeditions he was assisted in his naturalist studies by Ladygin, Kaznakov, Chetyrkin, Napalkov, Chernov, Pavlov, and his wife Kozlova-Pushkareva. Kozlov explored and described the fauna of many parts of Inner Asia: the eastern Tien-Shan (1883, 1890, 1893); much of Chinese Turkestan (1884, 1890, 1893–1895), mainly the oases of Khotan, Niya, Cherchen; the Lukchun depression, Lake Lopnor; the large tracts of Dzungaria (1893, 1895); Mongolian Altai, the northern, central and southern Mongolia (in his last four journeys); the Alashan range (1901, 1908); Lake Kokonor (1884, 1894–1895, 1900); Tsaidam (1884, 1893–1895, 1900–1901); the northwestern part of Gansu Province (1909); Amdo (1894, 1908–1909); northeastern and, partly, inner Tibet (1884, 1894, 1900–1901); and southeastern Tibet (Kham, 1908–1909).

The natural collections brought by Kozlov from his six journeys and transferred by the Geographical Society to the Zoological Museum of the Russian Academy of Sciences and the Botanic Gardens are an important resource for zoologists and botanists. His zoological collection consists of over 1,400 samples of mammals, some of which are either rare or unique, such as those of the wild camel (*Camelus bactrianus ferus*), wild yak (*Poëphagus mutus*), Tibetan bear (*Ursus pruinosus*), Thorald's deer (*Przevalskium albirostre*), Chinese rock deer (*Gervus affinis*), Chinese musk-deer (*Moscus sifanicus*), antelopes of the Gazella genus, and the Tibetan monkey (*Macacus tibetanus*). His collection of wildcats includes some new forms, such as *Felix anastasiae* Satun., *Felix chutuchta* Birula, *Otocolobus manul mongolicus* Satun. Among the hoofed animals Kozlov's expeditions discovered a new sub-species of Manchurian deer (*Cervus Canadensis alashanicus* Bobr. et Flerov), typical of the Alashan range, and Persian gazelle (*Gazella subgutturosa reginae Adlerberg*), found in Western Tibet, as well as the Tibetan antelope (*Pantholops hodgsoni*). Among small mammals two new genera of jerboa (*Cardiocranius paradoxus* Satun. and *Salpingotus Kozlovi* Vinogr.) are also noteworthy.

---

63 Ovchinnikova 1964, 73–75; Murzaev 1947, 3–23.

The ornithological collections were even more abundant (over 5,000 samples, including Himalayan griffons (*Gyps nivicola*), lammergeiers (*Gypaëtus barbatus*), golden eagles, falcons and other birds of prey). Kozlov also obtained several new species of birds, some of which bear his name (*Tetraogallus Kozlovi* Bianchi, *Emberiza Kozlovi* Bianchi, *Accentor Kozlovi* Bianchi). Particularly remarkable were two birds of a new genera, *Kozlovia roborovskii* (Przev.) and *Kaznakovia kozlovi* (Bianchi).

The reptiles and amphibians collected by Kozlov totaled about 750 samples, some of which were totally new forms; fish made up only 300 of this total, a reflection of the paucity of Inner Asian basins. As for insects, the expeditions of Przhevalskii (the last one), Pevtsov, Roborovskii and Kozlov accumulated no fewer than 80,000 specimens, of which 200 were new species and over thirty new genera. Many were given Kozlov's name.[64]

Kozlov's botanical collections totaled about 100,000 specimens, representing hundreds of new species and dozens of new genera. The results of the extensive floristic studies of Kozlov and other Russian travelers were summarized by B.A. Fedchenko.[65] Although Kozlov was not a professional botanist, he had a good understanding of the geographical distribution of plants and of the important vegetation types of Inner Asia and he tried to determine the borders of the major floristic regions in the region.

One last thing. Kozlov was also environmentally aware. His major efforts as an environmentalist were directed toward preservation of the unique wildlife at the Ascania-Nova nature reserve. He visited the place several times between 1913 and 1927, and was seriously concerned about the damage caused to Ascania during the civil war and tried to repair it. Kozlov's activities urged the Soviet Ukrainian government to bestow the status of the 'state-protected steppe reserve' on Ascania in 1921. In 1927, Kozlov drew up an inventory of the entire animal stock at Ascania, upon which he submitted a report to the Soviet authorities, calling them to "take the utmost measures" to protect the few extent aurochs from complete extinction.[66]

As a lover of wildlife, he particularly admired the Bogdo-Ula (Mong. *Bogd Uul*) forest refuge in the environs of Ulan-Bator, the first one in the world, which he called "the precious natural setting" of the Mongolian capital. In 1924 Kozlov published an essay in which he described minutely the Bogdo-Ula's fauna and flora and referred to a decree by one of the Grand Lamas of Mongolia, according to which the mountain was proclaimed a protected or sacred

64 Semenov-Tian-Shanskii, A.P. 1937, 131–137.

65 Fedchenko 1940, 669–678.

66 Gnatiuk 2014, 32–44.

area in the eighteenth century. Hunting and woodcutting on its slopes was strictly prohibited under penalty of death and Bogdo-Ula became accessible to Mongol ascetics only.[67] The travel diary of Kozlov's last expedition to Mongolia has numerous references to the 'sacred mountain' which served as a model for human protection of nature and a source of constant inspiration for him.

67 Kozlov 1924.

CHAPTER 5

Bronislav Liudvigovich Grombchevskii (1855–1926)

Mikhail K. Baskhanov

The Formative Years: Life and Service on the Borders of the Empire

In the history of Russian geographical study of Inner Asia, a prominent place belongs to the name of Bronislav Liudvigovich Grombchevskii (Polish: Bronisław Grąbczewski). In the five years between 1885 and 1890, he had completed three expeditions to Eastern (Chinese) Turkestan, Pamir, Kanjut (modern Hunza) and northwestern Tibet. These expeditions brought Grombchevskii fame and advanced his name amongst the number of distinguished travellers and explorers of Inner Asia.

Grombchevskii's research activity was inextricably linked with political events in Central Asia at the end of the 19th century – the Anglo-Russian rivalry for political, military and economic domination, otherwise known as the 'Pamir crisis', as well as Russia's delimitation with Afghanistan and China. Grombchevskii became involved in the events of the Great Game – Britain and Russia's struggle for supremacy in the region. Grombchevskii's geographical research was never entirely scientific, it always seemed to have a political undertone, relating to Russia's foreign and defence policy in the Inner Asian region.

Grombchevskii was born on January 15, 1855 into an aristocratic family in the Kovno *guberniia* (province), in what is known today as the northern half of modern Lithuania.[1] He was born Roman Catholic and obtained his elementary education at the Warsaw Classic High School. In April 1873 he joined the military service as a volunteer for the Kexholm Grenadier Regiment, stationed in Warsaw. In 1874 Grombchevskii entered the Warsaw Infantry Junker School, where between 1864 and 1867, Lieutenant N.M. Przhevalskii had taught geography.[2] In July 1875 he graduated from the School and joined the Kexgolm

1 Biographical data provided by archival document – Rossiiskii gosudarstvennyi voenno-istoricheskii arkhiv (RGVIA), Moscow, F. 400. Op. 12. D. 18583. ll. 48–55: A brief record of service of acting commandant of Osh district, listed as Army Infantry Lieutenant-Colonel Bronislav Ludvigovich Grombchevskii.

2 The reference that Grombchevskii studied at the Academy of Mines, but had not completed the course (see Postnikov 2005, p. 435), does not comply with his official record of service.

 | DOI 10.1163/9789004376267_007

FIGURE 5.1
Captain Bronislav Liudvigovich Grombchevskii, 1890.

Grenadier Regiment in November of the same year as warrant officer. In March 1876, he was enlisted in the 14th Turkestan Line Battalion. From this point onwards, Grombchevskii's 'Turkestan period' begins, encompassing almost 20 years of service.

Grombchevskii arrived at the 14th Turkestan Line Battalion, in Tashkent, in late spring – early summer of 1876. Grombchevskii's start date in Turkestan coincided with the conquest of the Khanate of Kokand. Russian military operations in the territory of the Khanate were completed by the end of January 1876 and already by February 19 an Imperial Order was issued about the inclusion of the territory of the former Khanate of Kokand to the Russian Empire with the formation of a separate administrative unit – the Ferghana *oblast'* (region) as part of the Governor-Generalship of Turkestan. The first appointed military governor of Ferghana *oblast'* was Major-General M.D. Skobelev. In July 1876 Grombchevskii participated in the Alai military expedition, undertaken on the order of Commander-in-Chief of the Turkestan Military District, Adjutant General K.P. von Kaufman. The aim of the campaign was to bring into submission the Kyrgyz people in the mountainous parts of Ferghana *oblast'* and have them accept Russian authority, and to conduct reconnaissance of this little-known mountainous area up to and including the Trans Alai Range.

Russian troops remained in the Alai valley until August 28. This time General Skobelev conducted reconnaissance near the Chinese border in order to determine the border line between China and the former Khanate of Kokand, the troops quickly developed wheel way from Alai to Gulcha frontier post through the Taldyk Pass. In the course of the Alai expedition, so-called Alai military and scientific expedition, work took place and held an important significance for the study of the geography of Alai and the Pamir. Lieutenant Grombchevskii among other officers of his battalion took part in the expedition. During the expedition astronomical points were determined, route surveys were carried out, botanical and zoological collections were assembled. The course of the work and the results of the Alai military and scientific expedition were presented by General Staff Captain L.F. Kostenko in his works.[3]

After his return from the Alai expedition Grombchevskii was appointed acting adjutant of the battalion and for a brief time was seconded to the 3rd Orenburg Cossack Regiment. The Russo-Turkish War of 1877–78 considerably strained Anglo-Russian relations. A number of representatives of the Russian military elite insisted on the need for a preventive strike on India. In June 1878, Grombchevskii as adjutant of the 14th Turkestan Line Battalion, was posted to the Main Samarkand Detachment, which had been deployed to the Afghan frontier in case of an armed conflict with Britain. The Russian troops did not cross over the Amu Darya River and on July 19, 1878, following the Supreme decree, the Main Samarkand Detachment was disbanded and troops were returned to permanent deployment.

For Grombchevskii, participating even in an assumed, march on India, had its own specific consequences. The anti-British atmosphere of those days in many respects contributed to creating this image of Britain as the main enemy of the Russian Empire in Central Asia. Fighting this British influence in Asia in a way became Grombchevskii's *idée fixe*, imperative to all his subsequent activities in Turkestan.

Grombchevskii with his extreme views on Imperial foreign policy was not the only man with such views at the time. On the 'front' side there were numerous true 'soldiers of the Empire' who were prepared to defend the interests of Great Britain in the Asian expanse till their last breath. Almost in parallel to Grombchevskii – both in time, in irreconcilability of views and occupation, the formation of another prominent member of the Great Game in British India was occurring, the British counterpart to Grombchevskii – Francis

3 L.F. Kostenko produced two major works on the Alai expedition: 'Ekspeditsiia v Alaiskie gory', in *Russkii Invalid*, 1876, nos. 206, 211, 224, 235, 239, 244, and 'Voenno-nauchnaia ekspeditsiia na Alai i Pamir', in *Voennyi Sbornik*, 1879, no 4. An extract from Captain Kostenko's account was later translated and published by the Royal Geographical Society, see Robert Michell, The Russian Expedition to the Alai and Pamir in 1876, in *The Proceedings of the Royal Geographical Society of London*, London, Vol. 21 (1876–77), pp. 122–140.

MAP 5.1 The routes of B.L. Grombchevskii. Compiled by Mikhail Baskhanov on the basis of the map published in: *Pamir, Khunza i Kashgariia v ekspeditsionnykh fotografiiakh B.L. Grombchevskogo* (*1888–1890*), Ed. by M.K. Baskhanov, A.A. Kolesnikov, M.F. Matveeva and A.I. Glukhov, Moscow: Pelikan, 2017.

a n
issar
Eastern Turkestan
Taklamakan
Yarkand
Khotan
Niya
Polu
Tibet

Younghusband. Their meeting would occur several years after the events described here, in the uninhabited uplands of High Asia at the height of the Anglo-Russian struggle for the Pamir.

In less than three years of service in Turkestan, Grombchevskii had developed three dominant tendencies for travel, politics and administration. This explosive mixture was to determine the rest of his fate.

In 1880 Grombchevskii submitted a report to the battalion commander requesting permission to continue service with the *voenno-narodnoe upravlenie.*[4] His transfer into the new service took place on the basis of the Supreme order dated June 15, 1880. During his service in Ferghana, Grombchevskii was in a civil marriage with a Kyrgyz woman, who had a daughter, Maria.[5] The daughter was never officially adopted, and was listed as his foster child. After the death of his wife, Grombchevskii sent the girl to an all-girls boarding school in St. Petersburg. He briefly mentioned his "baby orphan child" in a travelogue of his expedition to Hunza of 1888.[6]

In St. Petersburg on February 12, 1881, a treaty was signed between the Russian and Chinese Empires regarding the Ili Region, according to which Russia would hold dominance over the western part of the Ili region and the rest of the land would be restored under Chinese rule. The parties determined the boundaries of the divided land as established by the Protocol of Chuguchak, signed September 25, 1864.[7] Furthermore, both sides appointed by the Boundary Commission, had to conduct differentiation between the Russian Semirechye, Ferghana *oblast'* and Chinese Turkestan. To accomplish the entrusted task, the Boundary Commission had geodesists, surveyors, translators, guides and guards at their disposal. On March 9, 1882 Major-General V.Iu. Medinskii was appointed as the Russian plenipotentiary commissar to the Boundary Commission, and assigned to the delimitation of Turkestan general-governor-

4 *Voenno-narodnoe upravlenie* (Military-Popular Government) – a system of administration of the Russian borderlands in the Caucasus, Siberia and Central Asia. The provincial governors under this system were military of senior ranks, and senior civil servants were army officers seconded from their units to perform civil administrative duties. The distinctive feature of this system was the division between the military administration and the local self-government by the natives (*tuzemtsy*).

5 See The National Archives (TNA), Kew, FO539/56/118: Mr. Eliot to Sir R. Morier, February 29, 1892.

6 Nauchnyi arkhiv Russkogo geographicheskogo obshchestva (NA RGO), St. Petersburg, F. 45. Op. 1. D. 2. l. 68.

7 The Protocol of Chuguchak (25 September 1864) defined the state border between Russia and China from Sayan to the Tian Shan mountain ranges.

ship with West China. In August 1882 Grombchevskii was appointed as Medinskii's assistant to help with work on the boundary commission.

On May 22, 1884 in Novyi Margelan, Major-General Medinskii and the Chinese border commissioner Sha, signed the protocol on the delimitation of the Russo-Chinese boundary from the Bedel Pass (Kokshal Range) to the Uzbel Pass (the Northern Pamir, Sarykol Range).[8] With work completed, Medinskii returned to his duties as assistant to the military governor of Ferghana *oblast'*. As for Grombchevskii, for his work on border delimitation he was promoted in rank. On April 30, 1885 he was promoted to assistant to the Osh district commandant. He stayed in this post for a mere six weeks before obtaining the post of senior civil servant on special duty to the military governor of Ferghana *oblast'*.[9]

Wanderlust: The First Trip into Kashgaria (1885)

From this point on, Grombchevskii would begin his expeditionary period, which would establish his name and reputation. Although this period was brief, it was very intense and created a lasting impression. Grombchevskii's first independent expedition took place in 1885 into neighbouring Kashgar, adjacent to Fergana. In fact, this was not an expedition in the truest sense of the word but rather an official trip. In accordance with the Margelan Protocol on the demarcation of the Russo-Chinese frontier of 1884 (Article 5), by mutual agreement of both sides, annual inspections of border marks were carried out by officials, who were responsible for their safety.

Grombchevskii's trip to Kashgar took place between August 10 and November 24, 1885. He was accompanied by five Cossacks from the 6th Orenburg Cossack Regiment and two *jigits*.[10] He received 300 rubles for the expenditures. 'I did not take an interpreter with me', – Grombchevskii wrote in his report – 'as there was no one who knew Chinese [in the Fergana region], whilst I know the Turkic languages well and can speak freely. In communicating with the Chinese, I used translators belonging to the Chinese officials, and explained myself in Turkic.'[11]

8 For the document, see *Russko-kitaiskie dogovorno-pravovye akty*. 1689–1916, Moscow: Pamiatniki istoricheskoi mysli, 2004, pp. 160–163.

9 There are some references in literature that Grombchevskii was appointed to the post of 'diplomatic officer', this is however not true, because such a position under the military governor of Ferghana *oblast'* did not exist.

10 *Jigit* – from Turkic, a skilfull and brave equestrian.

11 Grombchevskii 1886, p. 1.

On August 20 Grombchevskii arrived into Kashgar and stopped off at the Russian Consulate. Here occurred his first encounter with N.F. Petrovskii, a Russian Consul, an expert of the East and one of Russia's most prominent diplomats and administrators. During Grombchevskii's travel the Consul assisted him as much as possible, from obtaining the right to travel to the Chinese territories to providing monetary assistance. Petrovskii was an outstanding personality, who possessed many talents. In addition to the office, his interests included – politics, diplomacy, trade, as well as history, archaeology and translating from Oriental languages. He was a member of many Russian and foreign scientific societies. Petrovskii assisted many Russian and foreign expeditions who arrived into Kashgar. Grombchevskii had at his disposal consulate political documents as well as Petrovskii's personal library, all of which was offered to him by Petrovskii himself.

From August 28 to September 19, Grombchevskii carried out reconnaissance of the Russo-Chinese border. Together with the Chinese representative, he refined the location of border signs and drew up protocols of verification. Work progressed quickly, however, a misunderstanding arose around a frontier mark in Irkeshtam. The Chinese had insisted on transferring the border mark, which required returning to Kashgar and raising the matter with the Chinese governor of Kashgar. As governor-general of the Xinjiang province, needed to make a decision on the Irkeshtam issue, the Chinese suggested Grombchevskii take 30 days break. Grombchevskii readily accepted this proposal.

> Daotai's[12] offer was accepted with complete willingness, staying for longer in Kashgar, and, moreover, at the invitation of Chinese authorities, opened up opportunities for reconnaissance work in Southern Kashgaria. All of my free time ... was spent on acquainting myself with the Kashgar administrative systems and its military forces, I took advantage of every opportunity to attend military training of the Chinese troops.[13]

Grombchevskii set himself the task of gathering as much material as possible, focusing on two main issues – the provincial administrative system and the military forces of Kashgar. Documents from the consular archive, as well as Petrovskii's official consular report of 1885[14] provided much needed valuable

12 Daotai – from Chinese, a local governor of sub-provincial level.

13 Grombchevskii 1886, p. 19.

14 [Petrovskii N.F.], *Otchet Imperatorskogo Rossiiskogo konsul'stva v Kashgare*, as manuscript, not for publication, n.p., [1886]. For a copy of this document, see Rossiiskii gosudarstvennyi arkhiv drevnikh aktov (RGADA), F. 1385, Op. 1, D. 466, ll. 114–155.

information, Grombchevskii only needed but supplement this with information from personal acquaintances with Chinese troops both in Kashgar and beyond.

October 10, 1885 Grombchevskii left Kashgar and set out on his first independent expedition. Within two weeks, he traveled from Kashgar to Yangi Hisar, Yarkand, Karghalik, Guma, Khotan, 515 *verst*[15] in 13 marches. As noted by Grombchevskii himself, the intense speed at which he travelled strongly influenced the completeness of the information collected. Throughout the expedition he conducted route surveys, visited Chinese garrisons, composed descriptions of roads, settlements, water sources, etc. On October 24, Grombchevskii reached Khotan, the furthest point of his route, and spent three days in the city. He returned back to Kashgar via Sanju, Kargalyk, Yarkand and Yangi Hissar (515 *verst* in 11 marches).

In his report on the three and a half month Kashgar expedition, Grombchevskii reported of works accomplished, which included completing 1,796 *verst* (of which – 480 through mountainous terrain), writing up descriptions of three routes, creating drawings and plans of several fortifications and conducting route survey from Irkeshtam to Khotan through Guma and Sanju (total 900 *verst*, scale: 1 inch equals 5 *verst*) sumarising his work Grombchevskii concluded:

> The most valuable work undertaken can be considered to be the route survey from Kashgar to Khotan, as our maps up until now have been supplemented by English sources. I have undertaken the most pleasant assignment, to be the first Russian to visit the country and to connect the surveys of Kashgar with Colonel Przhevalskii's surveys,[16] who had arrived not long before me in Khotan from the east, and went to the north, whilst I made my way from the west.[17]

At the beginning of 1887, based on his surveillance materials, Grombchevskii published an article entitled 'A sketch note on the Chinese troops and fortifications in Chinese Turkestan'.[18] The article was largely a compilation and contained sufficiently known information; however the mere fact that it was

15 *Versta* (*verst*, pl.) – is an obsolete Russian unit of length, one *versta* equals to 1.0668 kilometres (0.6629 miles). Hereafter, we will use *versta* as a unit of length.

16 This refers to the fourth journey of N.M. Przhevalskii in Inner Asia (September 1883 – October 1885).

17 Grombchevskii 1886, p. 19.

18 Grombchevskii 1887.

published in several issues of the official military publication strongly raised Grombchevskii's authority among the Russian military.

In Tashkent, most were well aware of the importance of what was accomplished by Grombchevskii: the assembled military-statistical data and route surveys significantly expanded the possibilities of war planning for the Staff of the Turkestan Military District. Well-deserved awards followed. During the Kashgar trip Grombchevskii was promoted to Staff Captain, 'for distinguished conduct', which was followed by an Imperial award of 1,000 rubles 'for useful work in Kashgaria'.

Towards 'The Robber's Nest': Expedition to Hunza (1888)

During the trip to the Southern Kashgaria, Grombchevskii for the first time came into contact with so called 'Kanjut question'. Whilst working with the political affairs in the archive of the Russian Consulate in Kashgar, Grombchevskii focused on Consul Petrovskyii's reports, in which he briefed the authorities that Kanjut's[19] ruler attempted to establish relations with Russia. Soon he witnessed the arrival of Gulzor Khan, envoy of the Hunza's ruler Ghazan Khan, into Kashgar. Grombchevskii saw new opportunities in the area of high politics and diplomacy. It was a dangerous precedent, which assumed the beginning of his irresponsible, if not adventurous, foray into Russian political interests in the east. The ambitious Grombchevskii could not pass on such an opportunity.

There was, however, one sensitive matter – in Kashgar political contacts as well as political issues in general, were entirely under the jurisdiction of the Russian Imperial Consul. However, Consul Petrovskii was absent from Kashgar at this time, and Grombchevskii decided to create his own diplomatic 'party' – to meet with Hunza's envoy unauthorised. The meeting took place not long before Grombchevskii's departure from Kashgar. During the discussions, Gulzor Khan mentioned to Grombchevskii, the domestic and external position of Hunza and many a time repeatedly mentioned Ghazan Khan's dislike towards the British. Gulzor Khan's reports gave Grombchevskii grounds to assume that the 'Hunza direction' was underestimated by Petrovskii, and that further developing the 'Hunza theme' to a state of 'a problem', meant that one could make a name for themselves.

19 Present-day Hunza. Hereafter, we will use the modern name of this former independent possession in northern India.

The Hunza princedom was located on the southern slopes of the Hindu Kush Mountains in the region where it joins the Karakoram mountain range. To the north, the princedom bordered with Afghanistan and China, to the east with the Chinese holdings in Raskam, to the south its holdings were limited to the Karakoram ridge and the Hunza River, in the west with Gilgit. About three fourths of the territory of Hunza is covered in high mountain peaks and glaciers, whilst intermountain valleys in Hunza are mostly river valleys. The only communication means lie in the river valleys, overcoming the watershed crossings of up to 5,000 meters. The mountainous country, with a total area covering 8,000 sq. km, had the population of 56,000 inhabitants, who occupied 55 settlements. A majority of the population were the Yeshkun people (up to 80%) as well as the Wakhi people. The population was primarily Shia Muslims, Ismailism was widespread. Communal feudal landownership existed in Hunza. Power was presented by important feudal lords and was passed on by birthright. The favourable geographical position of Hunza, with its straight roads leading out to the Southern Kashgaria and the Pamir, led to its strategic value. Nowadays the strategically built Karakoram Highway runs through Hunza, connecting China with Pakistan.

In mid-1885, with the expansion of British possessions in northern India, the princedom became more and more the object of increased attention for the Indo-British authorities. The Hunza lords tried to maintain independence through political intrigue against British India, Russia and China. An important source of income for the rulers of Hunza, that is until the establishment of the British control, were raids on neighbouring territories, robberies of trade caravans and the slave trade. At the time of these events, the ruler of Hunza acknowledged vassalage to China.

In 1886, Colonel W.S.A. Lockhart's British political mission, known as Lockhart's Chitral Mission, under the pretext of continuing the work of the Afghan Delimitation Commission, was sent to the Hindu Kush principalities – Chitral, Hunza, Badakhshan and Wakhan. Important military geographic data was accumulated during the mission, regarding the political situation of the Hindu Kush principalities.

Lockhart's Mission was a prelude to the policy of the British authorities in incorporating the principalities into Indo-British possessions. In 1887–88, Colonel A. Durand conducted a mission to Chitral and Gilgit, set with the task of reinstating British political power in Gilgit as well as reconnaissance of the territories, which lead from Chitral to the Pamir. Colonel Durand's Mission was seen as a complete success – he secured British political influence in Chitral and restored political power in Gilgit by deploying a large number of British garrisons there. Gilgit was turned into a major political and military centre on the northern frontiers of British India.[20]

Soon after Colonel Durand's visit to Hunza a palace coup occurred in the principality. During the riots, Ghazan Khan was assassinated by conspirators, including Safder Ali Khan, his son, who seized power in the principality. After brief diplomatic overtures with the British, he continued his father's line in establishing close connections with Russia, in case of further deterioration of relations with British India and struggles with neighbouring principalities. This political game eventually led to the British occupation of Hunza.

The thought of completing a large expedition into the distant and unknown Hunza completely consumed Grombchevskii. The trip to Kashgaria brought Grombchevskii a certain amount of fame, nevertheless, it clearly showed his limits for new riskier enterprises. The main issue was his lack of experience managing large geographical expeditions and lack of practice of topographic surveying and astronomical observations. Having studied Przhevalskii's experience along his journeys, Grombchevskii came to the conclusion that without this basic knowledge of geodesic surveying, he could never really fulfil the principal goal of any geographical expedition, that is to say, to determine the precise geographical location of positions, without which it was impossible to proceed with mapping the explored regions.

In order to further improve his knowledge in the field of geodesy, Grombchevskii took leave and travelled to St. Petersburg and attended lectures on astronomy and geodesy at the Physics and Mathematics Faculty of St. Petersburg University, hoping to then pass the practical training at the Pulkovo observatory. Early December 1887, Grombchevskii arrived in St. Petersburg and began the course, which lasted for four months (until April 1888). Between lessons Grombchevskii acquired the necessary skills to use different tools including an aneroid barometer and hypsothermometer, used to determine heights and topographical relief, as well as portable astronomical tools including chronometers, universal instruments, and others, used to determine astronomical positions. He also acquainted himself with the basic skills of carrying out triangulation surveys, meteorological observations and photographic works.

During his stay in St. Petersburg Grombchevskii persistently searched for opportunities to present to the Council of the Imperial Russian Geographical Society (IRGO), his plan for an expedition to Hunza. In February 1888 he

20 For more details on these events, see W.S.A. Lockhart, and R.G. Woodthorpe, *Confidential Report of the Gilgit Mission, 1885–6*, London: Eyre and Spottiswoode, 1889; A.G.R. Durand, *The Making of a Frontier*, London: John Murray, 1899; G.J. Alder, *British India's Northern Frontier, 1865–95*, London: Longmans, 1963; J. Keay, *The Gilt Game. The Explorers of the Western Himalayas, 1865–95*, London: John Murray, 1979.

managed to secure a meeting with the Chairman of the Council of the IRGO, Grand Duke Konstantin Nikolaevich of Russia. Making reference to his merits to the Society, Grombchevskii turned to him for assistance and gave a brief note on the purpose of the expedition. Grand Duke Konstantin Nikolaevich on behalf of the Council of the IRGO turned to the War Minister P.S.Vannovskii with a letter outlining Staff Captain Grombchevskii's proposal, requesting support of the Geographical Society to assist with the Hunza expedition.

The letter triggered the official correspondence, which continued for a considerable amount of time starting in February till July 1888. The War Minister, believed that these kinds of questions should be agreed upon with the Governor-General of Turkestan, who Grombchevskii serviced under, and who was also in charge of relations with neighbouring countries, handed the matter over to Adjutant General N.O. Rozenbach.[21] In his reply addressed to Adjutant General N.N. Obruchev, Chief of Main Staff, the Governor-General of Turkestan pointed out that sending an officer to Hunza would not only be dangerous for his personal safety but could also cause a significant amount of harm 'to the recently established relationship with the British cabinet and the newly formed feelings of friendship and trust'.

Referring to the political risks, associated with Grombchevskii's planned journey, Adjutant General Rozenbach noted:

> From the recent report from our Consul in Kashgar, it seems, that the presence of Russian traveller Grum-Grzhimailo[22] on Kashgar territory, who had crossed our boundary at a point not indicated in the Treaty of Saint Petersburg (1881), instilled a certain level of trust in the Kanjut ruler, who deemed him to be none other than a messenger from the Emperor himself. It is therefore possible to predict the bombshell effect among the locals and deep concern from the Viceroy's Executive Council, if they where to see a Russian officer within the limits of Kanjut or Chitral, if he were seen to carry out interrogations or route surveys.[23]

21 Nikolai Ottonovich Rozenbach (1836–1901) – Adjutant General, General of Cavalry, Governor-General of Turkestan and Commander-in-Chief of the Turkestan Military District (1884–89).

22 The Adjutant General Rozenbach had meant a journey to the Pamir undertaken in 1887 by a full member of the IRGO naturalist G.E. Grum-Grzhimailo and his brother M.E. Grum-Grzhimailo, an officer of Imperial Guard 2nd Artillery Brigade. The travellers intended to reach Hunza through Sarykol and the Taghdumbash Pamir, but after crossing the Chinese border in Sarykol were stopped by the Chinese authorities in Tashkurgan. The ruler of Hunza considered detained travellers envoys of the Russian Emperor.

23 RGVIA, F. 401, Op. 4, D. 46 (1888), l. 5.

Adjutant General Rozenbach proved to be a seer all had happened precisely as he had predicted.

We can assume that in the Military-Scientific Committee of the Main Staff, the main military administrative body, which at that time was in charge of co-ordinating military intelligence, held considerable sentiment towards Grombchevskii's project. Within the Military-Scientific Committee, the supernumerary member of the committee was Major-General Przhevalskii known for his views on the need for an offensive policy in Asia. At the time, Przhevalskii was developing his concept of research priorities in Inner Asia, built on identifying several regions, which would represent the greatest scientific interest. Among the areas he singled out was 'the Pamir with parts of the Hindu Kush and Karakoram'.[24] The authoritative generals A.N. Kuropatkin (General on special duty to the Chief of Main Staff) and L.F. Kostenko (Head of the Asiatic Department of the Main Staff), could both have assisted Grombchevskii, as they took part in the Alai expedition and held a natural interest in the proposed area of travel. Kuropatkin had even conducted reconnaissance of Kashgar, whilst Kostenko had studied neighbouring Dzhungaria and conducted solid military-statistical survey of the area.[25]

Whilst the War Ministry and the office of the Governor-General of Turkestan corresponded on the matter, Grombchevskii stayed in St. Petersburg. In the meantime, he was tightening his ties within the Geographical Society and was often present at various functions. On April 12, 1888, Grombchevskii on the recommendation of I.V. Mushketov, A.V. Grigoriev and Iu.M. Shokal'skii, all prominent Russian geographers, was elected a full member of IRGO.[26]

In April, 1888 the Chief of the Main Staff reported to the War Minister that Przhevalskii 'spoke in favour of assigning' Grombchevskii to explore Hunza via the Mustagh Pass. I.A. Zinoviev, Head of the Asiatic Department of the Ministry of Foreign Affairs, also supported this expedition. He suggested the need to 'beautifully present' the reason for Grombchevskii's trip beyond the Hindu

24 RGVIA, F. 401, Op. 4, D. 53 (1888), l. 23.

25 L.F. Kostenko, Dzhungaria. Voenno-statisticheskii ocherk, in *Sbornik geograficheskikh, topograficheskikh i statisticheskikh materialov po Azii*, Vyp. XXVIII, St. Petersburg: Voennaia tipografiia, 1887.

26 Zhurnal zasedaniia Soveta IRGO, 12 April 1888, in *Izvestiia IRGO*, T. XXIV. 1888, Vyp. VI, St. Petersburg: tipografiia A.S. Suvorina, 1889, p. 513. At this Council meeting simultaneously with Grombchevskii the other two officers from Turkestan, famed for their geographical and ethnographic research in Central Asia, were elected as a full members of the Geographical Society – Lieutenant-Colonel G.A. Arendarenko, Samarkand district commandant, and A.I. Skassi, a class topographer of the Military-Topographical Department of the Turkestan Military District.

Kush. He suggested that Grombchevskii travel with a letter from Emperor Alexander III himself, expressing his gratitude to the ruler of Hunza for his letter which he had sent with traveller Grum-Grzhimailo and in it explaining that Grum-Grzhimailo was not a special envoy, but rather a regular traveller.

In mid-May 1888 matters regarding the forthcoming expedition beyond the Hindu Kush was completed. The expedition was carried out under IRGO auspices, and with the direct support of the War Ministry. At the end of May 1888 Staff Captain Grombchevskii left St. Petersburg and headed to Turkestan, in order to prepare for his first solo expedition. The Council of the IRGO allocated almost 1,500 rubles towards the exhibition, to which an extra amount of 300 rubles in gold[27] was donated by Count I.I. Vorontsov-Dashkov.[28] The War Ministry paid for the travelling allowance, a sum of 300 rubles, from St. Petersburg to Osh. Another 300 rubles was allocated to the expedition by the Governor-General of Turkestan.[29] In the course of the expedition itself Grombchevskii received a loan from the Russian Imperial Consulate in Kashgar for another 500 rubles, which was later covered by the IRGO.[30] The Hunza expedition was supplied with decent scientific instruments: an aneroid barometer, hypsothermometer, photo camera (from the Geographical Society), a small universal instrument by Kern, and two pocket chronometers (from the Military-Topographical Department of the Main Staff). The expedition convoy included four Cossacks of the 6th Orenburg Cossack Regiment and three employed *jigits*. It

27 About 450 credit rubles according to the exchange rate of that time.

28 Regarding the origin of this amount, there is no reliable information. Grombchevskii himself gives inconsistent information about this money. In one of the first reports on the results of the expedition of 1888, he informed that he had received from the Cabinet of His Imperial Majesty '300 gold rubles specifically for the purchase of gift items for the ruler of Hunza'. Elsewhere he speaks of 300 gold rubles 'donated to the expedition by Count I.I. Vorontsov-Dashkov, the Minister of the Imperial Court'. It is obvious that the amount of 300 gold rubles could not have been both personal funds of the Imperial family and the private gift of Count Vorontsov-Dashkov.

29 These two amounts are not mentioned in the official cost of the expedition.

30 Grombchevskii often mentioned that he travelled to Hunza on paltry means and by the end of expedition he would have 'only a few tens of rupees in his pocket, no horses and food supplies'. To compare Przhevalskii's 4th Inner Asian expedition (the total budget was 59,670 rubles, 21 people, and duration of 25 months) and Grombchevskii's expedition (3,050 rubles, 5 people, four and a half months), it becomes obvious that the expenditure rate per member for Grombchevskii's expedition is higher than the same indication related to Przhevalskii's expedition (135 against 113 rubles). It must be noted that Przhevalskii's expedition had a significant number of 'expensive' members – a General Staff Colonel, two company officers and two civil experts.

seems however that no specific program was spelled out either by the IRGO or the Main Staff. Besides obtaining military-political information on Hunza and neighbouring territories, Grombchevskii's main task was to collect military-geographical information – surveying the major routes leading from Russian Turkestan to the northern frontier of India, rough route and basic instrumental surveying, determining heights and geographic coordinates.

Grombchevskii's expedition set out from the town of Margelan on July 12, 1888 and on September 1, arrived into the village of Baltit, the capital of Hunza. On September 7 he departed from Hunza, via the Ming Teke Pass and reached the Chinese territory on the Taghdumbash Pamir. Then he proceeded via Raskam and Sarykol into Kashgar, returning to Margelan on November 27. On his return he briefed the Governor-General of Turkestan on the results of the expedition, informing that the expedition members travelled 2,808 *verst* (2,284 *verst* beyond the Russian frontiers) on horseback, produced route surveys (scale: 1 inch equals 5 *verst*) over the distance of 1,383 *verst*, made astronomical determinations of 14 points, carried out 158 hypsographical observations, and constantly conducted meteorological observations along the route. Additionally, small zoological, ethnographical and rock sample collections were assembled. Grombchevskii also compiled a Kanjut dictionary of up to 2,000 words and a Wakhan dictionary numbering several hundred words. In addition to the ethnographical material comprised he compiled a photographic collection of views along sections of the route and ethnographic subjects. Amongst other achievements Grombchevskii mentions that the expedition "discovered huge reserves of jade, lead and copper ore, rock salt deposits, etc".[31]

Grombchevskii's linguistic field materials were transferred for processing to well-known Indologist I.P. Minaev, who unfortunately was unable to commence his work having passed away shortly thereafter (June 1, 1890). Thus the materials were handed over to K.G. Zaleman, the Director of the Asiatic Museum of the Imperial Academy of Sciences. It remains unknown; to what extent these materials were processed or used by Zaleman.[32]

Based on Grombchevskii's astronomical observations, Major-General K.V. Sharngorst, Professor of the Geodesics Department of the Nicholas General Staff Academy, carried out calculations of latitude and longitudes of 15 points, located between the Ak Baital River, village of Baltit and the Tom Kara

31 RGVIA, F. 401, Op. 4, D. 46, l. 34: Memorandum of Staff Captain Grombchevskii, a senior civil servant on special duty to the military governor of Ferghana *oblast'*.

32 Currently, the language materials collected by Grombchevskii in Hunza are held in *Sankt-Peterburgskii filial Arkhiva Rossiskoi akademii nauk* (SPbF ARAN), Fond 87, K.G. Zaleman.

FIGURE 5.2 Hunza, a view from the orchards of Baltit towards the gorge of the Hunza River. PHOTO BY GROMBCHEVSKII, 1888.

area (on the road from Tashkurgan into Kashgar).[33] Sharngorst praised the results of the astronomical determinations of the 1888 expedition, noting that Grombchevskii 'managed to determine his route on the basis of astronomical positions'. Thirteen points in Hunza and Raskam were calculated on the basis of Grombchevskii's observations, and were included in the *Full catalogue of astronomical positions of the Turkestan Military District and adjacent territories*,[34] indicating a high level of geodetic work. Grombchevskii's barometric observations, which were carried out over the course of the expedition, proved to be somewhat less successful.[35] A brief note on the meteorological observations is

33 Sharngorst 1889, pp. 437–439.

34 See P.K. Zalesskii, *Polnyi katalog astronomicheskikh punktov Turkestanskogo voennogo okruga i prilezhashchikh k nemu zemel'* (Полный каталог астрономических пунктов Туркестанского военного округа и прилежащих к нему земель), Ташкент: Туркестанский отдел ИРГО, 1914, pp. 218–227.

35 A review of the barometrical determinations made by Grombchevskii was undertaken by General Staff Lieutenant-Colonel D.D. Gedeonov, an assistant to the head of the geodesic section of the Military-Topographical Department of the Main Staff. See Gedeonov 1889, pp. 440–448.

published by Grombchevskii himself, reporting some details of the climate in different sections of the expedition route.[36] These four notes, in fact, limited the entire enumeration of the expedition's published scientific results.

Grombchevskii's astronomical determinations and route surveys, which he had prepared during the 1888 expedition, held significant importance for the mapping of the region. Grombchevskii's expedition routes to the Pamir specifically in Hunza, Raskam and Sarykol intersected with the routes of several British and Russian explorers, specifically General Staff Captain D.V. Putiata and brothers G.E and M.E. Grum-Grzhimailo. Immediately after the completion of the expedition, the Geographical Society, declared that

> … the 'cartography of Hunza and approaches from both the Little Pamir and the Taghdumbash Pamir, had become available thanks to Grombchevskii's heroic efforts, thanks to who, the areas between the rivers Raskam Darya to the east, Taghdumbash Darya to the west and Yarkand Darya to the north were also mapped out, as well as the discovery of the giant peak of Charkum, a rival of the majestic Mustagh Ata.[37]

This assertion was, however, only partly true, and mainly in relation to the Russian side. At that moment neither the Main Staff nor the IRGO had accurate information on the results of the work of the British expeditions in the region, many of which were conducted in secrecy. Members of Lockhart's Mission (1886) carried out a detailed survey of the entire Hunza, the Kalik Pass and upper Wakhan Darya. The valleys of Sarykol and Tagharma were investigated thoroughly by Colonel T.E. Gordon and his topographers, during Sir T.D. Forsyth's mission to Kashgar in 1873–74. The peak of Charkum, or North Mustagh Ata as it was named by Grombchevskii (modern peak of Kuksai), which was allegedly 'discovered' by Grombchevskii, had in fact already been located on Colonel Gordon's map of Kashgaria.[38] The above said area was extensively surveyed by the pundits (or 'native explorers'),[39] having been sent there over the

36 Grombchevskii 1889b, pp. 449–453.

37 See 'Karta Kandzhuta, Raskema i Sarykola', in *Izvestiia IRGO*, T. XXV, 1889, Vyp. VI, p. 524.

38 See 'Preliminary Map of Eastern Turkestan to illustrate The Report on Sir Douglas Forsyth's Mission to Kashgar, 1873–74'. Scale 1 inch = 40 miles. Compiled by Captain H. Trotter, Deputy Superintendent Great Trigonometrical Survey of India. January 1875.

39 Pandits (from the Sanskrit: paṇḍita, wise, learned man) – specially trained surveyors from the local tribes of the northern territories of India; widely employed by the Great Trigonometrical Survey of India for the exploration of remote and politically unstable areas of Inner Asia.

years by the Great Trigonometrical Survey of India.[40] The area was also surveyed during Ney Elias' expedition to Eastern Turkestan in 1885–86.[41] On the subject of what parts of Raskam and Sarykol were explored by Grombchevskii for the first time, is not entirely known. This would have to be subject to separate research, in which Grombchevskii's routes in this region be compared with the routes of British expeditions (until 1888).

Grombchevskii loved to compare himself with Przhevalskii. In his memories he proudly noted that no blood was spilled, no life was lost during the expedition. Although true, the comparison is not quite fair, taking into consideration the difference between the two in the number of expeditions undertaken, number of participants involved, and more importantly, distance travelled from the Russian borders. Grombchevskii studied Przhevalskii's accounts in great detail, the influence of the latter is visible both in his diary writing and in certain expeditionary techniques. He had acquired from Przhevalskii the practice of naming geographical features – peaks, glaciers, ridges, rivers, etc. A tradition, which did not evoke sympathy amongst members of Geographical Society, as the large number of renaming in Inner Asia had caused an outcry amongst members.

During his stay in Hunza, a reception was held for Grombchevskii by Safder Ali Khan, the ruler of the principality. Whilst in Hunza, Grombchevskii learned that he was not destined to become the first European to visit this remote possession of Inner Asia, credit belonged to British Colonel Lockhart. Grombchevskii failed (or rather he chose not) to ensure his status as a simple traveller who was to deliver a letter from the Russian authorities with the expression of gratitude to the ruler of Hunza and his willingness to serve Russia. Safder Ali Khan refused to believe that before he stood a simple traveller and not the envoy of the Great White Czar. It is possible to assume that the reception and the situation surrounding Grombchevskii's arrival, strongly warmed his vanity and he did not fight the part the ruler of Hunza had seen in him. 'I had to accept it' – he later wrote in his memoirs. Grombchevskii thereby violated the orders of

40 For the pundits' surveys in the area, see OIOC/L/P&S/20/238: Selections from the Records of the Government of the Punjab and its Dependences. Confidential Series – No A X, Report on Yarkand. By F.B. Lahore, 1880. A wealth of information on the pundit's exploration of the area can be found in D. Waller, *The Pundits. British Exploration of Tibet and Central Asia*, Lexington: The University Press of Kentucky, 1990.

41 See N. Elias' confidential report, his route surveys prove that he was fully acquainted with Sarykol and Raskam areas; for details of Elias' surveys, see OIOC/L/P&S/20/A27: Report of a Mission to Chinese Turkestan and Badakhshan in 1885–86. By N. Elias, Political Agent on Special Duty. With a Route Map. Calcutta: The Superintendent of Government Printing, 1886.

FIGURE 5.3 A group of Kanjutis musicians, 1888. PHOTO BY GROMBCHEVSKII, 1888.

the Russian Ministry of Foreign Affairs, which were specifically put in place to avoid any political issues. General Staff Captain Z.I. Zaichenko, an expert on Inner Asian issues, presented a more critical opinion on Grombchevskii's trip to Hunza.

> It is difficult to decide whether Captain Grombchevskii was given orders, but there is a reason to believe that the named traveller, arrived in Hunza and was met with a brilliant reception, extinguishing any hopes of an alleged incorporation of Hunza into the Russian possessions or assistance in the case of a conflict with the British. It is was clear that the painful suspicion of the British and the way Grombchevskii was warmly met in Hunza, was the main reason behind the deplorable fate that befell the unfortunate Khan.[42]

42 Z.I. Zaichenko, *Pamiry i Sarykol. Ocherk vozniknoveniia i posledovatel'nogo razvitiia i sovremennogo polozheniia pamirskogo voprosa* (Памиры и Сарыкол. Очерк возникновения и последовательного развития и современного положения памирского вопроса), Ташкент: типография штаба Туркестанского военного округа, 1903, p. 48.

FIGURE 5.4 Hunza, fortress of Kala-i-Altit, the second capital of the domain. PHOTO BY GROMBCHEVSKII, 1888.

Consul Petrovskii held a similar view, saying that Grombchevskii's trip to Hunza, his solemn reception there, conversations with the Safder Ali Khan, instigated the suspicion of the British and eventually led to the final occupation of Hunza.[43]

It must be noted that on the eve of the trip Grombchevskii poorly visualised Hunza as a country. His knowledge of the country, in the period of preparation for the expedition was fragmented, and his hopes of creating a political party at the common border of the three empires – naive. Having arrived into Hunza, he had come to face with reality. Whilst on his way to Hunza, he came across a Hunza saying – 'it is easier to trek down into our country than it is to leave'. On August 12, 1888, after his first impressions Grombchevskii wrote in his diary, 'today we entered the robber's nest'.[44]

Grombchevskii found himself in the realm of poverty and incivility, amongst cattle raiders, slave-traders and parricides (about whom he writes in detail in

43 *Tsentral'nyi gosudarstvennyi arkhiv Respubliki Uzbekistan* (TsGARU): F. 2, Op. 2, D. 52, l. 17: A dispatch of the Imperial Russian Consul in Kashgar to the Russian Imperial Minister in Peking.

44 Grombchevskii 2015, p. 108.

his travelling diary). Grombchevskii's short stay in Hunza was full of humiliation, where he was forced to tolerate obediently, fearing for his life. In a way, he was Safder Ali Khan's hostage and was forced to perform humiliating services – treat and feed his subjects, and give property, including that which was government issued. He would write in his diary, how 'entirely disappointed' he was with the ruler of Hunza, and found his 'begging unbearable'. He had but one desire and that was to 'to get out fast', which becomes the burthen of his entire stay in the Hunza domain.

Grombchevskii often mentioned in his diaries, that although he was the unofficial face, things changed once he arrived in Hunza, he instantly realised that his 'new' status of Russian envoy was his only guarantee of safety. Contrary perhaps to his will and desire he had to play out this role, which perhaps he had originally not thought of taking up. But the logic of events and their pace was such that he had no room to maneuver. This is evidenced by an entry in his diary:

> I am moving with the assistance of the [Hunza] government, tolerating so many failures, what would happen if they end up disappointed in me. Here's hoping, I get to leave, even it were to be by foot.[45]

In British India at the beginning of the 1880s, within the framework of forward policy, political and military control was established over the northern territories adjacent to the frontiers of India, between the Hindu Kush and Karakoram. The British needed Grombchevskii. The emergence of a Russian officer, unaware of the intricacies of policy in the Hindu Kush, was a gift of fate, a chance that the British could not afford to miss. If Grombchevskii had not existed, he would have had to be invented. Lord Curzon identified him as the 'stormy petrel of Russian frontier advance'. The British needed an excuse to go north and Grombchevskii presented this occasion after appearing in Hunza in military uniform, accompanied by Cossacks.

The fate of an independent Hunza was predetermined. In 1889–91 the British accumulated their military forces in Gilgit and accelerated the construction of roads to the borders of Hunza. In the anticipation of a fast march by British troops, Safder Ali Khan sent envoys to the Governor-General of Turkestan, requesting protection from Russia, and urging the dispatch of weapons whilst referencing Grombchevskii's 'promises'. Safder Ali Khan's political mission to Tashkent proved to be unsuccessful. Russia refused to interfere in the affairs of Hunza primarily from fear of aggravating relations with Britain. Safder Ali

45 Ibid., p. 120.

Khan appeared doomed and his fall imminent. It did not take long, on December 1, 1891 a strong expeditionary force of British troops with artillery under the command of Colonel Durand invaded Hunza. After a series of brief skirmishes, the major stronghold of Hunza was seized, and on December 20 the entire territory came under British control. Safder Ali Khan and his entourage fled to Kashgar.

With the conquest of Hunza, the British significantly strengthened their political and military influence in the territories adjacent to the Pamir, the principalities of the Hindu Kush and West China. At their disposal were the shortest and most convenient routes to Kashgar and the Pamir. Seizing Hunza made Russian authorities concerned for the safety of Russian Turkestan and they undertook urgent efforts in protecting Russian interests in the Pamir. These political and military conditions led to the emergence of the Pamir crisis 1891–95, which yet again only aggravated Anglo-Russian relations in Inner Asia.

On November 26, 1888 the Governor-General of Turkestan sent a wire to the Chief of the Main Staff of the safe return of Grombchevskii's expedition. Adjutant General Obruchev sent a wire to Tashkent, requesting that Staff Captain Grombchevskii be assigned to St. Petersburg along with his collected expedition materials.[46] At the end of December 1888, Grombchevskii arrived in St. Petersburg and met with the War Minister. Vannovskii, seemed to be pleased with Grombchevskii's report results. It is possible that during this meeting Grombchevskii raised the prospect of a new expedition beyond the Hindu Kush. Grombchevskii's stay in St. Petersburg was extended until May 1889 in order to process field data. In March of the same year, he was promoted to the rank of Captain for his 'excellence' in service.

At the beginning of 1889 Grombchevskii's name becomes sufficiently well known in the circles of the Geographical Society. On January 10, Grombchevskii together with the Governor-General of Turkestan Rozenbach were present at the Council of the IRGO meeting. Vice-Chairman of the Society P.P. Semenov put together a brief summary of Grombchevskii's work during the expedition in Hunza and noted that the officer 'carried out the entrusted mission brilliantly', having himself been 'convinced from conversations with the traveller'. The Council decided – 'whilst waiting for the results of the materials obtained by Grombchevskii, to award him for the successful completion of his mission a small gold medal'.[47]

---

46 RGVIA, F. 401, Op. 4, D. 46 (1888), l. 32: A copy from a wire of Chief of the Main Staff to the acting Commander-in-Chief of the Turkestan Military District, 26 November 1888.

47 Zhurnal zasedaniia Soveta IRGO, 10 January 1889, in *Izvestiia IRGO*, T. XXV, Vyp. I, 1889, p. 5.

Thus, the award from the Geographical Society was an advance on the basis of what Grombchevskii had reported to Semenov of his works and route surveys. However, he perceived this award as undeservedly small, taking into account the risks of the expedition and the volume of the accomplished work. It should be noted that Society's small gold medals were in status slightly higher than silver medals.

During his trip to St. Petersburg Grombchevskii was actively involved in the life of the Geographical Society. On February 1, 1889, he reported the results of his expedition to Hunza at the general meeting of the Geographical Society.[48] His report was accompanied by objects that were part of the Hunza ethnographic collection, which he would later donate to the museum of the IRGO, as well as the collection of photographs of Hunza. Grombchevskii's zoological collection was donated to the Zoological Museum of the Imperial Academy of Sciences. Commenting on the collection, the Zoological Museum Director Academician A.A. Strauch said that it 'proved to be very interesting with perfectly preserved specimens of lizards and rabbits skin. The birds were affected, unfortunately, as they weren't carefully laid; therefore it is hard to say whether we can succeed in preparing them for research'.[49]

Recently selected as a full member of the IRGO, Grombchevskii was already recommending new candidates to the Society. Together with V.I. Roborovskii, he proposed to the members of the Society the election of Ia. Liutsh, a secretary of the Russian Imperial Consulate in Kashgar. Grombchevskii met Liutsh during a duty journey to Kashgar in 1885, Liutsh had helped him organise the trip to Khotan. It would only be a matter of time before Grombchevskii, having ended up in a desperate situation, would again request Liutsh's help.

Strolling on the Roof of the World: The Second Inner Asian Expedition (1889–90)

During his stay in St. Petersburg, Grombchevskii was already developing a new plan for an expedition beyond the Hindu Kush. Now, on his side were fame, personal connections in the Geographical Society and the Main Staff as well as high patrons of noble aristocratic families. The main lobbyist for the new expedition was P.P. Semenov, Vice-Chairman of the IRGO. In a letter addressed to

---

48 The list of items in the collection is contained in the minutes of the Council of the IRGO, 30 January 1889. For details, see *Izvestiia IRGO*, T. XXV, Vyp. I, 1889, pp. 16–17.

49 Zhurnal zasedaniia Soveta IRGO, 13 May 1889, in *Izvestiia IRGO*, T. XXV, 1889, Vyp. I, 1889, p. 58.

the Chief of the Main Staff, he pointed out that the main purpose of the future expedition would be to 'study the mountain valleys of the Hindu Kush and in particular the independent holdings of Kafiristan, which are located on its northern slopes, as well as the frontier territories which are located within the influence of Russia, India and Afghanistan, areas which previously had not been uncovered by a single European'.[50] The willingness of the Geographical Society, in the face of its authoritative figures to yet again be drawn into dubious affairs is simply staggering.

It should be noted that the Geographical Society's Central Asian direction was falling on hard times. Przhevalskii, had recently passed away, and so with his passing ended the golden age of exploration in Inner Asia, as a result, the Society's credibility both in Russia and overseas put into question. The Society desperately needed an influx of fresh forces, a new name, which could continue traditions laid down by Przhevalskii.[51] Grombchevskii's arrival at the IRGO, in these hard times was seen by the Society's senior executives as a gift from above, a chance they could not pass. Semenov was enchanted by Grombchevskii, his personal and professional qualities – 'it is Grombchevskii who possesses this rare combination of qualities in one person that are needed in order to achieve this goal'.[52] These hypnotic qualities so compelled the once experienced Semenov, that he no longer recognised the clear signs of folly in

50 RGVIA, F. 401, Op. 4, D. 46 (1889), ll. 2–2 ob.: A confidential letter of the Vice-Chairman of the IRGO to Chief of the Main Staff, 26 February 1889.

51 The sudden death of N.M. Przevalskii revealed the absence of figures, on par with Przhevalskii in personal and professional qualities, capable of undertaking a large-scale exploration of Inner Asia. A document of great interest which perfectly reflects this 'personnel crisis', is P.P. Semenov's letter to the Adjutant General N.N. Obruchev in which he mentions that the Council of the IRGO, having not enough candidates, delegated military authorities a right to find a suitable candidate for the role of leader of the abortive 5th Inner Asian expedition of Przhevalskii. Semenov outlined: '<...> It is essential, that the head of the expedition should be from the military, and at the same time be a person of academic knowledge with a wide experience in travelling, directing expeditions and undertaking surveys as well as astronomical observations. There is no doubt that our War Ministry has a number of many gifted servicemen, those who meet the most stringent requirements. The Council of the IRGO with a deep sense of respect recalls the outstanding geographical works and contributions to the study of Inner Asia, both by officers of the General Staff and the Army, such as A.N. Kuropatkin, N.I. Grodekov, A.V. Kaulbars, M.V. Pevtsov, and many others. The Council of the Society welcomes the appointment of a person of such qualities as head of the expedition'. See RGVIA, F. 401, Op. 4, D. 53 (1888), l. 2 ob. – 3 ob.

52 Ibid. Grombchevskii never concealed P.P. Semenov's crucial role in his career, who 'has discovered him to be well prepared for a long-distance reconnaissance'.

Grombchevskii's new enterprise. It is unclear, why the Geographical Society began implementing this new large-scale expedition with such haste at a time when the previous expedition had yet to be concluded – materials had not yet been developed and official reports not published.

There was another influential patron acting behind the scenes, prior to Grombchevskii's expedition to the Hindu Kush – Prince Esper E. Ukhtomskii, a prominent Russian diplomat, Orientalist and publicist. Prince Ukhtomskii was fascinated by Buddhism, Oriental mysticism and the idea of Russia's special mission in the East.[53] He was rightfully called one of the first Russian ideologists of Eurasianism. Prince Ukhtomskii was close to Tsesarevich[54] Nikolai Alexandrovich, heir to the Russian throne, and held a certain amount of influence over the Tsesarevich. Prince Ukhtomskii greatly contributed to organising the Tsesarevich's trip to the East in 1890–91, The Great Eastern Journey, having personally accompanied the future Russian Emperor on the trip and published a richly illustrated travelogue as a result of his travels.

Grombchevskii's subsequent meeting and acquaintance with Ukhtomskii would be significantly important in advancing his proposed project and expedition to the Hindu Kush. This connection highlighted the Imperial family and the heir's involvement in assisting with equipment and financing of the Grombchevskii's expedition. This also explains the reason for how quickly matters were finalised and approved by the Ministry of Foreign Affairs and the War Ministry. Grombchevskii's ambitious plan was to prove Russia's power and influence on the Indian borders, which surprisingly coincided with the heir's views and aspirations for Russia. Prince Ukhtomskii supported the idea that should the expedition prove to be successful, it should be named after the heir. This request was completely unexpected for Grombchevskii, but nevertheless changed the general disposition of organising the expedition in his favour.

Grombchevskii's second expedition beyond the Hindu Kush was organised in a very short space of time. He insisted on carrying out the expedition on such a tight schedule because of the 'successful unfolding political circumstances' – the Afghans had retreated from Badakhshan and the Pamir principalities, of Wakhan, Shugnan and Roshan. Grombchevskii heard news of this whilst still in Hunza and had tried to move out in the fall of 1888. Whilst in Kashgar, however, he did not receive permission from Consul Petrovskii to do

53 For a good summary of Prince Ukhtomskii' views and his role in forming of the Russian Eastern Policy, see David Schimmelpenninck van der Oye, *Toward the Rising Sun. Russian Ideologies of Empire and the Path to War with Japan*, DeKalb: The Northern Illinois University Press, 2001.

54 Tsesarevich – the title of the heir apparent or presumptive in the Russian Empire.

so. Rumours of the Afghans retreating had circulated in Kashgar and were associated with the ongoing turmoil in Afghanistan – the rebellion of Ishaq Khan in Northern Afghanistan in the summer of 1888. However, Grombchevskii had no reliable source of information as to the real situation in the region along the route of his future expedition. Whilst discussions took place in St. Petersburg regarding Grombchevskii's second expedition, Ishaq Khan's revolt was suppressed and the political situation in Afghanistan stabilised. The power of Abdur Rahman Khan, Emir of Afghanistan, had not been fully restored in the Badakhshan Province of Afghanistan, moreover Afghan outposts had advanced far into the Pamir principalities of Wakhan, Shugnan and Roshan.

Grombchevskii's objective for the second expedition was Kafiristan (modern Nuristan province of Afghanistan), an Afghan vassal domain on the southern slopes of the Hindu Kush.[55] Grombchevskii's desire to become the first European to venture into Kafiristan, was consistent with his character, but, just as in the case of Hunza, his actions were motivated by vanity and were not in anyway based on reality. The first European to visit Kafiristan was in fact British surveyor William McNair of the Great Trigonometrical Survey of India. In 1883 he conducted a secret reconnaissance to and from Kafiristan and mapped a large portion of its territory, gathering important political and geographical information.[56] In the summer of 1885 Colonel Lockhart's reconnaissance party visited the eastern part of Kafiristan.

Whilst planning the new expedition, Grombchevskii was aware of how difficult it would be to achieve his plan, a result of a set of fortunate circumstances. Hence his phrasing – 'to slip over the Hindu Kush'. Had he faced any political complications en route, it would have resulted in the complete collapse of the whole expedition, which had already taken up so much of his personal efforts as well as public funds. It was an enterprise that was backed by the Imperial family, the Geographical Society, the War Ministry, as well as Russia's political prestige in Asia and they all depended on its success. As a result, Grombchevskii decided to play it safe and plan out two alternative routes for the expedition, a primary one and a backup just in case. This multivariance is already noted in Semenov's letter to the Chief of the Main Staff, in which he states that in such case that

55 Kafiristan became an integral part of Afghanistan in 1896 after boundary delimitation between British India and Afghanistan ("The Durand Line").

56 See OIOC, PFI/39: William MacNair. Confidential Report on the Explorations in part of Eastern Afghanistan and in Kafiristan during 1883; enclosures 15 and 19, India, 8 January 1884.

> … advancing into Kafiristan is totally impossible, Grombchevskii's expedition could still greatly benefit science by exploring the northern slopes of the Hindu Kush, by circumventing Roshan through the Alichur Pamir and Great Pamir to the Baroghil Pass and from there through the Taghdumbash Pamir to the sources of Raskam Darya, keeping as close as possible to the main ridge of Mustagh.[57]

Grombchevskii was aware of the changing military-political situation in Afghanistan and the Pamir before his departure from St. Petersburg.

> The task with which we are faced ahead is difficult, shortly before my departure from St. Petersburg we received news that Ishaq Khan had been defeated and had fled to Bukhara, and Emir Abdur Rahman Khan had arrived into Mazar-i-Sharif accompanied by a strong force. These news caused a great deal of concern for Russian authorities, for the security of the Bukhara borders, and for the protection of Bukhara [Russian] troops were redeployed from Turkestan.[58]

How in such circumstances the Ministry of Foreign Affairs and the Main Staff agreed on Grombchevskii travelling through Afghan possessions and into the Hindu Kush, remains a mystery. Cancelling the expedition plans would have seemed like the logical thing to do, especially against the backdrop of tensed military and political conditions. However, in St. Petersburg it was considered inconvenient to call off the expedition, which was supported by the Imperial family. In fact, the authorities drew into the unpredictable adventure. On March 3, 1889, the War Minister, presented Alexander III a report on Grombchevskii's 'mission into Kafiristan'. The following day a Supreme order assent for the expedition was issued.

Grombchevskii reported on the cost of the expedition in his diary.

> In total, the expedition was allocated 5,000 rubles by the Geographical Society and in addition to this, Count Vorontsov-Dashkov, a minister of the Imperial Court of Cabinet of His Imperial Majesty allocated 100 pieces of 5 rubles gold coins for the acquisition of gifts for the expedition.[59]

57 RGVIA, F. 401, Op. 4, D. 46 (1889), l. 2 ob. – 3.

58 RGVIA, F. 401, Op. 4, D. 46 (1888), l. 63.

59 Grombchevskii 2015a, p. 186.

The expedition expenses should also include allowances, both Grombchevskii's 2,184 rubles per year as well as his men's allowances.

Another thing not made public – the Geographical Society in their official documents understood that 'in the case of success' the expedition is to be named after the Tsesarevich. The fact the Geographical Society concealed these two interesting details regarding the finances and naming of the expedition shows the unfortunate consequence of Grombchevskii's ambition for the Hindu Kush project.

The expedition was expected to last between seven months to a year. The scientific work of the expedition consisted of carrying out route surveys and compiling descriptions of covered routes, determining geographical coordinates and altitudes, conducting meteorological observations and assembling zoological, geological and ethnographic collections. Besides Grombchevskii, the expedition included seven Cossacks, three employed *jigits* and a preparator. Including the latter was not a coincidence, Grombchevskii needed an experienced specialist, as his previous experience compiling zoological collections proved to be not so successful. As his assistant, Grombchevskii elected M.L. Konrad, who was supposed to take part in Przhewalskii's failed 5th Inner Asian expedition (1888). Among the Cossack convoy was *uriadnik*, or subaltern officer, Fedor Koziakaev, who had accompanied Grombchevskii into Hunza in 1888. In order to equip the expeditionary unit, Semenov appealed to the War Ministry with a request to allocate twelve Berdan dragoon rifles with 2,500 rounds of ammunition and twelve revolvers with the same number of rounds.[60]

The Hindu Kush expedition was not only well set financially and technically, but had a wide range of political, military, and geographical tasks. It even held a symbolic value – its purpose was to demonstrate the increased power of Russia in Asia and the ability of the Empire to stand up for its interests. Grombchevskii's fortuitous hour had come.

> March 29 I had the honor of meeting in person the Emperor and the Tsesarevich, with me I brought my photographic album, taken during the expedition to Hunza and Raskam.[61]

Grombchevskii's Second Inner Asian expedition took place from 1 June 1889 to 20 October 1890. After leaving Margelan, Grombchevskii tried crossing the Trans Alai Range via the western point of the Alai Valley in order to reach the Pamir. He failed to do so due to the high snow in the Trans Alai Range. He

60 RGVIA, F. 401, Op. 4, D. 46 (1889), l. 17: A memorandum of the deputy of General of Artillery (General Feldzeugmeister) to Chief of the Main Staff, 30 March 1889.

61 Grombchevskii 2015a, p. 186.

therefore ventured through the Bukhara possessions of Karategin and Darwaz and carried on through the Panj River valley. In early July, he arrived into the capital of Darwaz, Kalai-Khumb, and there learned that Shugnan, which he intended to cross was occupied by the Afghans. It is here that Grombchevskii realised that it was impossible to cross over into the Hindu Kush. The political situation in the region was in fact very different to what Grombchevskii reported to the Council of the IRGO. From Darwaz, the expedition passed along the right bank of the Panj River to the boundaries of Roshan where Grombchevskii sent a letter to the Afghan authorities requesting permission to pass into Kafiristan.

The Afghan frontier officials responded that they themselves do not have the authority to let the Russian traveller through; his request was forwarded to Kabul, where it would be examined by the Emir. At the same time, the Afghans strengthened their posts along the expedition's assumed route. The presence of the expedition in Roshan, under the condition of open hostilities between Syed Akbar Shah, a local ruler, and the Afghan force, was very dangerous. The encampments, located at the confluence of the rivers Vanj and Panj, were in clear visibility of the Afghan forces and could potentially be subject to a surprise attack at any moment. Furthermore, problems of a political nature arose.

In an effort to avoid a conflict with the Afghans as well as any possible political complications, Grombchevskii decided to bypass the territory of Shugnan which was occupied by the Afghans, and enter from the Eastern Pamir.

> The hostile attitude of the Afghan commander towards the expedition should not be questioned. Anticipating the possibility of an armed conflict with the Afghans and not wanting to cause trouble for our Foreign Ministry, I carried on forward into the depth of the Vanj River Valley for the Bukhara stronghold of Kalai-Rokhorf.[62]

Continuing on through the Sytargi Pass, the Hingob River Valley, Karategin, the Great Alai Valley and a number of other crossings, on August 14 the expedition arrived at the Kudara River source, the summer pastures of the Pamir robber Sahib Nazar.

In the Pamir, Grombchevskii found himself between a rock and a hard place – to the west his actions were observed by the Afghans, whilst Chinese posts were located to the east. A representative of the commander of the Chinese troops arrived at their camp, located near Lake Bulunkul in the Pamir and advised them that they were in fact beyond the line of the Afghan posts in the

62 NA RGO, F. 45, Op. 1, D. 6, l. 4.

FIGURE 5.5 A road along the gorge of the Panj River. PHOTO BY GROMBCHEVSKII, 1889.

Western Pamir and should proceed on into Kashgar. The expedition moved on through the Hargush Pass and down into the valley of the Pamir River, where they received news of movement of the Afghan force in their direction. Grombchevskii reported on the political condition to his superiors in St. Petersburg.

> We were in a difficult position not knowing whether the Afghans would allow us to proceed into Kafiristan, or whether we would have to carry out further works within the limits of China, either way, I had to avoid a conflict with the Afghan and Chinese troops at all costs, so as not to provoke either party. Meanwhile, with recent heightened frustrations and heated passions, as a result of recent military actions, a meeting with the groups could result in us having to use self-defense.[63]

In St. Petersburg, of course, the officials had no idea that Grombchevskii, who was in fact set off to travel beyond the Hindu Kush for purely 'scientific purposes', was in fact drifting dangerously and unpredictably between the lines of

63 RGVIA, F. 401, Op. 4, D. 46, l. 74 ob.

military and political conflict, and that Russia's whole policy in Central Asia at this time was balancing on the edge of war and peace.

Grombchevskii spent almost one and a half months waiting for a response from the Afghan Emir, maneuvering between Chinese and Afghan posts. At the beginning of September, the expedition moved to the Aksu River source, where Grombchevskii received news of Emir Abdur Rahman Khan's military campaign in Kafiristan and the occupation of the upper reaches of the Karakash River and the Chinese stronghold of Shahidulla Hoja by the British (the latter proved to be false). From Aksu, Grombchevskii sent a letter to the ruler of Chitral Aman ul-Mulk requesting to pass through his possessions in Kafiristan, but was rebuffed under the pretext of inability to ensure safety. On October 1 Grombchevskii received a letter from the governor of Afghan Badakhshan elucidating his refusal to let him into Kafiristan. In this letter Grombchevskii was advised to 'cease strolling on the frontiers of Afghanistan' and return home. As a result, continuing the expedition according to plan became impossible and Grombchevskii had nothing other to do than to fall back on his original mission, that is to study the Raskam Darya basin. Grombchevskii reported in a letter to P.P. Semenov:

> For failing to fulfil my mission, I have resorted to exploring the western marginal lands of the Pamir, after that I set off to travel east to the Taghdumbash Pamir and on to Raskam Darya to complete what was left unfinished by [my] previous expedition.[64]

It should be noted that at this stage there was still no notion of moving into northwestern Tibet in Grombchevskii's plans. The next alternative of the expedition was for him the least desirable. The scenario would potentially deprive him of creating a name for himself in the exploration of Kafiristan. The change of the initial expedition itinerary affected the political significance of his undertaking and ruined his dream of associating the expedition with the Tsesarevitch's name. In fact, already the beginning of October 1888 Grombchevskii's expedition suffered a complete fiasco and he had no choice but to save his reputation. The entire subsequent history of his Hindu Kush expedition will be a series of chaotic, almost senseless, running around in search of a 'golden' route, which if he were to pass, he would be making a brilliant discovery. But the bitter truth was that he had no choice but to trek an already sufficiently known region, known as the result of the work of British

64 Ekspeditsiia kapitana Grombchevskogo na Pamiry, in *Izvestiia IRGO*, T. XXV, Vyp. V, 1889, p. 426.

expeditions and was far from *terra incognita* of geographical science.[65] He seemed trapped, but had no choice but to continue.

The Council of the IRGO had no choice either, the promising expedition which had so brilliantly started, now seemed doomed to turn into a disaster. The Society decided to pretend as though nothing really serious had happened.

> Such a turn of events was foreseen by the Society whilst planning Grombchevskii's expedition, therefore taking into account that B.L. Grombchevskii made every possible effort to achieve the outlined goal; the Council expressed its sincere appreciation for his labours and allocated him new finances to continue the expedition along an alternative route. The works of the expedition cover the area next to the right flank of Colonel Pevtsov's[66] exploration area.[67]

Grombchevskii loved the method of 'linking explorations', and would frequently in his expedition plans make reference to the need to link his routes with those of other explorers (Przhevalskii, Pevtsov and British travellers). On November 8 in the defile of the Bazar Dara River, Grombchevskii found traces of Colonel Pevtsov's expedition. Here he made the decision to link his own surveys with those of Pevtsov's expedition. From this point on, according to Consul Petrovskii, Grombchevskii began aimlessly 'chasing Pevtsov'.[68]

After passing the Beik Pass in the Taghdumbash Pamir, Grombchevskii began preparing for the winter journey along the Raskem Darya basin. In early October, British officers, Major C.S. Cumberland and Captain H. Bower[69] while

65 The area of Grombchevskii's proposed route has been examined in sufficient detail during Sir T.D. Forsyth's British mission to Eastern Turkestan (1873–74), and the travels of British political agent Ney Elias (1886), Captain Francis Younghusband (1886), and French explorers Gabriel Bonvalot, Guillaume Capus, and Albert Pépin (1886–87).

66 Herein the Tibetan expedition under the command of General Staff Colonel M.V. Pevtsov (May 1889 – January 1891). At that time Pevtsov's expedition had progressed to Khotan.

67 Vesti iz ekspeditsii B.L. Grombchevskogo, in *Izvestiia IRGO*, T. XXVI, Vyp. I, 1890, p. 85.

68 A letter from N.F. Petrovskii to baron F.R. Osten-Saken, Kashgar, 14 May 1892, in N.F. Petrovskii, Turkestanskie pis'ma (Туркестанские письма), Москва: "Памятники исторической мысли", p. 226.

69 Major C.S. Cumberland and Captain H. Bower visited Kashar in 1889. Captain Bower, an officer of the Intelligence Department for the British Indian Army at Simla, gained fame thanks to the discovery in Kucha of one of the oldest extent books, dating from the 5th century A.D. Written in Sanskrit in the Brahmi alphabet, it was named the Bower manuscript, and paved the way to a new field of linguistics. Major Cumberland at the time of the events was a retired officer.

FIGURE 5.6 A hunter from the Pamir. PHOTO BY GROMBCHEVSKII, 1889.

awaiting the arrival of Younghusband's expedition, visited the Russian expedition camp in a recessed area along the Chatyr Tash Valley. The British officers told Grombchevskii of the imminent arrival of Captain Younghusband. On receiving this news, Grombchevskii cordially invited his English rival to dine with him in the Russian camp. On October 11, 1889, in the narrow valley of Kaindy Auzy[70] on the Taghdumbash Pamir a meeting of two Captains occurred, two heroes of the Great Game in High Asia.

> Captain Younghusband travelled from India to Hunza. This young man made his name by daring to journey from Peking through all of China to Kashmir. He was accompanied by a small escort of Bengal soldiers, pundits and numerous servants. Our encounter was friendly and since my expedition arrived the night before Captain Younghusband, he was our

70 On British contemporary maps – Khaian-aksai.

> guest for almost three days. Both expeditions were an eclectic mix of 20 nationalities.[71]

Younghusband was on a secret mission to Chinese Turkestan and the Taghdumbash Pamir and was informed that somewhere in the Raskam Darya Valley, which separates the Pamir and Kunlun, a Russian officer was conducting topographic surveys. Grombchevskii mentions Younghusband's acquaintance in his diary.

> He was fully acquainted with my expedition, and showed me Veniukov's letter printed in the Proceedings of the Royal Geographical Society in March of this year, in which Veniukov reports all the details of my trip, extracted from my letter to him. This was a great indiscretion on Veniukov's part, especially as I had specifically asked him not to publicise the confidential content of the letter.[72]

During the joint stand, Russian and British officers exchanged views on frontiers issues and Anglo-Russian political relations. Many of Grombchevskii's frank opinions would be later used by Younghusband to justify British policy towards the Pamir. Peter Hopkirk, the popular British journalist, author and historian, in his work *The Great Game* describes several days of British and Russian joint stand at Kaindy Auzy. Based on Younghusband's memoirs, Hopkirk noted that the 'slightly tipsy' Grombchevskii openly spoke of Russian plans to invade India and offered Younghusband a secret Russian map of the Pamir and adjacent territories, based on data from the latest Russian reconnaissance and surveying.[73]

---

71 Doklad kapitana B.L. Grombchevskogo o puteshestvii v 1889–1890 gg., in *Izvestia IRGO*, T. XXVII, Vyp. II, 1891, p. 108.

72 NA RGO, F. 45, Op. 1, no 6, l. 412. Here Grombcheskii admits inaccuracy. An article published by M.I. Veniukov gave accounts of Grombchevskii's travel into Hunza and did not include his new expedition. It did not contain anything new, except what had been published in publications of the IRGO. Therefore, Grombchevskii's reference to certain confidential information contained in the letter, which had allegedly been disclosed by Veniukov, is not entirely accurate. In addition to the British publication, Veniukov published several short articles on Grombchevskii's travel in the Bulletin de la Société de Géographie (Paris), which were also based on publicly available information borrowed from the Russian press. For the French publications, see '*Bulletin de la Société de Géographie*', T. 11, 1890, p. 47; T. 12, 1891, pp. 417–420; T. 13, 1892, pp. 406–407.

73 Hopkirk 1992, p. 455.

Hopkirk's version of events, was perceived by Russian researcher A.V. Postnikov, who was studying the history of the Pamir boundary delimitation, without criticism. Having retold Hopkirk's semi caricatural story about the 'significant amount of vodka drunk by Grombchevskii together with the Englishman'(?!), and concluded that

> ... Grombchevskii committed an act which highlighted his lack of understanding towards the nature and importance of Younghusband's mission and the significance of the cartographic documents, as he gave the English officer the most up to date map of the upper Amu Darya with handwritten marks and corrections, used <...> by the British as a basis for the design of the Pamir partition between China and Afghanistan, to the detriment of Russia's interests. Let us note that Grombchevskii did not mention handing over this map to Younghusband in his diary.[74]

There is an opinion that Grombchevskii had committed a malfeasance, bordering on treason. However, the map that was handed over by Grombchevskii had been openly published by the IRGO in 1886, more than two years before the described events. This piece of cartography was published under the title *The Map of the Upper Amu Darya was compiled in 1885 by the Military-Topographical Department of the Main Staff according to latest available data.*[75] The British no doubt knew about this map and had it at their disposal as they were closely monitoring all cartographic material on Asia that was published in Russia. Moreover, the IRGO regularly sent all of its publications to London and exchanged materials with the Royal Geographical Society. A careful study of Grombchevskii's map published in Postnikov's work, proves this map does not relate to Grombchevskii's 1889–90 expedition. These were routes from his previous expeditions in 1885 and 1888, which too had been published by the IRGO. Neither the map nor handwritten notes were a secret, and were given to Younghusband as a sign of attention and courtesy. This practice was the unique etiquette, the unwritten code of conduct amongst travellers from different countries who met on no-man's-land of High Asia. It should be noted that Younghusband also presented Grombchevskii with a British map of the Taghdumbash Pamir, which Grombchevskii references to in his reports. At Grombchevskii's request, Younghusband made astronomical determinations of Kainda Auzy using his own chronometer as Grombchevskii's had failed.

74 Postnikov 2005, pp. 237–238.

75 See *Izvestiia IRGO*, T. XXII, Vyp. II, 1886, prilozhenie (supplement).

Grombchevskii had both the Russian and British expeditions photographed together. This miraculously preserved photograph would later on become almost a cult image and a classic photographic chronicle of what Count Karl Nesselrode, Imperial Russian Foreign Minister, called 'tournament of shadows' in Asia. British historian John Keay characterised this as the most distinctive and memorable image of the entire Great Game, 'a remarkable record of a celebrated and desolate encounter with which little can compare; imagine the excitement if a photograph of Stanley's meeting with Livingstone ever came to light'.[76]

Before the meeting with Younghusband, with winter on its way, Grombchevskii intended to set out to Kashmir to spend the winter months there, and by early spring cross through the Tibetan Plateau and be on the way to Polu in Southern Kashgaria. During his meeting with Younghusband, Grombchevskii expressed his intention of marching to Kashmir. Younghusband was surprised by such a plan and expressed his doubts as to its feasibility. Nevertheless, he advised Grombchevskii to write a letter to the British resident in Kashmir, Colonel Parry Nisbet, requesting his expeditionary force be permitted to winter in Ladakh. Younghusband himself volunteered to hand deliver the letter to Ladakh, as he was headed there. As expected, Grombchevskii received a categorical refusal from the British resident.

During Grombchevskii's meeting with Younghusband another episode had occurred, which had dire consequences for the Russian traveller. In an effort to prevent the reconnaissance of a caravan road from Polu to Ladakh, Younghusband suggested a shortcut from Shakhidulla Hoja (Chinese outpost in Raskam) to Polu through the Tibetan upland. This direction greatly interested Grombchevskii since according to Younghusband, this region had not yet been surveyed. Younghusband not only knew this route was completely impassable in the winter months, but also convinced the local Kyrgyz to lie and confirm its existence to Grombchevskii. Younghusband notes this period, recalling his second meeting with Grombchevskii in August 1890.

> He survived some hard times, and traversed many adversities, for he had attempted the Karakoram Pass in the middle of December, and then, in the dead of winter passed on eastwards to the edge of the high Tibetan Plateau. When I recall how inadequately he and his party were equipped with camp equipage, and under what conditions they were travelling,

76 Keay 1979, p. 190.

> I cannot help admire the stubborn perseverance with which the Russian explorer fulfilled his task.[77]

However, Younghusband made no mention of his admiration for Grombchevskii in his report to his immediate superior, Colonel Nisbet. He reported that he directed the Russian along a 'path of absolutely no importance, leading from nowhere to nowhere, that passes over very elevated plateaus and mountains, without grass or fuel, which if he were to cross in winter would cause him extreme hardship and a loss to his party'.[78] Grombchevskii always felt sympathy towards Younghusband, and it seems he never did see the whole truth, and most probably hardly thought of who the real culprit was who had brought him to disaster and failure.

Grombchevskii attempted to penetrate a region, which was not part of his alternative route – northwestern Tibet. This feverish search for new routes proved to be very dangerous, forward planning was replaced with ill-considered improvisation. The winter trekking to northwestern Tibet proved to be a catastrophe.

> The expedition on December 14 set out directly east, intending to cross the desert of northwestern Tibet and reach its inhabited parts. On December 26, after passing the Karakash River, we climbed to the Tibetan Plateau at a height of 17,000 feet. Not only were there no signs of fuels but all the springs had completely frozen over. The hardships the expedition faced exceed any possible limits. It was minus 33–35°C during the day, how far temperature dropped by night – is unknown, as our low-reading thermometer ruptured. <…> In the last four days our horses had nothing to drink and were exhausted from fatigue. According to the information I had acquired by now we should have encountered warm ice-keys. We reached these on the fifth day, having walked from four in the morning to two at night, having lost 28 horses in the process. Circumstances forced us to dispose of our horse-loads and collections, in short, leaving only the bare necessities.[79] To round everything off a blizzard ensued. The desert was covered deep in snow which made it almost impossible to navigate.

77 Younghusband, Captain Frank E., *The Heart of a Continent: A narrative of Travels in Manchuria, across the Gobi Desert, through the Himalayas, the Pamirs, and Chitral, 1884–1894*, Delhi: Book Faith India, 1994, pp. 292–293.

78 Cited by Patrick French, *Younghusband. The Last Great Imperial Adventurer*, London: Harper Collins Publishers, 1994, pp. 77–78.

79 The members of the expedition managed only to take diaries, documents, photography, silver, and a small amount of ammunition and food. All packs were piled together and covered with felts and stones.

> Envisioning the inability to move forward without horses and supplies, I decided to turn back.[80]

Grombchevskii had come to realise the full tragedy of his situation.

> The thought, that my carelessness and desire for new experiences, for which men might die, men who had followed me in the selfless faith of my experience and foresight.[81]

February 10, 1890, the expedition, 'physically detuned and morally killed' (Grombchevskii's words) with incredible difficulties proceeded from Shahidullah Hoja into the village Kilian in the Southern Kashgaria. Grombchevskii reported in a letter to P.P. Semenov on his disastrous glimpse into northwestern Tibet.

> The situation in Kilian was truly deplorable, our financial resources had finally dried up; we had brought only eight horses, who could not cope with the road ahead, our expedition gear was either partially burnt or abandoned, or otherwise become unfit to wear over the last nine months of use. In short, it seems that the expedition is over, the only thing left to do is thank God if we eventually made it to Margelan.[82]

Grombchevskii realised the dire situation he was in, which could put an end to his entire military and scientific career. At the beginning of April 1890, his assigned expedition term, approved by the War Ministry, was near expiration. In addition to the devastating state of affairs and complete absence of money, Grombchevskii also had time constraints. Nevertheless, with his intrinsic energy and persistence, he desperately undertook all efforts to salvage the situation. He knew requesting the Geographical Society to prolong his expedition, would only be possible if he were to find an alternative source of funding. He proceeds to write a letter to Ia. Liutsh, secretary of the Russian Consulate in Kashgar, in which he retells the sad fate of the expedition and appealed for support. Unexpectedly, an answer arrived from Liutsh. He was willing to privately loan Grombchevskii 4,000 rubles (a huge amount at the time, equivalent to the initial expedition budget), which Liutsh himself had to borrow. This sur-

80 RGVIA, F. 401, Op. 4, Д. 46 (1889), l. 80 ob. – 81: A report of Captain Grombchevskii on the progress of the expedition into northwestern Tibet.

81 NA RGO, F. 45, Op. 1, no 6, l. 629.

82 Vesti ob ekspeditsii B.L. Grombchevskogo, in *Izvestiia IRGO*, T. XXVI, Vyp. IV, 1890, p. 330.

prise of a loan instilled a sense of hope in Grombchevskii in rectifying the situation he was in.[83]

He wrote a letter to Semenov, in which, after describing all of their misfortunes he asked for support to receive permission of the War Minister to extend his expedition assignment until January 1, 1891. For Semenov who was under Grombchevskii's spell and trust, a failed expedition could spell terrible consequences for his social standing. As a result, he had in a way become Grombchevskii's involuntary hostage. Semenov appealed to the War Minister and Chief of the Main Staff with a request to prolong his mission.

Without waiting for a decision from St. Petersburg, starting in Kilian, Grombchevskii set out to the east, heading to the foothills of the northern slopes of Kunlun Mountains, passing through to Polu, and then trekking back up to the Tibetan Plateau. In the village of Pudjia, 50 *verst* south-west of Khotan, he received news that Pevtsov's expedition was wintering in the town of Niya. Pevtsov's camp made a great impression on Grombchevskii.

> From K.I. Bogdanovich,[84] we learnt that the [Pevtsov's] expedition from early October 1889 had camped overwinter in the village of Niya, where they have been for almost five months and intended to remain until mid-April. If we add to this the two months the expedition spent in the Kuk Yar Mountains – the expedition had already been stationary for seven months. They were well set up having rented a big house and well adapted for the Europeans. Even tables and chairs were brought in. Yes! Unlike us, with only some of us having tents, in freezing temperatures below 33–35° C, at a height of 19,000 feet. It seems there are aristocrats and pariahs – coolie workers, everywhere; in the service of pure science ordinary workers were not always well regarded.[85]

Colonel Pevtsov described the meeting with Grombchevskii as follows.

> On March 10, well-known explorer B.L. Grombchevskii arrives in Niya with his associate Mr. Konrad with several Cossacks, they had trekked

83 In a later version of events Grombchevskii presented the case, that he was forced to turn to Liutsh for support only after having received an official approval notice of the Main Staff, – i.e. after 1 April 1890. In fact, Grombchevskii's request for an expedition extension came only after he had secured the money from Liutsh.

84 Karl Ivanovich Bogdanovich (1864–1947) – a geologist, member of Pevtsov's Tibetan expedition.

85 Grombchevskii 2015a, p. 326.

> down from Khotan at the end of February from the upper reaches of the Karakash River. Meeting with a compatriot, who arrived together with Bogdanovich was for us a truly joyful event in this remote country. Unfortunately, on the 15th of the same month we had to say our farewells to Grombchevskii, who was in a hurry and on his way to Polu, from where he intended to make his way towards the region which lies between the upper reaches of Keriya Darya and Yurungkash Darya, in the Tibetan Plateau.[86]

At Grombchevskii's request, Pevtsov tested his astronomical instruments and calibrated chronometer rate. Grombchevskii additionally linked his survey to an astronomical location determined by the Tibetan expedition in Niya and Keriya. He had examined gold mines near Niya, and on his way to Polu had come across jade deposits.

On April 1, 1890, whilst in Polu, Grombchevskii received a telegram from Chief of the Main Staff with confirmation of extending his mission for a further six months. He also found out that the mission was allocated an additional sum by the Council of the IRGO, a total of 2,000 rubles. All of April he spent preparing to move from Polu into the Tibetan Plateau. Meanwhile, the situation surrounding the Russian expedition intensified. The local Chinese authorities under the threat of an armed force demanded Grombchevskii to leave Polu. They in every way possible were trying to impede a Russian march into Tibet. This included destructing the only road through the Kurabe River Gorge and prohibiting the locals supplying the Russians. Grombchevskii began to lose patience and was trying by all means possible to force the locals to restore the destructed road. He no longer excluded the possibility of an armed conflict with the Chinese. The local Chinese administration inundated him with complaints from the Daotai of Kashgar. Liutsh requested that Grombchevskii exercise caution. Grombchevskii was enraged at such advice: 'Liutsh advised me to be cautious, and so on. Instead of following the platonic advice of being cautious, I consider the need to exert vigorous pressure on the Chinese as the only way to save the expedition'.[87] The nightmare of an accumulating debt (by which time Liutsh's debt already totalled 1,800 rubles) as well as

86 M.V. Pevtsov, *Puteshestvie po Vostochnomu Turkestanu, Kun'luniu, severnoi okraine Tibetskogo nagor'ia i Chzhungarii v 1889 i 1890 godakh* (Путешествие по Восточному Туркестану, Куньлуню, северной окраине Тибетского нагорья и Чжунгарии в 1889 и 1890 годах), С.-Петербург: типография М. Стасюлевича, 1895, p. 197.

87 Grombchevskii 2015a, p. 339.

the possibility of a 'disgraced return home' (as recorded in his diary) held Grombchevskii in suspense.

Replenishing supplies, Grombchevskii spent his last available funds on hiring guides and porters in Polu and without the permission of the Chinese authorities on May 6, set out upwards into the Tibetan Plateau. Grombchevskii had not carried out any previous reconnaissance therefore primarily relied on luck to get to Tibet. At the beginning of the expedition, several pack loads had slipped off into the abyss, the expedition lost part of its entomological collection and its entire stock of alcohol. Grombchevskii arrived too early to the Tibetan Plateau, it was still cold and there were no available water sources. All attempts to undertake reconnaissance of the main route resulted in failure, whilst any attempts to conduct astronomical observations were hampered by bad weather. To add insult to injury all the hired guides and porters had left and most of the sheep had died. On May 12 Grombchevskii ventured back to Polu, having spent at least a week on the expedition. The remaining 500 rubles, the expedition possessed, were wasted. Thus the exploration of northwestern Tibet had ended. In the official report to the Main Staff, on the progress of the expedition, Grombchevskii strangely worded his account of the trip into Tibet.

> At the beginning of May we went to Tibet, where the temperature was still freezing below 22 – 24° C. We were more sensitive to the temperature change, as just a few days ago we were in the midst of the heat in the Keriya oasis with temperatures reaching 31° C. Having inspected a substantial part of Tibet and supported the survey with astronomical positions, we ventured down to Kashgar at the beginning of June.[88]

Noteworthy is Grombchevskii's misrepresentation of the length of time spent on the Tibetan Plateau, and his mention that they had explored a 'substantial part of Tibet'.

Whilst in Polu, Grombcheskii received permission from the Chinese authorities to move into Kashmir, thanks to Liutsh's efforts. He had no means to do so however and the mountain passes were still covered in snow. He decided to change the area, from Kashmir to Sarykol. It is possible that Sarykol, which was the last remaining area where Grombchevskii could continue exploring without the need of agreeing on the route with the Chinese authorities. However, given how well Sarykol had previously been surveyed, Grombchevskii's stay

88 RGVIA, F. 401, Op. 4, D. 46 (1889), l. 81: A report of Captain Grombchevskii on the progress of expedition into northwestern Tibet.

was more about passing time than anything. In Sarykol, he moved around a lot and his movements had no clear purpose or visible plan.

In early June Grombchevskii set out from Polu into Khotan and then on to Sanju and Kilian. On their way, strange things happened.

> The population of Char Imam was very unfriendly and imprecise. We attained the right amount of food only by arresting three village elders and several local notables, half of them had to be flogged and sent for food whilst the others remained arrested. The supplies were delivered.[89]

On August 27 Grombchevskii arrived in Yarkand, here he learned that for the past week the town had already been home to Francis Younghusband. He soon received a letter from Younghusband requesting a visit to 'renew acquaintances'.

> Younghusband's expedition was equipped extremely well and rich. Apart from Captain Younghusband the expedition included Monsieur Beech, M. Lennard, and M. Macartney[90] (as their Chinese, Manchu and French interpreter), a Chinese secretary, an Indian doctor, three pundits and many servants, all in a various dress. We were warmly met. In six days Captain Younghusband was due to depart for the Pamir, where he would stay for two months and then spend the winter in Kashgar. Two of his companions were sent via Maralbashi into Yulduz, where they were supposed to hunt. With particular pleasure, I drank my glass of champagne. This was my first glass of wine in nearly two years. Captain Younghusband paid a return visit and we had fun until late at night. He was very interested in the Pamir and looked somewhat confused when I told him about that the telegrams from the *Northern Agency*[91] stating that he had left Simla for Yarkand on a secret mission.[92]

Grombchevskii found out from the Amban of Yarkand, that Younghusband's primary mission was the delimitation of the Pamir. Before his departure from

---

89 Grombchevskii 2015a, p. 343.

90 Sir George Macartney (1867–1945) – a British political agent at Kashgar (1890–1907), Consular-General (1908–1918). Arriving in Kashgar as Younghusband's interpreter, in 1890 he remained there as British political agent. Macartney was chief coordinator of British policy in Xinjiang and the main opponent of N.F. Petrovskii, the Russian Consul in Kashgar.

91 Herein *Severnoe telegrafnoe agentstvo*, the Russian weekly news bulletin (est. 1882).

92 Grombchevskii 2015a, p. 367.

FIGURE 5.7 Captain F.E. Younghusband's expedition whilst staying in Yarkand, Chinese Turkestan. Francis Younghusband is standing in the centre, George Macartney to his left, and Henry Lennard and Richard Beech to the right.
PHOTO BY GROMBCHEVSKII, 1890.

Yarkand Grombchevskii became acquainted with the British officer's working map.

> Younghusband was very interested in the details of the route, and as we both knew French poorly, I had requested the map. Younghusband hastily presented me with a detailed map of the Pamir and India, the sketchily outlined in red ink the proposed border between Afghanistan and China was shown on a ridge along the Aksu River. Thus, the plan of upcoming border delimitation [of the Pamir] was confirmed.[93]

Grombchevskii immediately realised that near the borders of the Russian Central Asian possessions, important events were unfolding. Yet again he felt the need to get involved, unable to pass on such a chance. Having no authority to do so, he moved into the sphere of foreign policy, having not thought about the risks involved or possible consequences.

93 Ibid., p. 368.

> Today we visited the Amban of Yarkand, a dimwitted, but kind man. He in great detail queried me about which part of the Parmir one should consider as Chinese, I hurried to assure him that the Afghans never owned the Pamir, just as the British or the Kashmiris of Shahidulla Hoja, therefore this land is exclusively owned by the Chinese. All this I confirmed with maps.[94]

Indicative in this respect, was another episode. Shortly before Grombchevskii's departure the Amban of Yarkand requested a copy of the map of the Russian Pamir. Grombchevskii refused to make a copy of the map himself; for fear that Younghusband might get suspicious of him intervening in politics. However, he allowed the Amban to make a copy of his own Russian map.

> In the evening the chief secretary came to me, I made him draw the outline of the Pamir and indicated the course of rivers, mountain passes, etc. The Chinese proceeded to decorate this with his own inscriptions, and I firmly hammered into his head that all the land to the right side of the Panj River was the integral property of China.[95]

Grombchevskii having taken it upon himself to arbitrarily and boldly cut up the boundary, had in fact just given away the Eastern Pamir to the Chinese. In fact, Younghusband's secret mission was to undertake this very same issue, to convince the Chinese of their ownership of the Eastern Pamir, which had always belonged to them. The British idea was simple, to pull together the borders of Afghanistan and China on the Pamir, thus eliminating Russia as a possible contender for this strategically important region. By boldly determining what belonged to whom in the Pamir, Grombchevskii played a very dangerous game.[96] Whilst in Yarkand he was not aware that in St. Petersburg, at the highest level, complex diplomatic and military efforts were underway to incorporate the Pamir under Russia. Only a year remained until Russian troops would march under the command of Colonel M.E. Ionov into the Pamir.

Between July and August 1890, Grombchevskii's expedition explored the water-system area of the Tiznaf River and the middle part of the Yarkand

94 Ibid., pp. 367–368.

95 Ibid., p. 368.

96 A.V. Postnikov is of the opinion that Grombchevskii played a very negative role in the initial phase of the Pamir crisis, for he conducted with the Chinese unauthorized talks, with the main idea to convince them of 'China's certain historical rights in the Pamir'. See Postnikov 2005, p. 233.

FIGURE 5.8 Musicians and actors of Kashgar, Chinese Turkestan. PHOTO BY GROMBCHEVSKII, 1890.

Darya, as well as the eastern slopes of the Kashgar Range. From Yarkand he set out to Yangi Hissar, then further on to Kashgar, arriving here on September 16. Grombchevskii very sparingly mentions his meeting with the Russian Consulate, merely mentioning that he was met by Consul Petrovskii. From Kashgar he departed on September 28 and further proceeded to the Russian border. On October 16 on the Chernyshev Bridge[97] a wire was received:

> The expedition arrived safely today in Osh. Having surveyed more than 7,000 *verst*, made determinations of 73 astronomical positions as well as 357 heights. Throughout the whole time meteorological observations were carried out, significant collections, rich in geographical, political and ethnographic material, were compiled, with more than 200 photos. Members of the expedition are cheerful and healthy. Captain Grombchevskii.[98]

97 Since 1862, the head office of the Imperial Russian Geographical Society was located in the building of the Ministry of Education at the Chernyshev Bridge.

98 RGVIA, F. 401, Op. 4, D. 46 (1889), l. 67: Captain Grombchevskii's wire to the Vise-Chairman of the IRGO.

Upon receiving news of the successful completion of the expedition, P.P. Semenov appealed to the War Minister requesting Captain Gombchevskii's presence in St. Petersburg for a personal report.[99] On October 21, 1890 Adjutant General P.S. Vannovskii reported to Alexander III, about the successful completion of Grombchevskii's expedition and its main results.

On December 1890 Grombchevskii arrived in St. Petersburg to personally report to the head of Military-Scientific Committee of the Main Staff on the results of the expedition. While in St. Petersburg, Grombchevskii submitted to the Military-Topographical Department of the Main Staff his route surveys of 7,200 *verst*. He also claimed that the major part of his surveys (5,000 *verst*) covered the area which had not been visited by Europeans.[100] The results of astronomical observations were transferred to Major-General K.V. Sharngorst for further calculations.[101] Hypsographic determinations were transferred to military geodesist, Major-General A.A. Tillo, for further calculations.[102] The meteorological journal was handed over to Professor A.I. Voieikov for processing. All collections (with exception of the geological materials) were transferred to the Imperial Academy of Sciences. The geological collection and nephrite samples were transferred to the Mineralogical Cabinet of the Academy of Mines. Photographs were processed by V.I. Kassen, a photographer of the Imperial Academy of Sciences. Grombchevskii was left with his travel diaries, with the help of which he had to put together his official report on the works of the expedition.

99 RGVIA, F. 401, Op. 4, D. 46 (1889), ll. 68–69 ob.: A letter from the Vice-Chairman of the IRGO to the War Minister, 17 October 1890.

100 This statement was, of course, ill-founded, taking into account the European researchers' state of knowledge of the territories visited by Grombchevskii. His original route survey sketches are preserved in the collections of RGVIA, F. 386, Op. 3, nos. 1295 and 1296.

101 Because of other employment Sharngorst abandoned processing the materials. Calculations were carried out by a military geodesist Lieutenant-General O.E. Shtubendorf and accordingly published, see Shtubendorf [1895]. In the same article on the basis of more precise determinations, obtained by military geodesist P.K. Zalesskii in the Pamir, he corrected the calculations of astronomical determinations previously made by Grombchevskii during the expedition to Hunza in 1888. Shtubendorf 's calculations were related to 52 points only, while in the official report Grombchevskii mentions 73 astronomical points. Another interesting detail is the fact that the astronomical positions identified by Grombchevskii during his second Inner Asian expedition were not included in the important reference work of P.K. Zalesskii, *Polnyi katalog astronomicheskikh punktov Turkestanskogo voennogo okruga i prilezhashchikh k nemu zemel'* (Полный каталог астрономических пунктов Туркестанского военного округа и прилежащих к нему земель), Ташкент: Туркестанский отдел ИРГО, 1914.

102 The calculations are presented in the following work, Tillo 1892.

Приложеніе къ описанію путешествій
КАРТА
ПУТЕШЕСТВІЙ
ПОДПОЛКОВНИКА ГРОМБЧЕВСКАГО
ВЪ ДАРВАЗѢ, НА ПАМИРѢ, ВЪ ДЖИТЫ-ШААРѢ, ВЪ КАНДЖУТѢ,
ВЪ РАСКЕМѢ И ВЪ СѢВ. ЗАП. ТИБЕТѢ,
въ 1885, 1888, 1889 и 1890 годахъ.
МАСШТАБЪ
въ англійскомъ дюймѣ 20 вер.
1/840000
Составлена и литографирована при Турк. Воен. Топогр. Отдѣлѣ въ 1895 году.
Подъ редакціей Подполк. Родіонова.
Ошъ
Дол Ирак
Оз. Кара-куль

MAP 5.2 A map complied by the Turkestan Military-Topographical Department on the basis of Grombchevskii's surveys, 1895.

At the beginning of January 1891, Grombchevskii was received by Vannovskii, and briefed him on the results of the expedition. On January 10, 1891 at a special meeting of the IRGO, Grombchevskii presented the results of his expedition. January 16, he had an audience with Alexander III, who listened with great interest to the report on the progress and results of the expedition. Upon the recommendation of the War Minister and on the supreme pleasure of the Emperor, Grombchevskii was awarded the rank of Lieutenant-Colonel[103] and a lifetime pension of 400 rubles annually. In addition, he was allocated 2,500 rubles 'to cover the loan borrowed in Kashgar'.[104] Various other awards were bestowed upon other members of Grombchevskii's expedition.

Soldier of the Empire: Grombchevskii's Call 'To Stop the British'

On March 14, 1891, Grombchevskii presented his lecture at the Nicholas General Staff Academy of *Our Interests in the Pamir. A Military and Political Sketch*.[105] The subject of the report was the Pamir, for which the Ministry of Foreign Affairs and the Military had already formulated new policies and indicated that Gromchevskii's focus of interest had shifted from Hunza and the Hindu Kush. He had never before spoken in public on matters such as geopolitics, foreign relations and strategy. In his lecture, Grombchevskii not only outlined the geopolitical value of the Pamir, but also noted the inevitable struggle with Great Britain for control over the Pamir, and called out 'to stop and resist the British'. It is hard to believe, considering the content of the lecture, that it was addressed to the Russian General Staff elite, as it was full of unprofessional judgments and superficial conclusions. For example, Grombchevskii sincerely believed that 'the occupation of Afghanistan and a mutual frontier with India would cause a considerable amount of concern and grief for the East Indian government'.[106] This assertion about the seizure of Northern Afghanistan completely lacked any rational ideas or strategic needs. Grombchevskii was unacquainted with Afghanistan and vaguely envisioned the political situation

103 Alexander III drew the War Minister's attention to the fact that 'promoting Grombchevskii to a higher rank of Lieutenant-Colonel, should be considered as a special exception to the law, and this act should not be a precedent for further promotions of officers of administrative service with the rank of Captain to a higher rank of Lieutenant-Colonel'.

104 RGVIA, F. 401, Op. 4, D. 46 (1889), l. 116 ob. – 117: A report of the Main Staff, 18 February 1891. To cover Grombchevskii's debts, the Council of the IRGO allocated 1,000 rubles in addition to the funds of the War Ministry.

105 The lecture text, under the same title, was published in 1891 in St. Petersburg and Novyi Margelan.

106 Grombchevskii 1891d, p. 2.

of the country and the specifics of its population. Russia's annexations of Northern Afghanistan unavoidably lead to a military conflict with Afghanistan and Great Britain for which Russia was not prepared for. Advancing beyond the Amu Darya could have had disastrous consequences for Russia's entire policy in Central Asia as well as for Russia's supremacy over Turkestan. Far-sighted Russian strategists tried to convince authorities and the public that such an action would have fatal consequences for the Russian state.

Grombchevskii's proposal to provide the ruler of Hunza with 100 Berdan rifles and two mountain guns, 'for the sake of Russia's prestige and charm', seemed like adventurism.[107] Grombchevskii was well aware that Safder Ali Khan was tied in an agreement with the Indian government, in which they acknowledged his vassalage. Safder Ali Khan was also responsible for continuous caravan robberies and the established slave trade. It is no coincidence that Grombchevskii called Safder Ali Khan 'the ruler of one of the wildest robber tribes which had survived in the wilds of Central Asia'.[108] The proposed action could result in a serious deterioration of relations between Russia and Britain, and could further contribute to the Hunza robberies along caravan routes.

Grombchevskii had little knowledge as to the goals and objectives of British policy in Afghanistan and the Pamir, he relied more on information received from the lords of the Pamir Khanates and *caravanbasis* (persons in charge of a caravan). His ideas on British politics in the regions of the Hindu Kush and the Pamir were conjectural in nature, he often relied on unverified information, as was the case with the British occupation of the Chinese frontier post of Shahidulla Hoja. His statements about the desire of the British to 'divide the Pamir and settle near the valley of the Great Alai' were just as hypothetical, having not taken into account the huge mountain spaces, the communication vulnerability, and most importantly, the practical importance of these remote and barren areas for the British defence policy.

Such surface judgment was characteristic when approaching the question of the Pamir's military significance, which Grombchevskii had considered extremely important for the defence of Turkestan, and for a military campaign against India. He claimed that the roads in the Pamir are convenient for moving large masses of troops with artillery and carts. However, the Russian Pamir campaigns (1891–94) have first hand experience how inconvenient the Pamir is for moving large military contingents. In order to move mountain artillery through the Alai Valley to the Pamir, military engineers had to resort to rock

107 On the details of this proposal see Grombchevskii 1891d, p. 26.

108 RGVIA, F. 401, Op. 4, D. 46, l. 59 ob.: B.L. Grombchevskii, A current political condition of the Pamir principalities and the border with Kashmir. Camp near Shahidulla fort, 23 January 1890. Autograph manuscript.

FIGURE 5.9 The Chinese General Huang and his staff, with Lieutenant-Colonel Grombchevskii, at a frontier post of Ulugchat, 1891.

blasting on the Ak Baital Pass, as well as blowing up cliffs in order to construct a trail along the Panj and Gunt Rivers. Grombchevskii himself was unable to cross the Trans Alai into Pamir at the height of summer due to large amounts of snow. Mountain passes in the Hindu Kush, according to Grombchevskii were 'so convenient, that moving troops, presented no issues', he did not however note that the descent and further movement into India were impossible without serious engineer works. After the lessons learnt in the Pamir, Russian General Staff came to the conclusion that no serious movements, except for demonstrative ones, could be carried out from the Pamir towards British India. It is strange that Grombchevskii, a military man, who was familiar with the territory, which he had previously described as a stage for future military conflicts, paid so little attention to matters of geography relating to strategy and military planning.

The entire chain of events to follow, in the spring of 1892, however, clearly indicated that Grombchevskii's high point as a traveller and geographer was in the past. During this time he placed himself as a key specialist on the 'Pamir affairs'. In 1891 Grombchevskii published a series of works on military geography and the international relations in Central Asia.[109] His final idea was to produce a series of works under the joint title *In the Event of War with India*. This included such essays as – *The Routes from the Pamir to India* and *The Great Alai Valley as an operational basis*, however they remained unpublished for political reasons.[110]

The Great Journey's End

From 1892 onwards, Grombchevskii becomes less interested in geographical research. During this period he also felt dissatisfied with his official position, and there is evidence that he probed the possibility of serving for the British.[111] In 1893 he re-entered the *voenno-narodnoe upravlenie* and obtained the post of *nachal'nik* (commandant) of the Osh district. Osh was a transit point for many Russian and foreign travellers on their way to Chinese Turkestan, the Pamir, British India and Tibet. Grombchevskii provided great assistance to these travellers who were grateful for the support.[112] At the beginning of April 1895, Grombchevskii was appointed junior civil servant on special duties to the Governor-General of Turkestan. During this period, his relationship with the IRGO was in crisis. The Society had not received an official report from him upon completing his expeditions to Central Asia. In 1895 the IRGO celebrated its 50th anniversary and the Council of the Society was extremely interested in publishing Grombchevskii's expedition works. The scientific processing of the expedition materials, which was conducted by a number of well-known Russian scientists, was completed by this time. That same year the Turkestan Military-Topographical Department completed the publication of an important

109 See the following works, Grombchevskii 1891b, c, and d.

110 Grombchevskii himself brought these articles up in a letter to Semenov: "Subsequently, as a result of this absolutely unjust accusation against me and my expedition into Hunza in 1888, which supposedly prompted the British to take this Khanate in 1892 and the subsequent troubles I have not published some of my essays". See NA RGO, F. 1 – 1894, Op. 1, D. 9, l. 47–47 ob.: A letter from B.L. Grombchevskii to P.P. Semenov, 12 August 1896.

111 Grombchevskii's interest did not go unnoticed by Major Elliot, the British military attaché at St. Petersburg, who personally met with Grombchevskii in 1891 during a visit to Ferghana. See TNA/FO539/56/118: Mr. Eliot to Sir R. Morier, February 29, 1892.

112 Amongst these travellers were St. George and Teresa Littledales who met Grombchevskii in Osh in early Spring of 1893 whilst on their way to Kashgar. Grombchevskii invited the Littledales to dinner with several Russian officers and their wives. He also organised their caravan as far as the Chinese border. See Clinch, Nicholas and Elizabeth 2008, pp. 135–136.

map, assembled on the basis of Grombchevskii's expedition materials, *A Map of Lieutenant-Colonel B.L. Grombchevskii's travels into Darwaz, the Pamir, Jetyshaar, Hunza, Raskam and Northwestern Tibet in 1885, 1888, 1889–90*. Work on the map started in the early summer of 1894 once Semenov requested this to Chief of the Main Staff, General Adjutant N.N. Obruchev. The map is an index map consisting of four map sheets on a scale 20 *verst* per inch and is based on Grombchevskii's route surveys and barometric determinations.

In 1896, Grombchevskii was appointed border commissioner of the Amur *oblast'* and as a result departed Turkestan once and for all, where he had served for twenty years. Before departing for the Far East, Grombchevskii presented his diaries and route surveys to the IRGO, which had remained in his possession upon completing the official reports on the expedition. For the next ten years, Grombchevskii held various administrative positions in the Far East – border commissioner of Iuzhno-Ussuriiskii krai, South Ussuri region, commissioner of civil administration of Kwantung *oblast'*. In September 1903 he was appointed as governor of Astrakhan *oblast'*. In the same year he married the daughter of an Irkutsk merchant, Vera Kompaneitseva, who was 26 years his junior, however, the marriage was short-lived. After leaving the military service with the rank of Lieutenant-General, Grombchevskii returned to the Far East as head of civil administration of the Chinese Eastern Railway (1906–10).

In 1910 Grombchevskii returned to Poland and settled in Warsaw. In October 1911, he appealed to the Council of the IRGO with the request to return the original expedition diaries for further review and publication. Grombchevskii's request was granted and he was returned the original expedition diaries, two folders with letters and other documents relating to the expeditions, which included linguistic materials for compiling Kanjut and Yazgulem dictionaries.[113] The outbreak of the First World War prevented the materials from being published and Grombchevskii's diaries were lost. With the Germans approaching Warsaw Grombchevskii migrated to St. Petersburg. After the Bolshevik revolution in Petrograd, he moved to Kuban, where he awaited the arrival of the Volunteer Army. In 1919, he made attempts to be arranged into the civil authority of General D.L. Khorvat, who in Harbin, held the post of high commissioner in the Far East. However, the defeat of the White Army in the Far East prevented Grombchevskii from returning to his activity. In 1920 he left the Far East through Japan and returned to Poland.

Details of his later life are so scant, that it would be impossible to build a complete picture. He held a position at the Meteorological Institute and lectured on military geography in Polish military schools. During these years, he

---

113 NA RGO, F. 1 – 1894, Op. 1, D. 9, l. 54.

was very sick and in dire need of money. He wrote about this in his letters to Francis Younghusband, his old friend and main rival in the period of the Great Game. Shortly before his death, he prepared and published in Polish a few works, including his memories of serving in the Russian Imperial Army. Bronislav Liudvigovich Grombchevskii passed away on February 27, 1926 in Warsaw. He was buried at Cmentarz Powązkowski in Warsaw, where the remains of many famous figures of Polish history and culture were laid to rest. Grombchevskii's contribution to the exploration of Inner Asia was significant, but despite the vast scale of his expeditions and the resonance they evoked with the Russian people at the time, his actual influence on the development of geographical knowledge about Inner Asia was limited. In essence, this occurred primarily due to the fact that no materials from his expeditions had ever been published. This huge mass of data remained out dated both in scientific and practical terms. Although the geographic information, astronomical definition and route surveys, were extensively used to map the Pamir and adjacent regions of China, the natural-scientific details were less valuable, especially in the botanical and geological collections. Unfortunately, Grombchevskii's assembled collections were not used to their potential by zoologists, botanists, ethnographers, and linguists.

Despite the fact that Grombchevskii produced high-quality route surveys of Sarykol and Raskam, they were not used by the War Ministry to their full extent. It seems that the military did not know of their existence, which could have been the case since Grombchevskii's originals were at his request sent from the Military-Topographical Department of the Turkestan Military District to the IRGO, where they were deposited. During the Sarykol crisis (1900–01), it suddenly occurred to the Main Staff that they had no precise data on some parts of Sarykol and Raskam, the very areas that had been surveyed by Grombchevskii during his expeditions. It took several years and a lot of resources before General Staff officers attached to the Russian Imperial Consulate-General at Kashgar – Captains L.G. Kornilov and Z.I. Zaichenko, Lieutenant-Colonel V.G. Lastochkin, and officers of the Tashkurgan military post, could accurately represent Raskam and Sarykol on Russian military maps.[114]

Grombchevskii made an important contribution to the study of the region located at the junction of the Kunlun, Hindu Kush and the Karakoram ridges. Thanks to his surveys, the outlines of the Kashgar Range and the Mustagh Ata group were revealed. He investigated the basins of Yarkand Darya, Raskam Darya and Tiznaf River. Grombchevskii discovered the powerful Yurungkash Ridge, which forms the watershed of the Yurungkash and Karakash Rivers.

114 See RGVIA, F. 1396, Op. 2, D. 2135: Routes in Xinjiang. Routes nos. 2, 3, and 4.

Grombchevskii's studies filled a significant gap in the cartographic image of the south-west part of Inner Asia, in the area between the Pamirs and Northern Tibet. Of great importance were his topographic and geodetic works in Hunza, which became, perhaps, the only Russian experience on the other side of the Hindu Kush at the origins of the Indus River.

During the two Inner Asian expeditions, Grombchevskii produced over 8,500 *versts* of route survey, determined the astronomical position of 87 points, and the height of more than 400 points. He made an important contribution to the photographic fixation of the nature of Inner Asia and its peoples, having prepared about 400 photographs on expedition routes. Grombchevskii deserves credit for the ethnographic study of the peoples of Inner Asia and their languages.

CHAPTER 6

Grigorii Efimovich Grumm-Grzhimailo (1860–1936)

Tatiana I. Yusupova

> We looked at our expedition through the eyes of Przhevalskii, who aptly spoke of his explorations in Inner Asia as "scientific reconnaissance". To some extent, we have replenished the investigations of this famous traveler; yet after him, as well as after us, there still remains much to be done in Inner Asia.
>
> G.E. GRUMM-GRZHIMAILO (1896)[1]

Some Preliminary Remarks

Grigorii Grumm-Grzhimailo ranks among the cohort of outstanding Russian explorers of Inner Asia of the 'Przhevalskian age'. His expeditions covered the territories of the Pamir, Eastern Tian Shan, Western China, Nanshan, Western Mongolia and Uriankhaiskii Krai (now Republic of Tuva, within the Russian Federation). Unlike Przhevalskii and other Russian travelers who belonged to the military caste, Grumm-Grzhimailo was a civilian, with a university degree in natural sciences. His major expedition to Inner Asia in 1889–1890 was organized by the Imperial Russian Geographical Society (IRGO) as a follow-up of Przhevalskii's pioneering work in the region. In this journey he was assisted by his brother Mikhail, who was an army officer, and therefore it is often called the 'Expedition of Brothers Grum-Grzhimailo'. Though a civilian, Grigorii Grumm-Grzhimailo regarded himself as a follower of Przhevalskii and his methods of field work. In his report of the journey he made it a point to say that "we looked at our expedition through the eyes of Przhevalskii".[2]

The entire exploratory activities of Grumm-Grzhimailo included: 1) his field lepidopterological work[3] in the European part of Russia during his studies at

1 Grumm-Grzhimailo G.E. 1896, IX.

2 Ibid.

3 Lepidopterology is a branch of entomology concerning the scientific study of moths and butterflies.

 | DOI 10.1163/9789004376267_008

FIGURE 6.1
Grigorii Efimovich Grumm-Grzhimailo, ar. 1900. NA RGO, R. 117, OP. 1, D. 94.

FIGURE 6.2
Mikhail Efimovich Grum-Grzhimailo, 1890. NA RGO, F. 32, OP. 1, D. 57.

the St. Petersburg University, in 1881–1883; 2) the four expeditions he led into the Pamir country, between 1884 and 1887, collecting samples of local fauna (primarily his favorite butterflies), making route surveys and describing the landscapes of the places he visited. Three of these expeditions, as Grumm-Grzhimailo would later say, became possible "only thanks to the material and moral support of the Grand Duke Nikolai Mikhailovich" (Romanov)[4], while the fourth one was sponsored by Count S.D. Sheremetev; 3) his Inner Asian expedition, in 1889–90, the most prominent one judging by its results and the vastness of territories it covered. The journey combined general geographic exploration with the making of various collections en route; and finally 4) several trips he made, between 1903 and 1914, to Mongolia, Uriankhaiskii Krai and the Far East, at the behest of the Ministry of Finance. The most well-known of these was his Mongolia-Uriankhai expedition of 1903.

One last remark should be made on the spelling of the name Grumm-Grzhimailo. After the 1918 orthographic reform in Soviet Russia, G. Grum-Grzhimailo added the second 'm' to his family name (Grumm) which was earlier spelled with one 'm', i.e. Грумъ-Гржимайло. (It is believed that he did so because he was a superstitious man and wanted to avoid having a 13-letter last name.) This new spelling with double 'm' is used nowadays only by his direct descendants. In the essay and Bibliography we used the new spelling – Grumm-Grzhimailo. His brother's surname is spelled Grum-Grzhimailo. Also, we used the latter spelling in references to both brothers (the Grum-Grzhimailos).

•••

The biography of Grigorii Grumm-Grzhimailo has not been properly researched thus far. The first to pay attention to Grumm's extensive travels was Mariia Lialina, a popular author and translator of books for children and youngsters of the late nineteenth – early twentieth century. Lialina retold, in 1901, the content of the formal report of the brothers Grum-Grzhimailo on their journey to Western China to acquaint with it a broad reading audience.[5] The first genuine biography of the explorer was written in 1947 by his son, Aleksei Grigorievich Grumm-Grzhimailo,[6] at the request of the Moscow Society of Naturalists (*Moskovskoe Obshchestvo Ispytatelei Prirody*). Aleksei Grumm-Grzhimailo used a wide range of sources – the family archive, the materials of

4 Bukhert 2006, 179–180. Letter from G.E. Grumm-Grzhimailo to Count S.D. Sheremetev, February, 1887.

5 Lialina 1901.

6 Grumm-Grzhimailo A.G. 1947.

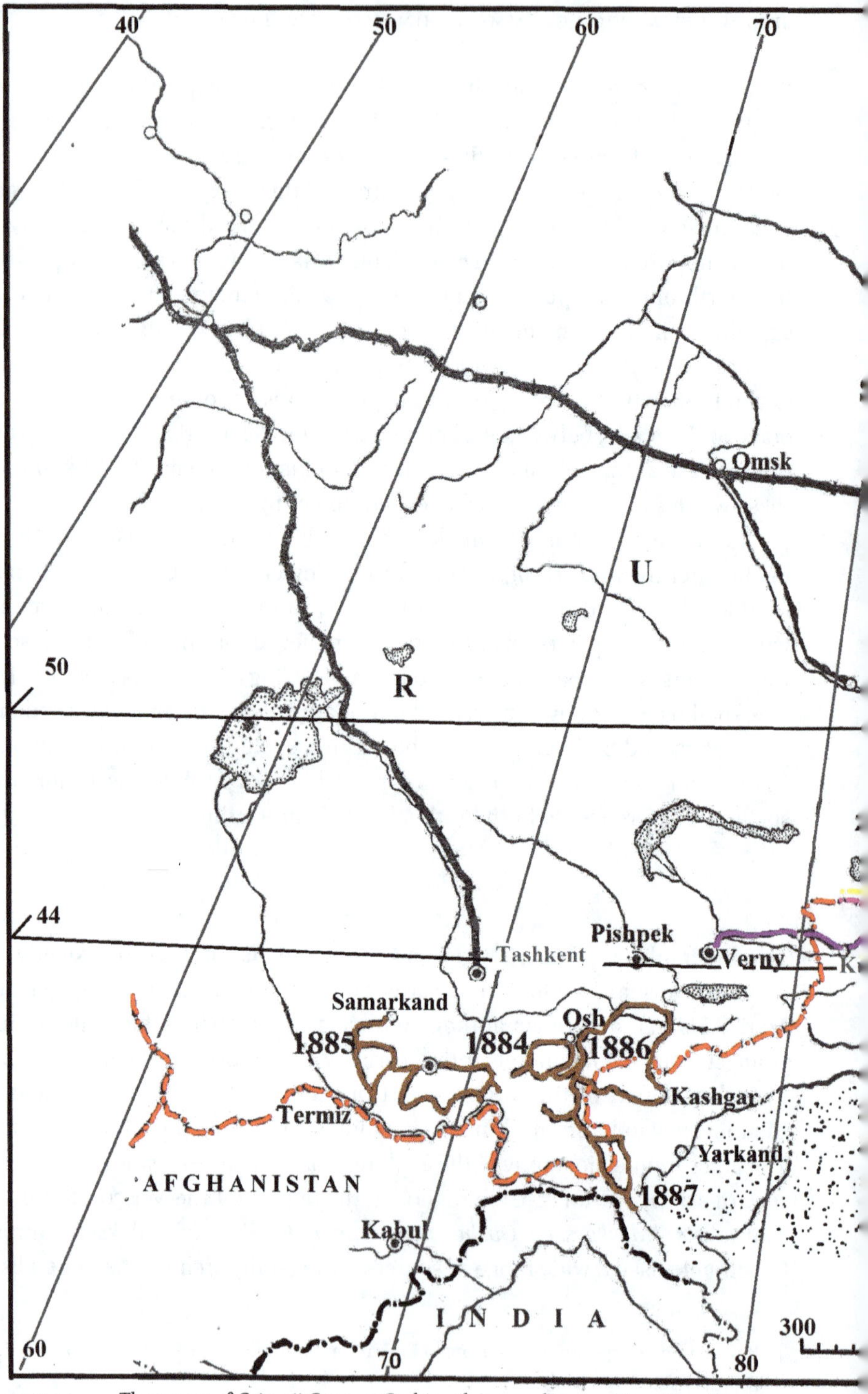

MAP 6.1 The routes of Grigorii Grumm-Grzhimailo's expeditions, 1884–1887, 1889–1890, 1903. Compiled by Tatiana Yusupova on the basis of the map from A.G. Grumm-Grzhimailo's *Dela i dni G.E. Grumm-Grzhimailo (puteshestvennika i geografa), 1860–1936*, 1947.

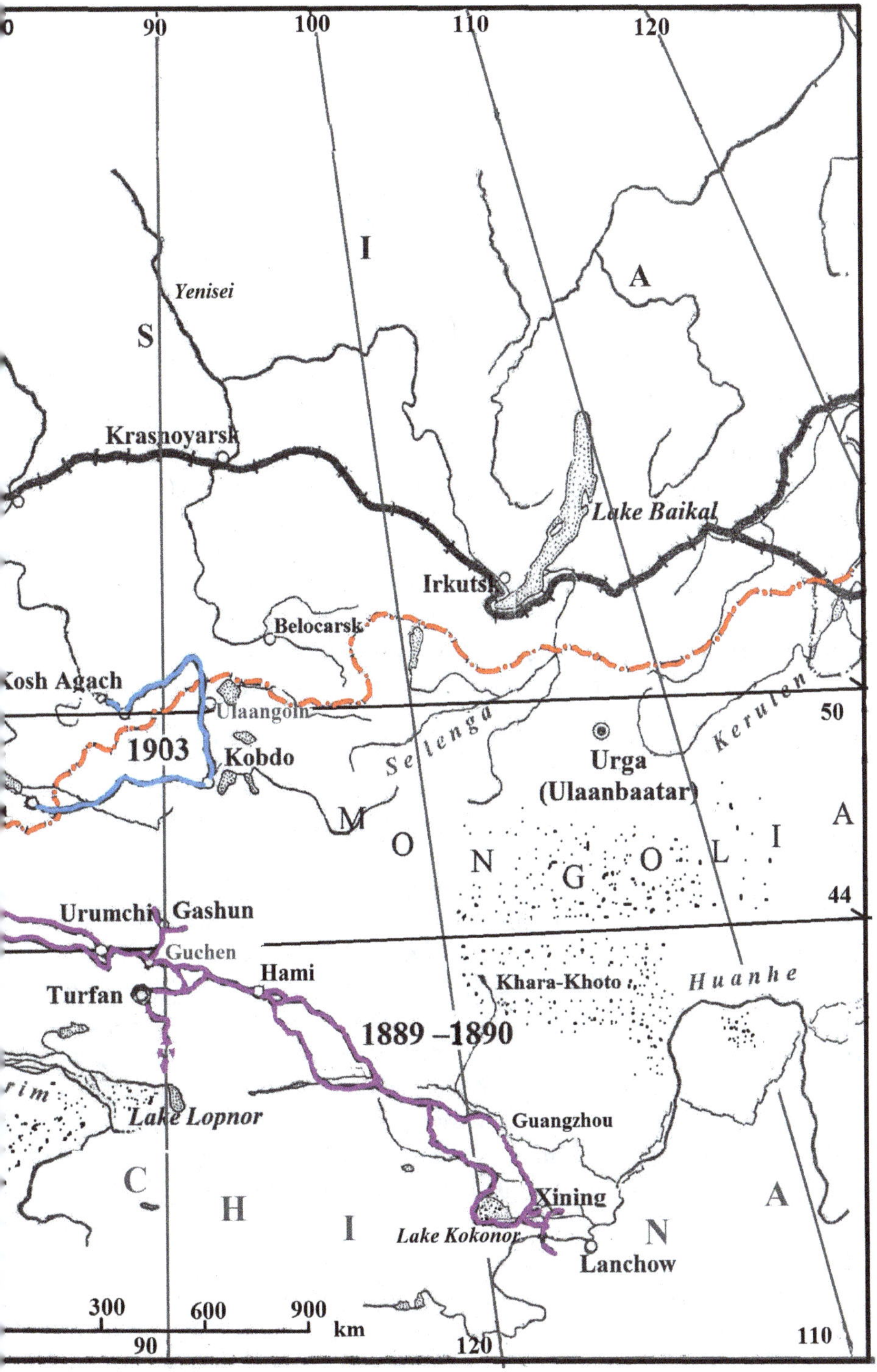

90
100
110
120
I
A
S
Yenisei
Krasnoyarsk
Lake Baikal
Irkutsk
Belocarsk
Kosh Agach
Ulaangom
50
Kerulen
Selenga
1903
Kobdo
Urga
(Ulaanbaatar)
M O N G O L I A
44
Urumchi
Gashun
Guchen
Hami
Turfan
Khara-Khoto
Huanhe
1889 –1890
Lake Lopnor
Guangzhou
C H I N A
Xining
Lake Kokonor
Lanchow
300
600
900
km
90
120
110

the Russian Geographical Society, including G.E. Grumm's documents and manuscripts handed over to the Society's Archive by his relatives after Grumm's death, as well as documents relating to his father he found in other archives in Leningrad. The book also included the full bibliography of G.E. Grumm-Grzhimailo's works and the list of publications devoted to the description of his naturalistic collections by specialists. It still remains the main source of information about the life and travels of his famous father.

New publications about G.E. Grumm-Grzhimailo have appeared only recently. One of these is a preface to the reprinted version of the above-mentioned report of the brothers Grum-Grzhimailo, written by A. Khoroshevskii.[7] Others deal specifically with G.E. Grumm-Grzhimailo's scholarly activities as a geographer, economist, entomologist and historian.[8] To these publications one may also add some short articles and entries in the encyclopedias (Brockhaus and Efron Encyclopedic Dictionary; The Granat; various editions of the Great Soviet Encyclopedia and the new Great Russian Encyclopedia).

This essay generally follows Aleksei Grumm-Grzhimailo's narrative, which in its turn was based on his father's detailed biography written as early as 1924.[9] We also used some new documentary sources (those of the RGO and the archive of the Russian Academy of Sciences), as well as the latest publications on G.E. Grumm-Grzhimailo.

Family Roots and Early Life

Grigorii Efimovich Grumm-Grzhimailo was born on 5 (17) February 1860 in St. Petersburg. His father Efim Grigorievich Grum-Grzhimailo (1824–1870) was an economist, a specialist in beet-sugar production, working in the Department of International Trade of the Ministry of Finance. He was of Polish descent, his ancestors belonging to the aristocratic Grzymala line, bearing the name *Grum-Grzymajlo.*[10] The Russian branch of the family came from a cornet (junior commissioned rank in cavalry) Luka Grum-Grzymajlo, who migrated to Russia in 1647. Grigorii's mother, Margarita Mikhailovna Grum-Grzhimailo (1838 – ?) was the daughter of military topographer, Major-General of the General Staff M.O. Bez-Kornilovich (also spelled Bezkornilovich) (1796–1862). He

7 Khoroshevskii 2014.

8 Latushko, Briantseva 2010; Koroliov, Murzin 2005; Yusupova 2016.

9 See NA RGO, F. 32, Op. 1, D. 73. Autobiography of G.E. Grumm-Grzhimailo, 1924; also SPbF ARAN, F. 50, Op. 2, D. 79. Correspondence between G.E. Grumm-Grzhimailo and A.A. Strauch, 1884–1885.

10 For Grzymala's coat of arms, see Gajl 2007.

was widely known as a historian and specialist in regional studies of Belorussia[11] (now Republic of Belarus'), where he spent much of his life on his estate Kornilovichi. Bezkornilovich's wife, grand-mother of Grigorii Efimovich, was named Nadezhda Aleksandrovna (nee Countess Dolgorukova).

The families of Grum-Grzhimailo and Bez-Kornilovich stayed in summer in a country-cottage in Ulianka, a southern suburb of St. Petersburg, next to the mansion of Count Dmitry N. Sheremetev (1803–1871), one of the most notable Russian aristocrats, known for his philanthropy.[12] The new-born Grigorii was baptized in Sheremetev's house church, his godfather being the sixteen-year old son of the Count, Sergei (1844–1918). The latter would become in future a prominent state and public figure, a philanthropist, as well as chairman of the Archeological Commission, and a member of several scholarly societies, including the Russian Geographical Society.

There were six children in Efim Grum's family – four sons: Grigorii, Mikhail (1861–1921), Dmitrii (1862–1912), and Vladimir (1864–1928), and two daughters – Margarita (1859–?) and Ekaterina (1869–?). The second son Mikhail graduated from the Mikhailovsky Artillery Cadet School and Academy (1886), and eventually attained the rank of Major-General; he was known as the inventor of the military technical devices and equipment. Mikhail took part, as mentioned before, in his elder brother's expeditions to the Pamir and Inner Asia, contributing considerably to their success. Grigorii's second brother, Vladimir, had a diploma of the St. Petersburg Mining Institute and became a scientist, one of the prominent Russian / Soviet experts in metallurgy, and a corresponding member of the Soviet Academy of Sciences (1927).

To support his large family, G. Grumm's father had to quit his service in the Finance Ministry which brought him little income. This occurred at the time when the public notary system was introduced in Russia. Having passed the necessary exams, Efim Grigorievich became a notary, one of the first in St. Petersburg. The new job helped him to improve substantially his financial situation. However the comfortable life of Efim Grum did not last long. In 1870 the head of the family died and his spouse Margarita with their six children remained practically without any means of subsistence. Luckily, she was given support by Sheremetev Jr. In particular, he paid for the education of his godson, Grigorii, who in 1871 joined the Imperial St. Petersburg College of Law. However Sheretev's assistance stopped in 1874, and Margarita had to send her son to the 3rd military gymnasium (later the Emperor Alexander II cadet corps), hoping that Grigorii would make a military career.

11 See, for example, Bez-Kornilovich 1855.

12 Sheremetev 2005, 212–213, 516 et al.

From an early age, Grigorii, as the elder son, assisted his mother in supporting the family. Being good at pencil drawing – a skill he would later use during his travels, he drew pictures from his mother's embroidery, who would then sell them. He also earned some money by tutoring and doing some work for the popular daily *Niva* Journal, such as writing columns and illustrating the journal's publications.

In 1879 Grigorii finished the military gymnasia. However, his mother's hopes that he would then join the army were not realized. This was because Grigorii suffered from short sight and because military service did not attract him at all. His real attraction was science and therefore he wanted to enter a university. Yet he could not do so as he did not attend the classical gymnasium which prepared pupils for university; in particular he did not know Latin and Greek which were not studied in military colleges. Therefore to fill the gap in his knowledge he had to master these languages on his own. It took him a whole year (from July 1879 to May 1880) to prepare himself for taking external exams for the full gymnasium course. Having overcome this major obstacle, Grigorii became a student, in September 1880, of the natural science department of the faculty of physics and mathematics at the St. Petersburg University.

In the course of his studies he attended the lectures of many outstanding naturalists, such as botanist A.N. Beketov (1825–1902), zoologist N.P. Vagner (1829–1907), soil scientist V.V. Dokuchaev (1846–1903), chemist N.A. Menshutkin (1842–1907), the future Nobel Prize winner, physiologist I.M. Sechenov (1829–1905). Grigorii was especially captivated by Vagner's lectures on the zoology of invertebrates,[13] and it was under Vagner's influence that he finally chose entomology as his future specialty with a focus on butterflies which had fascinated him from childhood.

The Early Entomological Field Work

Grigorii's studies at the university continued for slightly more than three years, instead of the regular four-year course. He combined his classes with the practical field investigations he conducted during summer vacations. The starting point of his professional career, in his opinion, was his work in the Phylloxera Committee, established at the Ministry of State Property, where he was invited, in the spring of 1880, to take up the post of secretary. This is certainly a remarkable fact as Grigorii was then just twenty and had not yet entered the University. He got this job thanks to his acquaintance with N.Ya. Danilevskii (1822–1885),

13 Grumm-Grzhimailo, A.G. 1947, 8.

a well-known Slavophile, publicist, one of the ideologues of the civilizational approach to history, and concurrently a naturalist. Danilevskii came to St. Petersburg from his estate in the Crimea to call upon the government to take urgent measures in order to fight the vine aphid (Fylloxera), insect pests which threatened to destroy the vineyards on the peninsular. Danilevskii was struck by Grigorii's knowledge of the subject so much that he immediately offered him to join the Committee.

The summer of 1881 Grumm spent on the estate of Danilevskii in the Crimea where he had an occasion to acquaint himself with the endemic lepidopterological fauna. As a result he published his first article in the *Horae Societatis Entomologicae Rossicae*.[14] The next year (1882) he was busy collecting the samples of lepidoptera, this time in Bessarabia and Podolsky Province in the middle course of the Dniester River, which he then transferred to the Zoological Museum of the Academy of Sciences. Here, in the Museum, Grumm would describe his collection and at the same time conduct a comparative analysis by collating his finds with a collection from a different region in the vicinities of Narva.[15] In the opinion of zoologists, these initial activities by Grumm showed him to be a genuine entomologist who had mastered the methods of zoogeographical investigations.[16]

His final students' vacation, in the summer of 1883, were spent in Saratov Province, where he had a chance to acquaint himself with the lepidopterological fauna of the Middle Volga region (Povolzhie). His collection caught the attention of Grand Duke Nikolai Mikhailovich Romanov (1859–1919) who also took much interest in entomology and played an important role in the development of the lepidopterological studies in Russia. An article by Grumm-Grzhimailo about his research was then published in the first volume of the many-volume edition of the *Mémoires sur les lépidoptères,* the editor of which was the Grand Duke himself.[17]

The acquaintance with N.M. Romanov helped G. Grumm to define the area of his explorations. It was the Grand Duke who proposed that he make a trip to the northern slopes of the Pamir for collecting entomological material. This territory was absolutely unknown to biologists so Grumm gladly accepted the offer. Nikolai Mikhailovich agreed to finance the journey, provided that the explorer left St. Petersburg before 1 March to be able to reach the Pamir at the

---

14 Grumm-Grzhimailo, G.E. 1881.

15 Grumm-Grzhmailo, G.E., Swiatsky 1882.

16 Koroliov, Murzin 2005, 37–38.

17 Romanoff 1884–1897.

best time for butterfly collecting. Another condition was that Grumm would hand over his entire collection to the Grand Duke.

For this reason Grumm pleaded with the administration of the university to allow him to take the external exams for the fourth course in the autumn of 1883, which they did. So he took the final exams ahead of time, and in January 1884, the Council of St. Petersburg University, taking into consideration his previous scientific work, awarded him the degree of Candidate of the Natural Sciences.

Having graduated from the University, Grumm left St. Petersburg, on 14 February 1884, to go on his first exploring expedition to the Pamir. This finally decided his fate – instead of a career as a university professor, he chose the path of a scholar-traveler, arduous and full of risks, yet rewarding him with geographical and other scientific discoveries.

The Pamir Expeditions (1884–1887)

The First Trip, 1884

G. Grumm's preparations for his first Pamir expedition did not take long. The Russian Geographical Society helped him chart his route and provided the necessary equipment. Meanwhile, the Council of the IRGO, acting in accordance with its Statute, elected him, on 6 February 1884, an associate member of the Society.[18] His expedition project was then discussed by several IRGO members familiar with the region, such as the geologist and geographer I.V. Mushketov (1850–1902),[19] the mining engineer D.L. Ivanov (1846–1924),[20] who had participated in one of Mushketov's earlier expeditions, and entomologist V.F. Oshanin (1844–1917).[21] The Society also supplied Grumm with recommendation letters, a permit to conduct field work (*otkrytyi list*), a surveying

18 *Otchet of the IRGO for the year 1884*, St. Petersburg, 1885, 4.

19 Ivan Vasilievich Mushketov undertook, in 1870s, several expeditions which covered practically all the mountainous part of the Turkestan country, from the Dzungar Alatau and Kuldja up to the northern Pamir, Gissar and the northern border of Afghanistan; he also explored the greater part of the Bukhara Khanate and the Kyzylkum desert.

20 Dmitrii Lvovich Ivanov – a mining engineer who participated in Mushketov's expedition to the Zaravshan glacier (1880), the Pamir expedition of Captain D.V. Putiata (1883); in 1899–1905, as head of the Irkutsk administration, he investigated the lower reaches of the Amur river and the island of Sakhalin.

21 Vasilii Feodorovich Oshanin – Russian entomologist, geographer and traveler, one of the explorers of Inner Asia.

compass, and, on top of that, Oshanin gave him his personal aneroid for making altitude measurements.

Grumm arrived in Osh, the starting point of the expedition, at the end of April. This Kyrgyz town located in the eastern part of the Fergana Valley, at the foot of the Alai Range, had been only recently annexed to the Russian Empire after Russia's conquer of the Kokand Khanate in 1876. Here Grumm equipped the expedition caravan and hired an armed escort. This consisted of eight men, four of which were Cossacks from the Orenburg Cossack troops, since travelling in this area was unsafe for Europeans. Finally, on 20 May 1884, Grumm's caravan took the road.

Moving along the planned route, Grigorii traversed the northern slopes of the Pamir, trekked down the Alatai Valley,[22] reached the river Muksu, and, having passed the Trans Alai Range, found himself on the shore of Lake Karakul in the northeastern part of the Pamir. This was the furthest point of his expedition, from which he proceeded to Karategin,[23] and, at the end of August, his caravan returned to Osh. During the three month period Grumm covered some 600 km, having surveyed the course of the river Muksu. His lepidopterological collection was huge – it consisted of about 12,000 samples of diurnal butterflies (Rhopalocera) of 146 species, 30 of which were totally new. Besides, G. Grumm collected a number of endemic vertebrates. The greater part of his collections (those of insects, with the exception of specimens presented to the Grand Duke, vertebrates, reptiles and amphibia) he donated to the Zoological Museum of the Academy of Sciences. He had been in close touch with the museum from his university days, his main contacts there being its Director academician Aleksander A. Strauch, and ornithologist M.N. Bogdanov, both acting as his consultants.[24]

In his letter to Strauch, Grumm reported proudly about his collection of insects saying that it included "practically everything that exists in Asia Minor and Armenia, where this group was studied most minutely. Definitely, I can say that the results surpassed all my expectations and the goal of the expedition has been achieved".[25]

22 Alai is a part of the Pamir, lying to the south of the Fergana Valley.

23 Karategin (lit. Black Mountains) is the historical region of Tajikistan, located in the mountains, in the middle course of the Vakhsh River, the right tributary of Amu-Darya.

24 Andreyev, A.I., Gnatiuk, T.Yu., Zoologicheskie kollektsii, in *Rossiiskie ekspeditsii v Tsentral'nuiu Aziiu*, in Andreyev 2013, 196–203.

25 SPbF ARAN, F. 50, Op. 2, D. 79, l. 20 ob. Letter from G.E. Grumm-Grzhimailo to A.A. Shtraukh, 17 September 1884.

G. Grumm's "Short report of the results of the expedition to Pre-Alai countries" was published in the *Proceedings of the IRGO* (*Izvestiia IRGO*),[26] while his preliminary description of his lepidopterological collection was included in the second volume of the *Memoires sur les lepidopteres* (1885).[27]

The Second Trip, 1885

The expedition began a year later, in spring 1885. This time it was initiated by Grigorii himself, who was fascinated with the unusual variety and novelty of the lepidopterological fauna of Alai and Pamir. From Osh he sent a letter to the Grand Duke Nikolai Mikhailovich with a proposal to continue his explorations in the Pamir region which had proved so successful by visiting Karategin and Darvaz. His aristocratic patron agreed and offered him a subsidy of 2000 rubles. The Geographical Society, for its part, fully supported G. Grumm's initiative and added to this sum 800 rubles more. The Society also petitioned the Governor-General of Turkestan, N.O. Rozenbakh, to attach to G. Grumm's party a topographer, Captain G.E. Rodionov, two assistants (preparators) to help him to collect and prepare naturalistic specimens, and eleven Cossacks as expedition escorts, which he did.

Grigorii was happy. On 26 March 1885, on the eve of his second journey, he wrote to Shtraukh optimistically:

> My first expedition was financed rather poorly. I had no hunter and no vertebrates. So I took only reptilia and amphibia, which were delivered partly to the Academy. Now I have a special assistant for vertebrates – Severtsov's[28] preparator and hunter. With my usual energy I hope I'll be able to gather huge collections which I will then present to the Academy. I am going to countries that have never been visited by zoologists and where no European trod before. I hope I will obtain many Himalayan species.[29]

The quotation shows the young Grigorii as a rather ambitious man, full of self-importance, yet also a great enthusiast for the kind of work he was doing.

---

26 Grumm-Grzhimailo, G.E. 1884.

27 Grumm-Grzhimailo, G.E. 1885.

28 Severtsov, Nikolai Aleksandrovich (1827–1885), zoologist and traveler, one of the pioneers of the ecological studies in Russia, an explorer of Middle Asia and the Pamir.

29 SPbF ARAN, F. 50, Op. 2, D. 79, l. 7. Letter from G.E. Grumm-Grzhimailo to A.A. Strauch, 26 March 1885.

G. Grumm left Samarkand with his team on 29 March 1885. In four and half months he passed through the mountain and steppe domains of Beys of the greater part of the Bukhara Khanate: Gissar, Kuliab, Bal'djuan, Kabadian, Shirabad, Karshi and others.[30] His routes linked up the earlier itineraries of N.A. Severtsov in the Khodjent area, those of the Turkestan expeditions of A.P. Fedchenko (1844–1873) in 1868–1871, and the Pamir ones of V.F. Oshanin (1844–1917), which provided a broad faunistic picture of the greater part of the Pamir and the adjacent areas.

The expedition's main trophy was its zoological collections, especially that of invertebrates. The entomological collection alone included over 20,000 specimens of insects, primarily lepidoptera, with a great number of new species and forms, unknown to zoologists. A part of these collections was described by Grumm himself,[31] the rest was done by other specialists. In recognition of his work, the Russian Entomological Society elected G. Grumm, on 4 November 1885, as a full member.

Grumm reported of the results of his two Pamir expeditions at the general meeting of the IRGO, on 4 December 1885.[32] And on 15 January 1886 he was elected as a full member of the Society, on the recommendation of its Vice Chairman P.P. Semenov and its secretary A.V. Grigoriev.[33]

The Third Trip, 1886

As soon as he was done with his reports, Grumm began to make preparations for his next journey – to the Cis-Pamir countries, being the southern parts of western end of the Tian Shan, the so-called Inner Tian Shan. His idea was to collect new zoological specimens to confirm the conclusions he had made during his first two trips to the Pamir.[34] The expedition was sponsored again by the Grand Duke Nikolai Mikhailovich. Its main object, apart from topographic surveying and collecting entomological specimens, was to conduct comparative zoogeographical explorations to ascertain the connections between the Pamir-Alai and Tian Shan faunas, on the basis of the lepidopterological material.

Grumm trekked from Osh – the expedition's starting point, up the River Naryn towards the undrained Lake Chatyr-Kul lying at the altitude of 3530 meters above sea level. His team then descended to Kashgar in Eastern (Chinese) Turkestan and thence, having crossed the Alai Valley, returned to Osh, in late

30 Grumm-Grzhimailo, G.E. 1885a.

31 Grumm-Grzhimailo, G.E. 1887, 397–406.

32 Grumm-Grzhimailo, G.E. 1886.

33 *Izvestiia IRGO*, 1886, XXII, 180.

34 *Otchet IRGO for the year 1886*, St. Petersburg, 1887, 40.

August 1886. In the course of this expedition Grumm established that "the lepidopterological fauna of the Pamir and of the Cis-Pamir countries is considerably different from the fauna of Tian Shan and is very similar to the fauna of the Cis-Hindukush countries".[35] This fact, in his opinion, evidenced that the Cis-Pamir and the Tian Shan regions lay separately at the time of the formation of the Tian Shan fauna. Grumm's conclusion provided an additional confirmation of Mushketov's geological theory according to which there had once existed a tectonic boundary between Pamir and Tian Shan – these mountain systems were separated and later tectonically pulled together.

The Fourth Trip, 1887

In early 1887 G. Grumm-Grzhimailo submitted to the Council of the IRGO a project for his next – fourth – expedition to the Pamir. His focus this time was the central part of Pamir, to the east of the mountain juncture Sel'tau. The expedition was organized by the IRGO, yet because of the Society's paucity of funds, Grumm had to look for sponsors. The one he found was Count S.D. Sheremetev who allocated 4000 rubles for his journey.[36]

He was accompanied on this trip by his brother Mikhail, then a lieutenant of the Second Guards' artillery brigade. The staff of the expedition consisted of 13 people – apart from both brothers, it included an interpreter, a preparator, three guides, and six Cossacks making the escort. The duties of the leaders were as follows – Grigorii was to conduct all exploratory field work, including the collecting of faunistic specimens, while Mikhail was in charge of all topographic surveys, the caravan and the safety of the expedition team.

The journey started again in Osh. Having reached the already investigated Lake Karakul, Grigorii planned to descend, down the river Kudar, to the town named Tashkurgan ('Stone Fortress'), located at the altitude of 3600 m; thence to cross the river Murgab and the Mardzanai Pass, and reach the Lake Yashilkul. From this spot the bothers intended to surmount the Pamir Ridge to get to Lake Zorkul and further to Lake Sarykov. Their next task was zoogeographical explorations in the Eastern Pamir.

However their plans were not realized fully. The brothers explored the basin of the river Kudar, where they discovered at its tributary Tanymas an unknown group of glaciers, which Grigorii named after his sponsor Sheremetev.[37] They

35 Ibid.

36 Meeting of the IRGO Council, 23 March 1887, Note by G.E. Grumm-Grzhimailo, see *Izvestiia IRGO*, 1887, XXIII, 369–371; *Otchet IRGO for the year 1887*, St. Petersburg, 1888, 6–8.

37 Subsequently, the major glacier was given the name of G.E. Grumm-Grzhimailo by decree of the Presidium of the Supreme Soviet of the Tajik Soviet Republic of 28 August 1948.

could not get to the Murgab Valley because of the flooding of Murgab and other mountain rivers. Thus the brothers had to constantly change their itinerary. To reach Lake Sarykol they followed a rather elaborate route – upstream Yark-and-Darya, over the Karakoram Ridge and further to the Aksu Valley, which eventually brought them to Chinese Turkestan. By doing so they violated the instructions of Governor-General N.O. Rosenbakh who strictly prohibited them to travel beyond the territories of the Russian Empire. Their emergence in the Western borderland of China was a real shock for the Russian consul in Kashgar, N.F. Petrovskii (1837–1908), who immediately sent Grumm an angry message through a Chinese patrol saying that his expedition should leave the foreign territory, as soon as possible, and return to Russia.

The incident spoiled Grigorii Grumm's relations with Petrovskii. The explorer put all the blame on the consul for not getting the approval of the Chinese authorities for the movement of his expedition well in advance. As a matter of fact, this was their second conflict – a year before, while in Kashgar, G. Grumm was largely discontented with Petrovskii's poor assistance given to his expedition.[38] And the consul, in his turn, complained in a letter to Przhevalskii about Grumm's "misbehavior" which caused him "a lot of trouble".[39]

Such conflicts were practically impossible when expeditions were headed by military officers. Unlike the civilians, they fulfilled the prescriptions of their superiors implicitly, knowing well that no scientific interests should interfere with Russia's geopolitical priorities in such places as Eastern (Chinese) Turkestan and Pamir.[40]

Grigorii Grumm returned to St. Petersburg, in early October 1887. He reported the results of his expedition to the IRGO, on 28 November, at the joint meeting of the Society's sections of physical and mathematic geography.[41] Despite all the problems he had faced, connected with his forced change of route and urgent return to Russia from the banks of the river Aksu, his expedition yielded fairly good results. It surveyed 2000 versts, in a 5-verst scale; determined the altitude of 20 points; discovered, as mentioned above, a whole new system of glaciers in the upper Tanymas river; conducted meteorological observations en

38 NA RGO, F. 32, Op. 1, D. 13, l. 72. Letter from N.F. Petrovskii to G.E. Grumm-Grzhimailo, 8 August 1886.

39 NA RGO, F. 13, Op. 2, D. 185, l. 22. Letter from N.F. Petrovskii to N.M. Przhevalskii, 25 November 1887.

40 See more on this in Postnikov 2005.

41 *Izvestiia IRGO*, 1887, XXIII, 787.

route; gathered information about the local population and their way of life by questioning natives; and finally took 150 ethnographic and landscape photos.[42]

Especially valuable were Grums' zoological collections. (The animals obtained were all shot by Mikhail, as Grigorii was unable to do any hunting.) One part of his trophies Grigorii donated to the Zoological Museum; and another part the Museum purchased from the traveler.

The results of his lepidopterological investigations and zoogeographic observations on his four Pamir expeditions G. Grumm described in a lengthy paper entitled *Novae species et varietates Rhopalocerorum e Pamir*[43] and in his major work *Le Pamir et sa faune lepidopterologique* (*Pamir and its lepidopterological fauna*).[44] In one of the chapters of the book the explorer described the ecological conditions of the flora and fauna of the Pamir (the lepidopterological fauna in the first place) and analyzed a variety of physiographic factors affecting their formation.

In the course of his four Pamir expeditions, Grumm the entomologist got interested in the borders of the geographical distribution of insects, and his interests in geography grew with every new journey. The experience of geographical research Grigorii had gained in the Pamir was to be useful in his major journey, the one to Inner Asia.

The Inner Asian Expedition, 1889–1890

Before starting on a new journey, Grigorii Grumm went to the Central Urals, in the summer of 1888, to verify his "new theories of the geographical distribution of some animal groups".[45] The collections he brought back he would later use for the comparative analyses when writing his works on the Pamir lepidopterological fauna.

In January 1889 Grumm was approached by the Grand Duke, who offered to support him in the continuation of his research of the Pamir and suggested that he make a journey to the southern slopes of the Pamir, into the Cis-Hindukush region, namely to Shugnan, Vakhan, Chitral and Kunzhut. This proposal concurred with Grumm's own plans, so he agreed. Grumm wanted to complete his zoogeographic exploration of the Pamir and the Cis-Pamir countries.

42 *Otchet IRGO for the year 1887*, St. Petersburg, 1888, 6–8.

43 Grumm-Grzhimailo, G.E. 1888.

44 Grumm-Grzhimailo, G.E. 1890.

45 Bukhert 2006, 185–186. Letter from G.E. Grumm-Grzhimailo to S.D. Sheremetev, 21 October 1888.

However, the Council of the IRGO recommended that he change his route "for political and other reasons"[46] and chose as an objective of new investigations the Eastern Tian Shan and the Nanshan. The "other reasons" were the Society's desire to fill the remaining 'blank spots' on the map of Inner Asia. The IRGO also wanted to link up the routes of Przhevalskii and Potanin in the Nanshan, which had been insufficiently explored by Russian travelers. (Przhevalskii traversed this region during his first and third Inner Asian expeditions (1870–1873; 1879–1881), and Potanin passed across its north-eastern outskirts on his way back from his Tangut-Tibetan journey of 1884–1886).[47]

The route of his expedition was elaborated together with P.P. Semenov and the well-known cartographer from the Military-Topographic department of the Main Staff, Colonel A.A. Bolshev (1828–1904). Grumm's expedition was to pass through Eastern Tian Shan and the Bogdo-Ula range, thence across the Dzungar Gobi up to Lopnor and, having passed over the Nanshan in its central-western part, to reach Lake Kokonor and the Valley of Huang He. This plan was approved by the Grand Duke Nikolai Mikhailovich who allocated the sum of 6000 rubles for the investigation of Western China – instead of the Pamir. The IRGO added to this sum 2000 rubles, which was not enough for Grumm's two-year journey into the depths of the continent, with an overall route of about 8000 km. The search for finances was a big problem for him this time. In his letter to Count Sheremetev G. Grumm wrote bitterly: "You cannot even imagine how much trouble and distress my every expedition gives me, and the present one is worse than all of them. ... Bowing and scraping, pleading, and paying all kind of compliments to people who do not deserve them, and finally resorting to flattery and things of that sort".[48]

Grigorii Grumm turned for help to the Sheremetev brothers asking them to donate 2000 or 3000 rubles for the sake of "the cause bequeathed by Przhevalskii". Yet neither his god-father S.D. Sheremetev nor the younger A.D. Sheremetev gave him any assistance on this occasion. His hopes to get financial support from the War Ministry were not realised either; the department had turned down earlier the request of the IRGO to retain Przhevalskii's fifth expedition after his death, saying that "it does not care about special research of Tibet and

46 Grumm-Grzhimailo, G.E, 1896, VII. Because of the renewed tensions in Anglo-Russian relations, the IRGO sent to the Hindu Kush L.B. Grombchevskii, see M.K. Baskhanov's essay on B.L. Grombchevskii in this volume.

47 Potanin 1893.

48 Quoted from Bukhert, 2006 189. Letter from G.E. Grumm-Grzhimailo to S.D. Sheremetev, 12 February 1889.

Kashgaria".[49] Similarly, the military showed no interest in the expedition of a civilian, even though he planned to visit a region of great geopolitical import. Moreover, when the IRGO petitioned the Emperor to send Lieutenant Mikhail Grum-Grzhimailo on "an official journey", the Tsar decided to grant him leave instead. However, it was noted that if his "work will be recognized as satisfactory and useful", his travelling time would be reckoned as "active service". The head of the General Staff N.N. Obruchev explained this decision by making a point that an officer could be sent on an official journey in the capacity of expedition leader, and not otherwise, while Mikhail Grum-Grzhimailo was only one of the expedition staff.[50] This meant that the latter was not entitled to get any allowance from the War Ministry. (It is worthy of note here that the members of the Pevtsov expedition mounted by the IRGO in 1889, following Przhevalskii's death, Lieutenants Kozlov and Roborovskii, were attached to the mission by the War Ministry and received double allowances from the department.)

Apart from Grigorii and Mikhail, the expedition staff included a seven-man escort, consisting of six Cossacks and one gunner. Unlike Lieutenant Grum-Grzhimailo they received their salaries and other allowances in advance, together with their military equipment and horses.[51] Besides, all the expenses required for the additional equipment of "low ranks" during their long march were met by the Governor-General of the *Stepnoi Krai* (Steppe Country), General G.A. Kolpakovskii, who was one of the key figures of the Russian conquest of Middle Asia.

The Zoological Museum of the Academy, on the other hand, provided G. Grumm with the accessories that he needed for collecting samples of fauna. These included gun-powder, small shot, spirits, instruments for preparing birds etc., plus a sum of 180 rubles. "The resources of all our [scientific] societies are so slender", Grigorii Grumm wrote to Sheremetev, "that they cannot give any help, despite their burning desire to do so".[52]

Thus the entire funding of Grigorii and Mikhail Grum's expedition came to about 10,000 rubles, which was ten times less than the sum donated by the Emperor, a year before, to Przhevakskii. Still the explorer hoped he would be

49 Quoted from Bukhert 2006, 187. Letter from G.E. Grumm-Grzhimailo to S.D. Sheremetev, 1 February 1889.

50 NA RGO, F. 32, Op. 1, D. 13, l. 5. Letter from N.N. Obruchev to P.P. Semenov, 14 March 1889.

51 NA RGO, F. 1–1889, Op. 1, D. 13, l. 19. Letter from N.N. Obruchev to P.P. Semenov, 3 May 1889.

52 Bukhert 2006, 189. Letter from G.E. Grumm-Grzhimailo to S.D. Sheremetev, 12 February 1889. Upon Grumm's return, the museum paid him 600 rubles more, since the collections he brought proved more valuable than he had originally expected. See Slepkova, 2014, 100–101.

FIGURE 6.3 The escort of the Grum-Grzhimailos expedition, 1889. NA RGO, F. 32, OP. 1, D. 57.

able to achieve no less substantial results than his great predecessor.[53] (His financial difficulties can perhaps be explained by the fact that the IRGO was mounting two more expeditions at that time, those under Pevtsov and Grombchevskii, to which the Society apparently gave priority). Luckily, the railway and shipping companies agreed to carry brothers Grum's luggage from St. Petersburg to Vernyi (now Alma-Ata, "the southern capital" of Kazakhstan), the expedition starting point, and back free of charge.[54]

In the middle of March 1889, brothers Grum-Grzhimailo left St. Petersburg for Verny where they began to organize and equip the expedition caravan. Here they were joined by other members of the staff – their escorts (Cossacks I. Komarov, A. Glagolev, I. Churkin, M. Komarov, P. Kolotovkin, M. Fateev and a gunner M. Zhiliaev), an interpreter, a Kalmyk N. Ananiev, and one more man,

53 Bukhert 2006, 189. Letter from G.E. Grumm-Grzhimailo to S.D. Sheremetev, 12 February 1889.

54 Grumm-Grzhimailo, G.E. 1896, VIII.

T. Issyman-Khodzhaev, a Sart by birth, who had participated in Grumm's Pamir expeditions. They were eleven people altogether.[55]

In April Grigorii Grumm received the Chinese travel passports from the Ministry for Foreign Affairs. However the documents did not guarantee the travelers a safe passage along their chosen route. According to Petrovskii, "the Chinese permitted the expedition to work in Kashgar Province rather reluctantly".[56] The consul realized that they could object to the exploration of the Tian Shan and Lake Lopnor, and therefore, recalling his earlier clashes with Grumm, he believed that the explorer might turn against him again and accuse him of non-cooperation with his (Grumm's) "scholarly undertaking". The head of the Asiatic Department also warned Semenov that the Chinese authorities could put obstacles in the way of brothers Grum's expedition and requested him to support Petrovskii, if the consul failed to settle the problems.[57]

On 27 May 1899 the expedition left the frontier Kazakh town of Dzharkent and crossed into Chinese territory. The brothers Grum spent the summer exploring Eastern Tian Shan and Southern Dzungaria. They collected a large amount of geographical data, within Eastern Tian Shan, where they discovered an uncharted mountain, the Dos-Megen-Ula, 6100 meters high. Then the Grums explored another mountain range, Bogdo-Ula, east of Urumchi. They were the first Europeans to do so. The Bogdo-Ula range divided the Turfan depression from the Dzungar Valley, into which they subsequently descended to hunt for the wild horse. The animal had been originally discovered by Przhevalskii, in 1881, and given the name of *Equus Przewalskii* by zoologists.[58] The Grums turned out to be the first observers of the horse in its natural habitat and their hunt was quite successful. They would deliver to the Zoological Museum four skins, three sculls and a part of the skeleton of several wild horses, both male and female, of different age, which were of much value to zoologists. The Museum's Director A. Strauch highly appreciated the gift and praised both brothers for their "special merit" – their remarkable energy "in pursuing their chosen goal".[59]

From the Dzungar Valley the expedition moved on to the oasis of Turfan. Some 250 km south-east of Urumchi the brothers explored the Lukchun

55 Grumm-Grzhimailo, G.E. 1896, VII. Grigorii Grumm-Grzhimailo originally wanted his younger brother Dmitrii also to take part in this expedition.

56 NA RGO, F. 1–1889, Op. 1, D. 13, l. 22–22 ob. Copy of Petrovskii's report to the Asiatic Department of the Foreign Ministry (MID), 12 May 1889.

57 NA RGO, F. 1–1889, Op. 1, D. 13, l. 21. Letter from the Asiatic Department to P.P. Semenov, 4 July 1889.

58 Poliakov 1881.

59 Slepkova 2014, 101.

FIGURE 6.4 The bivouac of the Inner Asian expedition of the Grum-Grzhimailo brothers at Djus-Agach in Dzungaria, 1889. NA RGO, F. 32, OP. 1, D. 57.

FIGURE 6.5 The Bogdo-Ula mountain ridge in the Eastern Tian Shan. In the foreground is Mikhail Grum-Grzhimailo, 1889–1890. NA RGO, F. 32, OP. 1, D. 57.

(Turfan) depression, having ascertained hypsometrically that it was located 130 meters below sea level.[60] Practically at the same time, the expedition of Pevtsov, working in the opposite (western) part of the depression, in the oasis of Toksun, attained the same result. The discovery of the Lukchun depression became one of the Grums' most important geographical discoveries and, quite naturally, it attracted the attention of geographers world-wide. The investigations of the depression continued in the subsequent years by V. Obruchev (1892–1894) and V. Roborovsklii (1893–1895).

From Turfan the expedition made its way via the oasis of Kham directly to Nanshan. Its route ran diagonally (from southwest to southeast) almost parallel to the routes of the expeditions of Potanin and Przhevalskii. This stretch of the Grums' journey was also crowned with much success, this time in the field of orography. The expedition discovered the Beishan upland, 2700–3000 m high, in a place where the maps showed the Hami desert. As G. Grumm was able to ascertain, Beishan provided a connection between the Tian Shan and the Nanshan. Three years later the geologist V.A. Obruchev would conduct a detailed geological investigation of the Beishan Mountains.[61] In the opinion of geologist Karl Ivanovich Bogdanovich (1864–1947), the discoveries of the Lukchun (Turfan) depression and of the Beishan Mountains by brothers Grum were very important theoretically for the understanding of the geological evolution of Inner Asia.[62]

In the spring and summer of 1890 the expedition explored the southern slopes of the Nanshan and thence descended into the valley of the Huang He river. Here a tragic incident occurred with Cossack Kolotovkin who wounded himself badly due to the mishandling of his gun. Since no medical assistance could be given to him, he died on 19 July and was buried on the bank of Lake Kokonor.[63]

The expedition proceeded further along the valley of the Heihe river (the upper and middle courses of Edzin-gol) towards Suzhou, and hence, again across the Nanshan and the northern foothills of the Tian Shan following their old route, to Urumchi. On 8 November the travelers arrived in Kulja, and a few days later (on 13 November) they were finally back to Russia.

60 According to the latest measurements, the lowest point of the depression makes 154 m below sea level. It is the third lowest part of the surface of Earth after the depressions of the Dead Sea in Israel and Lake Assal in the Republic of Djibouti in Africa.

61 Obruchev 1901, 483–535.

62 Bogdanovich 1907, 12–16.

63 Grumm-Grzhimailo, G.E. 1891, 180; Grumm-Grzhimailo, G.E., 1899, 3–8.

Incidentally, the year 1890 witnessed the termination of the three major expeditions to Inner Asia, all mounted by the Russian Geographical Society. While the focus of brothers Grum-Grzhimailo was the eastern part of the Tian Shan, the expeditions under the leadership of Pevtsov and Grombchevskii explored the northern outskirts of western Tibet, Kunlun (the former), and the Pamir (the latter). Considering the scientific importance of these journeys, the State Treasury Department allocated a sum of 24,000 rubles for the publication of the travel accounts by their leaders.

The results of the expedition were quite substantial. The travelers covered 7250 km, 6000 of which along the hitherto uncharted route. They made 140 hypsometric and aneroid measurements and determined the geographical coordinates of 30 points. Apart from these, the Grums conducted systematic meteorological observations. Especially valuable to geographical science was their discovery of the Lukchun depression and of the Beishan Mountains.

Their zoological collection included up to 114 specimens of large and middle-size mammals and 100 small ones; 1150 specimens of birds and 400 eggs with nests; 90 samples of fish, among which was a new genus Diptichus from the Huang He river, and 105 samples of reptiles and amphibia. In the opinion of A.A. Strauch, the expedition's faunistic material considerably broadened and even altered our "knowledge of the geographical distribution of Central Asian animals, primarily mammals".[64]

Grigorii's entomological collection consisted of over 35,000 insects, the botanical one of 800 herbarium sheets, and he also collected over 850 samples of rock and minerals.[65] These collections were then expertly described by zoologists from the Zoological Museum: E.A. Bikhner (mammals); F.D. Pleske (birds); A.P. Semenov and S.N. Alferaki (lepidoptera), assisted by G. Grumm himself; while A.M. Nikolskii from the St. Petersburg University studied reptiles, amphibia and fish. The botanical collection went to Director of the Botanical Gardens A.F. Batalin and the geological one to I.V. Mushketov.

The expedition also collected some ethnographic material and took many photos – of local people and of the lands they visited. The ethnographic photos were displayed, in 1891, at an exhibition set up by the Russian Technical Society and were awarded the bronze medal.[66] Also, Grigorii Grumm made a collec-

64 Slepkova 2014, 100.

65 Grumm-Grzhimailo, A.G. 1947, 32; Slepkova 2014, 100.

66 Archive of the RGO, F. 32, Op. 1, D. 82. Diploma of the medal, issued on 11 May 1891.

FIGURE 6.6
The Tangutans from the vicinity of Kokonor (Qinghai Lake). NA RGO, F. 32, OP. 1, D. 57.

tion of seeds, at the request of the Ministry of the State Property, which planned to use these for agricultural purposes in Russia.[67]

The explorer reported of the results of his expedition at the meeting of the IRGO which took place on 13 March 1891 in the Grand Hall of the State Duma. His report aroused much interest among those present, numbering about 600 people.[68] The scientific societies and institutions also highly appreciated the work done by the expedition. P.P. Semenov, for example, was of the opinion that "the expedition of brothers Grum-Grzhimailos is one of the most interesting and successful expeditions to Inner Asia in the past decade",[69] and he thought that Grigorii Grumm-Grzhimailo deserved one of the highest awards of the Society – either the Konstantin gold medal or the F.P. Litke gold medal.

67 Grumm-Grzhimailo, G.E. 1891, 180.

68 Grumm-Grzhimailo, A.G. 1947, 33.

69 Semenov 1892, 14.

However, the former had been already awarded, in 1891, to Mikhail Pevtsov, and the latter one to the explorer of the North Pole General A.I. Vel'kitskii. Therefore Grigorii Grumm received the newly-instituted Przhevalskii Prize (*Premiia Przhevalskogo*) for his explorations in Inner Asia. He was also awarded the order of St. Vladimir of the 4th degree by the Russian State "for his special service to the Fatherland", and, in May 1891, he was given the life pension of 600 rubles a year.[70]

The Society of Natural History, Anthropology and Ethnography elected Grigorii Grumm-Grzhimailo as its member, in 1892, and the Zoological Museum as its corresponding member, in 1895.

G. Grumm's explorations did not go unnoticed by Western scholarly community. The French Academy of Sciences (*Académie des Sciences*) on 4 December 1893 awarded him, the first among Russian scholar-travelers, the Chikhachev Prize (*Le prix Tchihatchef*) "for his research of Western China, the Pamir and the Tian Shan".[71] And the Royal Dutch Geographical Society (*Koninklijk Nederlands Aardrijkskundig Genootshap*), in 1895, elected him as the Society's honorary member.

As for his brother Mikhail, his work was also appreciated by the IRGO. The Society awarded him the Przhevalskii silver medal for his surveys and astronomic measurements. The same award was given to K.I. Bogdanovich, P.K. Kozlov and V.I. Roborovskii, the participants of M. Pevtsov expedition (1889–1890). Besides, the time Mikhail spent in Inner Asia (one year and nine months) was reckoned as his military service and he received his full salary for the period. Moreover, Mikhail Efimovich, like his brother, received the order of St. Vladimir of the 4th degree.

Medals and cash bonuses were also given to brothers Grums' companions. Apart from these, Dzhigit T. Khodzhaev, at the request of the IRGO, was given "the honorary *khalat* (robe) of the third degree" by the Governor-General of Turkestan.[72]

The full account of Grigorii and Mikhail Grum's expedition made three large volumes, including numerous illustrations, photographs, maps and other

70 NA RGO, F. 32, Op. 1, D. 13, l. 59. Letter from the Foreign Minister P.N. Durnovo to P.P. Semenov, 7 May 1891.

71 *Gauja* 1917, 289. This award was instituted in 1891, using the funds bequeathed to the French Academy of Sciences by the Russian geographer, geologist and traveler P.A. Chikhachev. Apart from G.E. Grumm-Grzhimailo, the prize was given to three more Russian explorers – V.A. Obruchev (1897 and 1925), Yu.M. Shokal'skii (1911), and P.K. Kozlov (1913).

72 NA RGO, F. 32, Op. 1, D. 13, l. 57. Letter of the Turkestan Governor General A.B. Vrevskii to P.P. Semenov, 1 May 1891.

appendices, published in 1896, 1899 and 1907.[73] This was his major work in the field of geography and his most prominent contribution to the study of Inner Asia.

The first volume Grigorii Efimovich dedicated "with a feeling of profound gratitude" to the Grand Duke Nikolai Mikhailovich. It was thanks to him, in 1884, that Grigorii Grumm started his career of scholar-traveler, an explorer of Inner Asia.

In the Service of the State, 1890–1917

The Inner Asian expedition finished the period of Grumm's active field work. One of the reasons was his marriage, in 1891, to his cousin Evgeniia Dmitrievna Bez-Kornilovich (1866–194?). They got to know each other as early as 1882, when Grigorii visited his uncle D.M. Bez-Kornilovich, on his estate near Kamenets-Podolsky (now Ukraine). Evgeniia was quite an extraordinary woman. She graduated from the Odessa Conservatory as a vocalist and concurrently took exams to obtain the certificate of secondary education at the Novorossiisk University at Odessa. The certificate allowed her to take up teaching, however instead she would devote herself entirely to the education of her own children and to assisting her husband's scholarly activities.

The Grumms had three children: two sons, Vladimir (1892–1915) and Aleksei (1894–1966), and a daughter Olga (1901 – ?). The elder son Vladimir died at the age of 22, when he was a student of a military college. The younger one, Aleksei, finished, in 1915, the Imperial Alexander Lyceum in Petrograd (as St. Petersburg was renamed after the outbreak of the 1st World War) and continued his education at the law faculty of Petrograd University, in 1918–1920, and at the Geographical Institute, where he attended lectures as an extern in 1921. Subsequently he worked in various research organizations, including the Russian Geographical Society. There Aleksei Grumm-Grzhimailo was in charge of the Society's library and he sorted out and described the personal archive of his father. He would eventually become the biographer of Grigorii Grumm-Grzhimailo.

To support his family Grumm had to abandon his exploratory field work and take up the job of a civil servant at the Ministry of Finance where he had moved from the Ministry of State Domains.[74] His vast erudition and analytical abilities found perfect use in the state service. In 1893 he was offered the post of

73 Grumm-Grzhimailo, G.E. 1896, 1899, 1907.

74 The Ministry of State Domains (1837–1894) controlled government-owned lands and any other government property in the Russian Empire.

secretary at the Ministry's Interdepartmental Commission for the development of trade in Middle Asia, headed by the prominent state figure D.F. Kobeko (1837–1917). And shortly after Grumm became head of the Asiatic section of the Trade Department in the same Ministry. In 1893 he compiled a summary of the "physiographic and economic conditions of the Amur region".[75] The Russian government badly needed the compendium for the construction of the Amur section of the Trans-Siberian railroad as well as for the discussion of measures for colonizing the Amur region and for the promotion of trade and industry therein.[76]

The concept of the work was elaborated, at the request of the Ministry, by P.P. Semenov, who personally wrote two chapters – on agriculture, industrial fishing and the trade of Russian colonists in the region. It took Grumm over one year to complete his summary. To this end, in the words of his son, A.G. Grumm-Grzhimailo, he collected and "passed through the filter of scientific criticism" practically all available Russian and foreign-language publications on the Amur region. The final result was a sort of encyclopedia, which included data on the history, geography, mineralogy, geology, meteorology, flora and fauna, ethnography and economy of the Amur country.[77] As Semenov noted in the preface to this work, Grigorii fulfilled "his obligation with his characteristic rapidity and talent".[78]

A need for material well-being of his family made Grumm to look for sources of additional income. In 1892 he began to collaborate with the editors of the Russian-language Brockhaus & Efron Encyclopedic Dictionary. Grumm wrote a number of entries for the edition devoted to the geography of Inner Asia (those on Eastern Tian Shan, Gansu, Tibet etc.).[79] And in 1896 he headed the Oriental section in a popular daily *Sankt-Peterburgskie Vedomosti* (St. Petersburg News) published by Prince Esper Esperovich Ukhtomskii (1861–1921), the well-known diplomat and publicist. For almost two years, in 1876–1877, the newspaper regularly published Grumm's notes discussing various economic, historical, geographic and political issues relating to Oriental countries and the colonial domains of the European powers.[80]

75 Semenov 1894, I.

76 Ibid.

77 See: Latushko, Briantseva 2010.

78 Semenov 1894, II.

79 This voluminous encyclopedic dictionary was published in Russia in 1890–1907 by the joint-stock company of F.A. Brockhaus and I.A. Efron based in St. Petersburg. Its full edition included 82 volumes and 4 additional ones.

80 For the bibliography G.E. Grumm-Grzhimailo's publications in *Sankt-Peterburgskie Vedomosti* see Grumm-Grzhimailo, A.G. 1947, 81–83.

Despite his heavy work schedule, Grumm succeeded in writing up and publishing, in the same period, two volumes of his travels, as mentioned above, entitled *Opisanie puteshesviia v Kitai* (The Account of the Journey to China). Apart from these, he published two major works on lepidopterology.[81] In these books, based on the study of his and other entomologists' collections made in the Pamir, Western Tian Shan, China, Altai, Himalayas, and other parts of the region, as well as in the Pyrenees, Labrador and in Canada, he singled out and described new Palearctic species of the lepidoptera.[82]

Grumm's interest in Inner Asia and its ongoing exploration by Russian and Western scholar-travelers did not wane, as he kept an eye on every new expedition into the region. When he learned about the publication of the travelogue of the American diplomat William W. Rockhill *The Land of the Lamas* (1891), describing his two journeys to Western China, Mongolia and Tibet in the 1880s, he welcomed the idea of having the work published in Russian and volunteered to write a preface and his comments for the edition. In the preface Grumm listed all the prominent explorers of Inner Asia, both Russian and Western, and he particularly highlighted the role of Przhevalskii, "the explorer of the outskirts of Tibet, who earned an immortal name in science",[83] and W. Rockhill, as one of the best explorers. He also mentioned Rockhill's later *Diary of a Journey through Mongolia and Tibet, 1891–1892* and made a point to say that extracts from his both works made a "secret edition" of the General Staff.[84]

G. Grumm's final entomological investigations appeared in the 3rd volume of his *Account of the Journey to China*, published in 1907. The chapters XV and XVI of the work were devoted to the zoographic analysis of the region based on the study of its lepidopterological fauna. In the preface of the volume Grumm stated that he regarded "his task of giving an account of his travels in Western China completed", by adding emphatically: "Feci quod potui faciant meliora potentes" (I have done what I could; let those who can do better). From that time on, Grumm conducted no further entomological research. His waning interest in entomology can perhaps be explained by the fact that his studies of that branch of science were underestimated by other Russian specialists.[85]

81 See Grumm-Grzhimailo, G.E. 1893; ibid. 1893a; ibid. 1893b; ibid. 1895.

82 Korolev, Murzin 2005, 44.

83 Grumm-Grzhimailo 1901, 3.

84 Grumm's preface to the Russian edition of W.W. Rockhill's travelogue, see Grumm, 1901, 5.

85 Quoted from Grumm-Grzhimailo, A.G. 1947, 54.

In the same period Grumm published a number of articles using the materials of his Inner Asian expedition. In these he touched upon the geographic, ethnographic and historical aspects of the regions he had explored.[86]

In 1899 Grumm got a position at the Department of Customs in the Ministry of Finance. It is this Department that would be linked with the final stage of his exploratory work, this time at the eastern and southeastern borderlands of the Imperial Russia. From 1903 until 1915, he made many business trips to various places – to Western Mongolia and the Uriankhaiskii Krai (1903), to the Amur and Cis-Amur countries (1908), to Kopet-Dag and the southern coast of the Caspian Sea (1911), to Lenkoran and along the entire Russian border with Persia and Turkey, including the Kars area (1912), to the Ussuri Valley and the northern slopes of the Little Khingan (1913), and, finally, to the Sayan Mountains and Uriankhai (1914), which, shortly before, had been placed under Russia's protectorate. In the spring of 1915, after his son's death, he went to the Far East again. His task this time was to speed up the delivery of a large quantity of goods, shipped from the USA to Vladivostok for the Russian army and fleet, to the central provinces of the country. Yu. M. Shokalskii, the new Vice-Chairman of the IRGO, after P.P. Semenov's death in 1914, assessing Grumm's work as a state official, remarked that although he was a functionary-bureaucrat ex officio, he "managed not to become one indeed".[87] He conveniently combined the assignments given to him by his superiors at the Ministry with the historico-geographical investigations of the places he visited.

G. Grumm's most successful trip, scientifically, was his first trip to Western Mongolia and the Uriankhai country. He made it at the behest of the finance minister Count S.Yu. Witte (1849–1915) to gather information on the conditions of trade with the Mongols and the situation at the Russo-Mongolian frontier. His itinerary was given in the foreign passport as Zaisan – Kobdo and Uliasutai, from which he was to return to Russia through the territory of Urianhai.[88] Accompannied by botanist M.M. Siiazov and a small group of Cossack escorts, Grumm crossed the Russian border on 2 June 1903. The party entered Mongolia through the basin of the river Saksai, reached the Lake Tal-nur, and thence proceeded to the Buiantu Valley and the town of Kobdo. From Kobdo the travelers struck northwards to the Uriankhai country from which, having crossed the Tannu-Ola range, they moved further to the Altai village Kosh-agach, on the Russian territory. Here started the newly built Chuiski Trakt (high road) which brought them to Biisk, the final point of their journey, on 20 August 1903.

---

86 The most significant of these are Grumm-Grzhimailo, G.E., 1894; idem 1898; idem 1898a.

87 Shokal'skii 1936.

88 NA RGO, F. 32, Op. 1, D. 69, G.E. Grumm-Grzhimailo's foreign passport, issued 10 April 1903.

Grumm reported on his journey at the meeting of the IRGO on 22 October 1903. The Society noted, in its annual report for that year, that his results were "fairly successful"[89] – Grumm-Grzhimailo surveyed 650 km of his route, determined the altitude of 42 points, conducted meteorological observations, made several collections of fauna and flora and gathered substantial material on the trade and ethnography of the native peoples. However, his greatest contribution was his major work *Zapadnaia Mongoliia i Uriankhaiskii Krai* (Western Mongolia and the Uriankhai Country). Its first volume, published in 1914, contained a detailed geological, geographic, climatologic, botanic and zoological description of the lands he visited.[90] Grumm began writing the book after the completion of his "Account of the Journey to China". While doing this, he deviated from the traditional scheme of narrative closely linked with the traveler's chosen route; instead he offered the readers a series of his personal observations providing a full and comprehensive picture of the regions he had explored. To this end he used all the available literature (more than 500 publications), thus summing up the contemporary knowledge of these countries. The well-known Russian geographer Leo S. Berg (1876–1950) noted in his letter to Grumm: "your *Western Mongolia* truly deserves to be called a classical work, which will bring you glory no less than your *Western China*. One can only regret that none of the regions of our country has a geographical description similar to the one you gave to Western Mongolia".[91]

The IRGO highly appreciated Grumm-Grzhimailo's explorations in Inner Asia by awarding him the Konstantin gold medal in 1907 for the sum total of his works on geography and particularly for the final volume of his *Account of the Journey to China*. And on 6 April 1910 the Society elected him as an honorary member, together with N.G. Potanin and P.K. Kozlov. In the opinion of the IRGO Council, these three explorers "completed the cycle of Przhevalskii's expeditions".[92]

After the completion of the Mongolia-Uriankhai expedition Grumm-Grzhimailo resumed his work on his other magnum opus, *Western Mongolia and Uriankhaiskii Krai*. Its second and third volumes would be published with some delay, after the 1917 revolution.

89 *Otchet of the IRGO for the year 1903*, St. Petersburg, 1904, 8.

90 Grumm-Grzhimailo, G.E. 1914.

91 NA RGO, F. 32, Op. 2, D. 35, ll. 1–2. Letter from L.S. Berg to G.E. Grumm-Grzhimailo, 14 January, 1927.

92 *Izvestiia IRGO*. 1910, XLVI, 7.

Activities in Soviet Russia, 1917–1936

The Russian revolutions in 1917 disrupted Grumm's habitual way of life. First and foremost, Grigorii Efimovich lost his permanent job. What helped him to survive in the severe conditions of hunger and economic devastation in the country, was his continued cooperation with the RGO (which was not the Imperial Geographical Society any more), where he was a member of the Society's Council, and his close contacts with the new Chairman of the entity Yulii Mikhailovich Shokalskii (from the autumn of 1917). Thus, for example, Grumm was among the Society's five most privileged members who received additional food rations, provided by the special Relief Commission for Scientists.[93] He was also licensed to keep his own lodging by the *Narkompros* (People's Commissariat for Public Education). In 1920 Grumm was elected Vice-Chairman President of the RGO, and he retained the position until 1931, when serious illness made him retire.

The civil war that followed the Bolshevik Revolution affected G. Grumm's family most dramatically. His brother, Major-General Mikhail Grum-Grzhimailo, who had participated twice in his expeditions, now served in the anti-Bolshevik White Army in the south of Russia. In September 1920, when moving to Moscow, he was arrested and put to the Butyrki prison, where he died on 8 May 1921.

Judging by his post-revolutionary career, G. Grumm was quite loyal to the new Bolshevik authorities. Moreover, like many other representatives of Russian liberal scholarly community he shared some of the Bolshevik ideas, such as the right of nations to self-determination.[94]

Despite his connections with his counter-revolutionary brother Mikhail, Grumm easily secured himself a place in the new Soviet system. What helped him were his high scientific credibility as an expert on the history and geography of Mongolia and China, the countries of paramount geopolitical importance to the Soviets, as well as his professional knowledge as an official of the Department of Customs at the Ministry of Finance. In the spring of 1921, on orders of the Institute of Economic Research at the Narkomfin (Commissariat of

93 NA RGO, F. 32, Op. 1, D. 87, l. 1. RGO Certificate, 10 April 1919. The Committee in question was set up by the writer Maxim Gorky in 1919 to improve the standard of living of the scholarly community. The Committee was supported by many international organizations which sent various food-stuffs to Soviet Russia.

94 NA RGO, F. 32, Op. 2, D. 8, l. 3 ob. Letter from G.E. Grumm-Grzhimailo to B.B. Baradiin, 15 February, 1931.

Finance), he wrote a "Report on the economic situation in modern Mongolia",[95] in which he made recommendations on the trade with this country.

Simultaneously, the People's Commissariat for Foreign Affairs requested him, in 1921, to write a "History of the Kalmyks." This was occasioned by the desire of the administration of the Kalmyk Autonomous Region to have a history of the Kalmyks published in the Kalmyk language, for which purpose they had approached the Soviet leader, Vladimir I. Lenin.[96] In order to meet the tight schedule, Grumm had to send his work in parts to Astrakhan where they were translated into Kalmyk. However an epidemic of typhus broke out in the region, the translator died, and the greater part of Grumm's manuscript was burnt.[97] Unfortunately, no copy of the text was made either by Grumm himself, or by the editor of the book, the well-known Mongolist V.L. Kotvich (1872–1944). And after Kotvich departed for Poland in 1923, the project went off the agenda.[98]

It was at this time that Grumm switched to a new sphere of activity. He began teaching a course of regional geography at the newly established Petrograd / Leningrad Institute of Geography (in 1921–1924), to become the Faculty of Geography at the Leningrad State University in 1925. In 1922–1930 he also lectured at the Institute of the Living Oriental Languages on the history of Mongolia, the ethnography of the Mongolian peoples and the regional geography of Asia.

At the end of 1917 Grumm completed the second volume of his *Mongolia and Uriankhaiskii Krai*, which focused on the history of these countries, from ancient times till nowadays.[99] However, he was unable to publish this rather voluminous work of his (being 900 pages long), as the RGO lacked the necessary funds. The book was published only in 1926 in Leningrad, thanks to the efforts of the executive secretary of the Mongolian Scientific Committee (*Uchkom*), Tseben Zh. Zhamtsarano (1881–1942), a Buryat scholar who had migrated there. The book presented a considerable interest for the Mongolian Uchkom so the entity, despite their paucity of funds, managed to print it, as evidenced by the note on its title page.

95 Grumm-Grzhimailo, A.G. 1947, 62.

96 The Kalmyk Autonomous Region, or The Autonomous Region of the Kalmyk Working People within the Russian Soviet Federated Socialist Republic (RSFSR), was established in November 1920. Today it is Republic of Kalmykia, within the Russian Federation.

97 NA RGO, F. 32, Op. 1, D. 20. Printed copy of G.E. Grumm-Grzhimailo's *History of the Kalmyk People*, 144 ps.

98 NA RGO, F. 32, Op. 2, D.8, l. 3 ob. Letter from G.E. Grumm-Grzhimailo to B.B. Baradiin, 15 February 1931.

99 Grumm-Grzhimailo, G.E. 1926.

In this work Grumm analyzed over 1000 publications of 700 authors, in the Russian and European languages, covering a broad period of Mongolian history. He was the first to produce a systematized and detailed account of the history of Inner Asia and to trace the ethnogenetic connection of the Mongols with the peoples who preceded them. Undoubtedly, his presentation of many historical events looks erroneous from today's point of view. Yet, on the other hand, there are topics which Grumm treated boldly in a rather innovative manner, such as his interpretation of the personality of Genghis Khan as a great military leader and the positive estimation of Genghis Khan's conquests for the evolution of the Mongolian statehood. For these views, running contrary to the Marxist approach, Grumm would be severely criticized by Soviet historians.

However, Grumm's *Historical review* was highly praised in Mongolia and he was offered an opportunity to write a textbook of Mongolian history for secondary school.[100] The writing took him more than a year, and by the end of 1928 the Uchkom received from Grumm a huge manuscript with genealogical and chronological tables appended. Zhamtsarano was delighted and made a decision to publish the textbook, yet its large volume created a serious problem for translation.[101] The work proceeded very slowly and then, in 1930, Zhamtsarano was suddenly removed from his post as a result of a change of the political leadership in the country. In his letter to Grumm he explained that the attention of the Mongolian leaders is directed entirely towards "the planned construction of socialism, so it is natural that the humanitarian work remains in the background, and we are unable to promote such capital works of yours as your course on the *History of the Mongols*".[102]

Having faced these difficulties with the publication of his book in Ulan-Bator, Grumm tried to have it published in Russian, in the USSR, the more so that some of the Soviet scholarly institutions, such as the Leningrad Oriental Institute, the Communist Academy and the Buryat-Mongol Institute of Culture, showed interest in the edition. Still, his *History of the Mongols* did not go to the printers, and in 1933 the idea of its publication was finally abandoned. This occurred, so it seems, for two main reasons – an acute shortage of printing paper in the country, but, more importantly, because of the author's ideological dissension with the official Soviet assessment of the events and political figures of

100 See Yusupova 2016.

101 NA RGO, F. 32, Op. 2, D. 61, l. 30. Letter from Ts. Zhamtsarano to G.E. Grumm-Grzhimailo, 22 December 1931.

102 NA RGO, F. 32, Op. 2, D. 61, l. 28. Letter from Ts. Zhamtsarano to G.E. Grumm-Grzhimailo, 8 August 1931.

Mongolian history. Thus both historical projects of Grumm – *History of the Kalmyks* and *History of the Mongols* – failed to reach the public.

In the midst of his writing the *History of the Mongols*, in September 1928, Grumm was awarded the title of 'Honored Worker of Science', one of the first in the USSR, on the initiative of the Council of the Geographical Society.[103] Now the Soviet government appreciated his contribution to the development of scientific knowledge of Inner Asia.

The year 1930 marked the 50th anniversary of Grumm's scientific career which had started with his work on the Filoxera Committee. In the same year the second part of the third volume of *Western Mongolia and Uriankhai Country*, sub-titled *Anthropologic and ethnographic description of these countries*[104] was published. (The first part was printed by the GO in 1926[105]). On this occasion the Geographical Society, in recognition of Grumm's long-time and extremely productive scientific activity, organized a gala evening at its quarters on 21 March 1930. Numerous Soviet scientific institutions and individual scholars forwarded their congratulations to the explorer. Grumm's foreign colleagues also welcomed him warmly and congratulated him on his scholarly jubilee, such as the German geographer Albrecht Penk (1858–1945), the French Orientalist Paul Pellio (1878–1946), the explorer and geographer Alfred Lacroix (1863–1948), the French mathematician Émile Picard (1856–1941), who was an honored member of the Academy of Sciences of the USSR, and others.

This meeting became a kind of summing up of the many accomplishments of the explorer. And shortly after, in the second half of 1930, Grumm fell seriously ill. He was diagnosed with uremia due to kidney failure. Yet what really made Grumm suffer and particularly depressed him, as Shokalskii noted, was not his ailing condition, but rather his "inability to work".[106] When he felt relieved, he resumed his studies, writing articles and book reviews for the *Izvestiia Geograficheskogo Obshchestva*. One of the reviews was devoted to the work of the new President of the Society, the well-known Soviet geneticist Nikolai I. Vavilov (1887–1943), who would die tragically in the wake of the political repression in the USSR.[107] In the summer of 1935 Grumm wrote his last article – a review of China for the *Great Soviet Encyclopedia*.[108]

103 NA RGO, F. 32, Op. 1, D. 91, l. 1. Letter from the Glavnauka of the Narkompros to the Presidium of the State Geographical Society, 18 October 1928 (Reference to the Sovnarkom decree of 24 September 1928, pr. № 55).

104 Grumm-Grzhimailo, G.E. 1930.

105 Grumm-Grzhimailo, G.E. 1926a.

106 Shokalskii 1936.

107 Grumm-Grzhimailo, G.E. 1931.

108 Grumm-Grzhimailo, G.E. 1936.

Grigorii Grumm-Grzhimailo died of uremia on the night of 3 March 1936. He was buried at the Volkovskoeo cemetery in Leningrad, at a place called *Literatorskie Mostki*, a special necropolis of the Russian and Soviet cultural elite.

In his obituary Shokalskii would write: "The last of the famous galaxy of explorers of nature in Inner Asia, whom the Geographical Society will always be proud of, has passed away".[109] Grumm's death and Petr Kozlov's death a year before finally closed the bright page of the Russian explorations in the region, so brilliantly started by Nikolai Przhevalskii.

Conclusion

Grumm-Grzhimailo had a particular quality which made him different from other Russian explorers portrayed in this book. In the opinion of Berg, this was the "perspicacity and erudition of the armchair scholar" combined with an excellent knowledge of the methods of field work.[110] His interests were much broader than the geographical research of Inner Asia. He had a profound knowledge of entomology (lepidopterology) and the history of Inner Asian countries as well as aspects of economy and trade in the eastern borderlands of the Russian Empire.

He started his career as an entomologist and a traveler to the Pamir, Bukhara and Western Tian Shan. From there he brought his first collections of the lepidoptera, which he presented to his sponsor, a great lover of entomology Prince Nikolai Mikhailovich Romanov. His personal lepidopterological collection he presented partly to the Zoological Museum of the Academy of Sciences, and partly exchanged with other collectors of those days or simply sold them to European museums. One of his buyers was the well-known British botanist and lepidopterologist Henry John Elwes (1846–1922). Grumm came to know him during his expedition to Western Altai in 1898. Subsequently, the collection of Elwes, together with the samples collected by Grumm, went to the British Museum of Natural History, where it belongs now.[111]

In his first major work based on his four trips to the Pamir, *Le Pamir et sa faune lepidopterologique,* Grumm provided data on 200 species of Rhopalocera belonging to 32 genera, and 92 species of *Heterocera,* including 80 species of

---

109 Shokalskii 1936.

110 NA RGO, F. 32, Op. 2, D. 35, l. 7. Letter from L.S. Berg to G.E. Grumm-Grzhimailo, 20 March 1930.

111 Korolev, V.A. *Galereia lepidopterologov Rossii,* URL: <http://archive.is/l4HI>. (30.08.2016).

Sphingidae (34 genera) and 12 species of *Bombycidae* (6 genera).[112] G. Grumm's further analyses of the formation and dissemination of the genera *Parnassius* and *Colias* throughout the world revealed the connection of the faunas of the Pamir and such remote territories as America.

However, Grumm's entomological studies were not given proper attention and recognition by his Russian colleagues. Nowadays his books are widely known and quoted by many entomologists specializing in the lepidoptera of the Tian Shan, Alai and the Pamir.

The peak of Grigorii Grumm's activities as an explorer was his Inner Asian expedition made jointly with his brother Mikhail. This journey stands out prominently as his greatest scholarly achievement and the most significant event of his life. Using Przhevalskii's method of 'reconnaissance' of uncharted lands, the brothers covered a vast territory and made two major geographical discoveries – of the Turfan depression and the Beishan Mountains, the latter providing an orographic link between Eastern Tian Shan and the middle Kunlun. These discoveries, according to geologist Karl Bogdanovich, had "a profound theoretical importance for understanding the evolution of Asia",[113] as mentioned above.

The geological interpretation of the territories explored by brothers Grum-Grzhimailo, especially his view of the continental – and not marine – character of the so-called Khanhai tertiary deposits,[114] were supported by the prominent Austrian geologist Eduard Suess (1831–1914).[115]

In addition, the expedition of the brothers Grum made an important contribution to the geographical study of the Nanshan, determined the southern borders of the Central Gobi, the Takla Makan desert, the shape of the source of Edzin-gol – the river Heihe, as well as of the Tarim and Lopnor.[116]

The bibliography of works written by Grigorii Grumm-Grzhimailo is extensive and varied thematically; it includes his books and articles in geography, entomology, economy, history and a wide range of bibliographic, political, ethnographic and other notes. Grumm's remarkable erudition allowed him, in the words of his son, Aleksei Grigorievich Grumm-Grzhimailo, to produce "prominent synthesizing works".[117] His accounts of his travels were not simply

112 Grumm-Grzhimailo, A.G. 1947, 25.

113 Bogdanovich 1908, 12–16.

114 These were the red-colored sedimentary Meso-Cenozoic deposits (sandstones and conglomerates), the precise age of which was then unknown. Nowadays this term is out of use.

115 Suess 1901, 213–234.

116 Shchukina 1955, 218, 222 et al.

117 Grumm-Grzhimailo, A.G. 1947, 3.

descriptions of his routes. His works were based on the analyses of the vast literature of his predecessors, which allowed him to compare his own work with that of other explorers. Grumm was the first among Russian geographers who turned his attention to the Chinese sources and he spoke of the need to transcribe Chinese geographical toponyms.

As we already mentioned, not all of his major works were welcomed by his colleagues – experts in special branches of knowledge. Thus, several of his geological theories were refuted by V.A. Obruchev and K.I. Bogdanovich, and his historical concepts by the Orientalist and historian V.V. Bartold (1869–1930). The latter did not agree with G. Grumm's historico-ethnographic views expressed in his *Historical Past of the Beishan* and reproached him harshly for his dilettantism.[118] However, Grumm, as he put it, was not afraid "to spread himself". He was convinced that "with the modern specialization of knowledge, works of synthesizing character are more needed now than ever before".[119]

In recognition of Grigorii Grumm-Grzhimailo's explorations in Inner Asia, his name was given to three geographical features – the largest glacier discovered by Grumm within the Sheremetev glacier group in the upper Tanymas river in the Pamir, one of the passes in the Sikhote Alin Mountains, and one of the glaciers of the Bogdo-Ula range in Eastern Tian Shan.

118 Grumm-Grzhimailo, G.E. 1898.

119 Quoted from: Grumm-Grzhimailo, A.G. 1947, 74.

Bibliography

Akramov, N.M. (Акрамов Н.М.). 1974. Voprosy istorii, arkheologii i etnografii Pamira i Pripamir'ia v trudakh B.L. Grombchevskogo (Вопросы истории, археологии и этнографии народов Памира и Припамирья в трудах Б.Л. Громбчевского), Душанбе: издательство Ирфон.

Aldan-Semenov, A. (Алдан-Семёнов, А.). 1965. Semenov-Tian-Shanskii (Семёнов-Тян-Шанский), Москва: Молодая Гвардия.

Andreyev, A.I. (Андреев, А.И.). 1997. Ot Baikala do sviashchennoi Lhasy: Novye materialy o russkikh ekspeditsiiakh v Tsentral'nuiu Aziiu v 1-oi polovine XX veka (Buryatia, Mongolia, Tibet) (От Байкала до священной Лхасы. Новые материалы о русских экспедициях в Центральную Азию в 1-ой половине XX века (Бурятия, Монголия, Тибет), Санкт-Петербург; Самара; Прага: ООО «Агни».

Andreyev, Alexandre. 2001. Russian Buddhists in Tibet, from the end of the 19th century – 1930, *Journal of the Royal Asiatic Society*, series 3, 11 (3), 349–362.

Andreyev, Alexandre. 2001. Indian Pundits and the Russian Exploration of Tibet: An Unknown Story of the Great Game Era, *Central Asiatic Journal*, 45 (2), 163–180.

Andreyev, Alexandre. 2003. The Tsar's Generals and Tibet: Apropos of some 'white spots' in the history of Russo-Tibetan relations, in A. McKay (ed.) Tibet and her neighbours: A History, London: Hans-Jörg Mayer, 167–168.

Andreyev, Alexandre. 2003a. Ia.P. Shishmarev – diplomat, puteshestvennik, issledovatel' Mongolii (Я.П. Шишмарёв – дипломат, путешественник, исследователь Монголии), *Mongolica-VI*, Санкт-Петербург, 118–124.

Andreyev, Alexandre. 2003b. Soviet Russia and Tibet: The Debacle of Secret Diplomacy, 1918–1930s, Leiden; Boston: Brill.

Andreyev, Alexandre. 2006. Tibet v politike tsarskoi, sovetskoi i postsovetskoi Rossii (Тибет в политике царской, советской и постсоветской России), Санкт-Петербург: Изд-во СПбГУ; Изд-во А.Терентьева «Нартанг».

Andreyev, Alexandre. 2013. Rossiiskie ekspeditsii v Tsentral'nuiu Aziiu: Organizatsia, polevye issledovaniia i kollektsii, 1870–1920-e gg. (Российские экспедиции в Центральную Азию: Организация, полевые исследования и коллекции, 1870–1920-е гг.), Санкт-Петербург: Нестор-История.

Andreyev, Alexandre. 2014. An Unknown Plan of Lhasa by Nikolai Przhevalskii (1879), *Tibet Journal*, XXXIX (1), 41–55.

Andreyev, Alexandre. 2016. Prirodoi ocharovannye stranniki – N.M. Przhevalskii i ego ucheniki (Природой очарованные странники – Н.М. Пржевальский и его ученики), *Priroda*, 4, 76–85.

Andreyev, Alexandre, Gnatiuk, T.Iu. (Гнатюк, Т.Ю.) (eds, comment.). 2007. Sredi liudei i ptits: Ornitolog i puteshestvennitsa E.V. Kozlova (1892–1975) (Среди людей и птиц:

орнитолог и путешественница Е.В. Козлова (1892–1975), Санкт-Петербург: Нестор-История.

Andreyev, Alexandre, Yusupova, T.I. (Юсупова, Т.И.). 2001. Istoriia odnogo ne sovsem obychnogo puteshesviia: Mongolo-Tibetskaia ekspeditsiia P.K. Kozlova (1923–1926 gg.) (История одного не совсем обычного путешествия: Монголо-Тибетская экспедиция П.К. Козлова (1923–1926 гг.)), *Voprosy istorii estestvoznaniia i tekhniki* (*Вопросы истории естествознания и техники*), 2, 51–74.

Andreev, N.V. 1962. Geograficheskie Trudy ofitserov General'nogo shtaba za 1836–1868 gg. i ikh znachenie v razvitii geografii Rossii, in Trudy Moskovskogo instituta inzhenerov geodezii, aerofotos'emki i kartografii, XLIX, 11–25.

Andrews, R.C. 1932. The New Conquest of Central Asia. A Narrative of the Explorations of the Central Asian Expeditions in Mongolia and China, 1921–1930, New York: The American Museum of Natural History.

Anuchin, D.N. (Анучин, Д.Н.). 1922. Aziia kak prarodina i uchitel'nitsa chelovechestva, eio nastoiashchee i budushchee (Азия, как прародина и учительница человечества, её настоящее и будущее), *Новый Восток*, 1, 233–249.

Arenskii, P.A. (Аренский, П.А.). 1931. Przhevalskii, ego zhizn' i puteshestviia (Пржевальский, его жизнь и путешествия), Москва; Ленинград: Молодая гвардия.

Baabar. 1999. Twentieth Century Mongolia. Cambridge: The White Horse Press.

Badmaev, P.A. (Бадмаев, П.А.). 2001. Za kulisami tsarizma. Vospominaniia, memuary (П.А. Бадмаев. За кулисами царизма. Воспоминания, мемуары), Минск; Москва; Харвест: АСТ.

Bailey, Scott C. Matsushita. 2008. Travel, Science and Empire: The Russian Geographical Society's Expeditions to Central Eurasia, 1845–1905. PhD Thesis: University of Hawaii.

Baer, K.M. (Бэр, К.М.). 1849. Ob etnograficheskikh issledovaniiakh voobstche i v Rossii v osobennosti (Об этнографических исследованиях вообще и в России в особенности), *Записки ИРГО* (*Записки ИРГО*), 1, 64–81.

Baker, J.N.L. 1931. A History of Geographical Discovery and Exploration, London; Bombay; Sydney: George G. Harrap & Co. Ltd.

Barger, Evert. 1944. Some problems of Central Asian exploration, *Geographical Journal*, CIII (1/2), 1–16.

Baskhanov, M.K. (Басханов, М.К.). 2005. Russkie voennye vostokovedy do 1917 g. A Biobibliographical Dictionary (Русские военные востоковеды до 1917 г. Библиографический словарь), Москва: Восточная литература.

Baskhanov, M.K. (Басханов, М.К.). 2015. K poslednemu perevalu: zhizn' i stranstviia Bronislava Grombchevskogo (К последнему перевалу: жизнь и странствия Бронислава Громбчевского), in: Dervish Gindukusha. Putevye dnevniki tsentral'noaziatskikh ekspeditsii generala B.L. Grombchevskogo (Дервиш Гиндукуша. Путевые дневники центральноазиатских экспедиций генерала Б.Л. Громбчевского), Санкт-Петербург: Нестор-История, 7–76.

Baskhanov, M.K. (Басханов, М.К.), Kolesnikov, A.A., Matveeva, M.F. (Басханов М. К., Колесников А.А., Матвеева М.Ф.). 2015. Dervish Gindukusha. Putevye dnevniki tsentral'noaziatskikh ekspeditsii generala B.L. Grombchevskogo (Дервиш Гиндукуша. Путевые дневники центральноазиатских экспедиций генерала Б.Л. Громбчевского), Санкт-Петербург: Нестор-История.

Baskhanov, M.K. (Басханов, М.К.), Kolesnikov, A.A., Matveeva, M.F. Glukhov, A.I. (Басханов, М.К., Колесников, А.А., Матвеева, М.Ф., Глухов, А.И.). 2017. Pamir, Khunza i Kashgaria v ekspeditsionnykh fotografiiakh generala B.L. Grombchevskogo (Памир, Хунза и Кашмир в экспедиционных фотографиях генерала Б.Л. Громбчевского), Москва: Пеликан.

Bassin, Mark. 1991. Russia between Europe and Asia: The ideological construction of geographical space, *Slavic review*, 50 (1), 1–17.

Bassin, Mark. 1997. Imperial visions: Nationalist Imagination and Geographical Expansion in the Russian Far East, 1840–1865, Cambridge: Cambridge University Press.

Bayan, O.A. (Баян, Р.А.). 1946. V.I. Roborovskii (В.И. Роборовский), in Pervye issledovateli Tsental'noi Azii (Первые исследователи Центральной Азии), Москва: ОГИЗ, 37–52.

Baziants, A., Kuznetsova, N., Kulagina, L. (Базиянц, А., Кузнецова, Н., Кулагина, Л.). 1969. Aziatskii Muzei – Institut Vostokovedeniia AN SSSR (1818–1968) (Азиатский музей – Институт Востоковедения АН СССР (1818–1968)), Москва: Наука.

Berg, L.S. (Берг, Л.С.). 1946. Vsesoiuznoe Geograficheskoe Obshchestvo za 100 let (Всесоюзное Географическое общество за 100 лет), Москва; Ленинград: Издательство АН СССР.

Bez-Kornilovich, M.O. (Без-Корнилович М.О.). 1855. Istoricheskie svedeniia o primechatel'neishikh mestakh v Belorussii s prisovokupleniem i drugikh svedenii, k nei zhe otnosyashihsia (Исторические сведения о примечательнейших местах в Белоруссии с присовокуплением и других сведений, к ней же относящихся). Санкт-Петербург: Типография III Собственной Е.И.В. канцелярии.

Blombergowa M.M. 2007. Bronisława Grąbczewskiego (1855–1926) opisy starożytności spotykanych w trakcie podróży po Azji, in Polacy w nauce, gospodarce i administracji na Syberii w XIX i na początku XX wieku, red. A. Kuczyński, Wrocłąw: Uniwersytet Wrocławski.

Blombergowa M.M. 2008. Bronisława Grąbczewskiego uwagi o rolnictwie, uprawach i sadach spotykanych w trakcie podróży po Azji Centralnej, *Zeszyty Wiejskie*, XIII, 116–139.

Blombergowa M.M. 2011. Jak Bronisław Grąbczewski organizował powstanie w Afganistanie, in *Non sensistis gladios*, LXX, 661–674.

Bogdanovich, K.I. (Богданович, К.И.). 1892. Geologicheskie issledovaniia v Vostochnom Turkestane (Геологические исследования в Восточном Туркестане) in Trudy

Tibetskoi Ekspeditsii 1889–1890 gg. pod nachal'stvom M.V. Pevtsova (Труды Тибетской экспедиции 1889–1890 г. под начальством М.В. Певцова), II, Санкт-Петербург: Издательство ИРГО.

Bogdanovich, K.I. (Богданович, К.И.). 1908. Otzyv o trudakh Grum-Gzhimailo (Отзыв о трудах Г.Е. Грум-Гржимайло), Otchet IRGO za 1907 g. (Отчет ИРГО за 1907 г.), Санкт-Петербург, 12–16.

Bogolepov, M. (Боголепов М.). 1928. Pamiati B.L. Grombchevskogo (Памяти Б.Л. Громбчевского), *Zemlevedenie,* XXX (4), 92–96.

Bronisław Grąbczewski zasłużony podróżnik po Dalekim Wschodzie, autor licznych dzieł podróżniczych i naukowych zmarł 26 lutego 1926 r. w Warszawie [nekrolog]. 1926, in *Wiadomości Geograficzny*, 4 (3), 34.

Brower, Daniel. 1994. Imperial Russia and its Orient: The Renown of Nikolai Przhevalsky, *The Russian Review*, 53, July.

Bukhert V.G. (Бухерт В.Г.). 2006. "Tol'ko blagodaria Vashei pomoshchi": Pis'ma G.E. Grumm-Gszhimailo grafu S.D. Sheremetevu, 1887–1889 gg. ("Только благодаря Вашей помощи…": письма Г.Е. Грумм-Гржимайло графу С.Д. Шереметеву. 1887–1889 гг.), *Istorichrskii Arkhiv* (*Исторический архив*), 2, 178–191.

Caban, W. 2011. Wkład oficerów Polakówi osób polskiego pochodzenia w rozwój rosyjskiej nauki w XIX wieku. Wybrane problemy, *Wrocławskie Studia Wschodnie*, XV, 99–119.

Carruthers, Douglas. 1913. Unknown Mongolia. A Record of Travel and Exploration in North-West Mongolia and Dzungaria, 1, London: Hutchinson.

Cegielski, Max, and Długosz, Zdjęcia Mikołaj. 2015. Wielki Gracz. Ze zmudzi na dach Świata, Kraków: Wydawnictwo Karakter.

Chekhov, A.P. (Чехов, А.П.). 1985. N.M. Przheval'skii (Н.М. Пржевальский), in Sobranie sochinenii v 12 tomakh (Collected Works in 12 vols), VI, Москва: Издательство «Правда».

Chen, Parker C. 1935. Alternating Lakes. Hullningsskrift tillagnad Sven Hedin, Stockholm.

Chistiakov, K.V., Yusupova, T.I., Gnatiuk, T.Iu. (eds.). 2014. Rossiiskoe izuchenie Tsentral'noi Azii: istoricheskie i sovremennye aspekty (k 150-letiiu P.K. Kozlova) (Российское изучение Центральной Азии: исторические и современные аспекты (к 150-летию П.К. Козлова) / The Russian Exploration of Central Asia in Historical Perspective and its Contemporary Aspects (In commemoration of the 150th anniversary of Petr K. Kozlov), Санкт-Петербург: Политехника-сервис.

Chłapowski, F. 1891. Podróż kapitana B. Grąbczewskiego w Środkowej Azji, in *Roczniki Towarzystwa Przyjaciół Nauk Poznańskiego*, XVIII (1, 2), 201–223.

Chulun, S., Yusupova, T. (eds). 2012. Mongol'sko-Rossiiskoe nauchnoe sotrudnichestvo: Ot Uchenogo komiteta do Akademii nauk (Монгольско-российское научное сотрудничество: От Учёного комитета до академии наук), Ulaanbataar: ADMON.

Ciechanowicz, Jan. 1997. Mikołaj Przewalski, in W bezkresach Eurazji: Uczeni polscy w imperium rosyjskim, Rzeszow, 271–284.

Ciechanowicz, Jan. 2001. Zdobywca Azji (Mikołaj Przewalski) (The Conqueror of Asia: Mikołaj Przewalski), Mołodeczno; 2nd ed. 2017: Warsaw: Private edition.

Clinch, Nicholas and Elizabeth. 2008. Through a Land of Extremes. The Littledales of Central Asia, Stroud: Sutton Publishing.

Dabbs, Jack A. 1963. History of the Discovery and Exploration of Chinese Turkestan, The Hague: Mouton & Co.

Denisov, P.V. (Денисов, П.В.). 1997. Zhizn' monakha Iakinfa Bichurina (Жизнь монаха Иакинфа Бичурина), Чебоксары: Чувашское книжное издательство.

Dostoevskii, A.A. (Достоевский, А.А.). 1928. Petr Petrovich Semenov-Tian-Shanskii, biograficheskii ocherk (Пётр Петрович Семёнов-Тян-Шанский, биографический очерк) in Petr Petrovich Semenov-Tian-Shanskii, ego zhizn' i deiatel'nost' (Пётр Петрович Семёнов-Тян-Шанский, его жизнь и деятельность), Ленинград: Издательство Государственного географического общества.

Dostoevskii, F.M. (Достоевский Ф.М.) 1984. Polnoe Sobranie Sochinenii (Complete Works), in 30 vols, XXVII, Ленинград: Наука.

Dubrovin, N.F. (Дубровин, Н.Ф.). 1890. Nikolai Mikhailovich Przheval'skii. Bibliograficheskii ocherk (Николай Михайлович Пржевальский. Биографический очерк), Санкт-Петербург: Военная типография.

Edinarkhova, N.E. (Единархова, Н.Е.). 2000. Mongoliia i mongoly v izdaniiakh Vostochno-Sibirskogo otdela RGO (1860–1880-e gg.) (Монголия и монголы в изданиях Восточно-Сибирского отдела РГО (1860–1880-е гг.)), *Vestnik Evrazii* (*Вестник Евразии*), 3 (10), 133–155.

Edinarkhova, N.E. (Единархова, Н.Е.) (ed.) 2001. Russkii konsul v Mongolii: Otchet Ja.P. Shishmareva o 25-letney deiatel'nosti Urginskogo konsul'sta (Русский консул в Монголии: Отчёт Я.П. Шишмарёва о 25-летней деятельности Ургинского консульства), Иркутск.

Engelgardt, M.A. (Энгельгардт, М.А.). 1891. Nikolai Przhevalskii. Ego zhizn' i puteshestviia (Николай Пржевальский. Его жизнь и путешествия), Санкт-Петербург; repr. in: [М.А. Энгельгардт] (M.A Engelgardt). Kolumb, Livingston, Stenli, A. Gumbol'd, Przhevalskii. Biograficheskie povestvovaniia (Колумб, Ливингстон, Стэнли, А. Гумбольдт, Пржевальский. Биографические повествования), Челябинск: Издательство «Урал», 1995.

Esakov, V.A. (Есаков, В.А.). 2002. Mikhail Ivanovich Veniukov, 1832–1901 (Михаил Иванович Венюков, 1832–1901), Москва: Наука.

Fateev, A.N. (Фатеев, А.Н.).1945. Istoriia odnogo prizvaniia: Puteshestvennik Przhevalskii (История одного призвания: Путешественник Пржевальский), Прага: Издание Русской Учёной академии.

Fedchenko, B.A. (Федченко, Б.А.). 1940. Rastitel'nost' Tsentral'noi Azii i rol' N.M. Przheval'skogo v eё izuchenii (Растительность Центральной Азии и роль Н.М. Пржевальского в её изучении), *Izvestiia Vserossiiskogo Geograficheskogo Obshchestva* (*Известия Всероссийского Географического общества*), 4–5, 669–678.

Fleszar, M. 1962. Rękopisy Bronislawa Grąbczewskiego w Towarzystwie Geograficznym ZSRR (The Manuscripts of Bronislaw Grąbczewski in the Geographical Society of the USSR), *Przegląd Geograficzny*, XXXIV (4), 733–737.

Fradkin, N.G. (Фрадкин, Н.Г.). 1958. P.P. Semenov-Tian-Shanskii i ego puteshestvie, 1856–1857 (П.П. Семёнов-Тян-Шанский и его путешествие 1856–1857), Introduction to: P.P. Semenov-Tian-Shanskii. Puteshestvie v Tian-Shan (П.П. Семёнов-Тян-Шанский. Путешествие в Тянь-Шань), Москва: Государственное издательство географической литературы, 3–38.

French, Patrick. 1994. Younghusband: The Last Great Imperial Adventurer. London: Harper Collins Publishers.

Gajl, T. 2007. The Polish Armorial Middle Ages to the 20th Century, Gdansk: L & L.

Gauja, P. 1917. Les Fondation de l'Académie des Sciences (1881–1915), Hendaye.

Gavrilenkov, V.M. (Гавриленков, В.М.) 1989. Russkii puteshestvennik N.M. Ptzhevalskii (Русский путешественник Н.М. Пржевальский), 2nd ed., Москва: Московский рабочий.

Gavrilenkova, E.P. (Гавриленкова, Е.П.). 2011. Neizvestnye stranitsy zhizni N.M. Przheval'skogo (Неизвестные страницы жизни Н.М. Пржевальского), 2nd ed., Смоленск: Свиток.

Gedeonov, D.D. (Гедеонов Д.Д.). 1889. Barometricheskie opredeleniia kapitana B.L. Grombchevskogo vo vremia puteshestviia v Kandzhut i Raskem (Барометрические определения капитана Б.Л. Громбчевского во время путешествия в Канджут и Раскем), *Izvestiia IRGO* (*Известия РГО*), XXV (6), 440–448.

Glinoetskii, N.P. (Глиноецкий, Н.П.). 1882. Istoricheskii ocherk Nikolaevskoi Akademii General'nogo Shtaba (Исторический очерк Николаевской академии Генерального штаба), Санкт-Петербург: Типография штаба войск гвардии и Петербургского военного округа.

Glushkov, V.V. (Глушков, В.В.). 2007. Istoriia voennoi kartografii v Rossii (XVII – nachalo XX v.) (История военной картографии в России (XVII – начало XX в.)), Москва: ИДЭЛ.

Gnatiuk, T. Iu. (Гнатюк, Т.Ю.) 2014. Prirodoochrannaia deiatel'nost' P.K. Kozlova: Askania-Nova (Природоохранная деятельность П.К. Козлова: Аскания-Нова), in K.V. Chistiakov, T.I. Yusupova, T. Iu. Gnatiuk (eds.) Rossiiskoe izuchenie Tsentral'noi Azii: Istoricheskie i sovremennye aspekty (K 150-letiiu P.K. Kozlova) / The Russian Exploration of Central Asia in Historical Perspective and its Contemporary Aspects

(In commemoration of the 150th anniversary of Petr K. Kozlov), Санкт-Петербург: Политехника-сервис.

Grąbczewski Bronisław [nekrolog] 1926, in *Wiadomości Geograficzny*, 4 (3), 33.

Grigoriev, A.V. (Григорьев, А.В.). 1898. Otzyv o trudakh V.I. Roborovskogo i P.K. Kozlova po issledovaniiu Tsentral'noi Azii na prisuzhdenie Konstantinovskoi medali (Отзывы о трудах В.И. Роборовского и П.К. Козлова по исследованию Центральной Азии на присуждение Константиновской медали), Otchet IRGO za 1897 g., Prilozhenie (Отчёт ИРГО за 1897 г., Приложение), 8–15.

Gritskevich, V.P. (Грицкевич В.П.). 1972. S Alaia na Karakorum (С Алая на Каракорум), *Literaturnyi Kirgizstan*, IV, 79–82.

Gritskevich, V.P. (Грицкевич В.П.). 1972. K izucheniiu biografii K.I. Bogdanovicha i B.L. Grombchevskogo (po materialam sovetskikh arkhivov) (К изучению биографий К.И. Богдановича и Б.Л. Громбчевского (по материалам советских архивов), in: Istoriia russko-pol'skikh kontaktov v oblasti geologii i georafii (История русско-польских контактов в области геологии и географии), Ленинград: Географическое общество СССР.

Grombchevskii, B.L. (Громбчевский Б.Л.). 1886. Otchet o komandirovanii v Kashgar i Iuzhnuiu Kashgariiu v 1885 godu starshego chinovnika osobykh poruchenii pri voennom gubernatore Ferganskoi oblasti poruchika B.L. Grombchevskogo (Отчет о командировании в Кашгар и Южную Кашгарию в 1885 году старшего чиновника особых поручений при военном губернаторе Ферганской области поручика Б. Л. Громбчевского), Новый Маргелан: типография Ферганского областного правления.

Grombchevskii, B.L. (Громбчевский Б.Л.). 1887. Ocherk kitaiskikh voisk i ukreplenii v Vostochnom Turkestane (Очерк китайских войск и укреплений в Восточном Туркестане), *Russkii Invalid*, XXII–XXIV, 32–34.

Grombchevskii, B.L. (Громбчевский Б.Л.). 1889. [Pis'ma iz Darvaza] ([Письма из Дарваза]), *Turkestanskie Vedomosti*, XXXII.

Grombchevskii, B.L. (Громбчевский Б.Л.). 1889a. Ekspeditsiia kapitana Grombchevskogo na Pamiry [vyderzhki iz pisem k vitsepredsedateliu IRGO] (Экспедиция капитана Громбчевского на Памиры [выдержки из писем к вице-председателю ИРГО]), *Izvestiia IRGO* (*Известия ИРГО*), XXV (5), 423–426.

Grombchevskii, B.L. (Громбчевский Б.Л.). 1889b. Zametki po klimatologii kapitana Grombchevskogo vo vremia puteshestviia v Kandzhut i Raskem (Заметки по климатологии капитана Громбчевского во время путешествия в Канджут и Раскем), *Izvestiia IRGO* (*Известия ИРГО*), XXV (6), 449–453.

Grombchevskii, B.L. (Громбчевский Б.Л.). 1889c. Quelques details sur un voyage dans les montagnes de l'Asie Centrale, *Comptesrendus de la Société de géographie*, IV, 89–91.

Grombchevskii, B.L. (Громбчевский Б.Л.). 1890. Vesti iz ekspeditsii [na Pamir] (Вести из экспедиции [на Памир]), Санкт-Петербург: типография А.С. Суворина.

Grombchevskii, B.L. (Громбчевский Б.Л.). 1891. Doklad o puteshestvii v 1889–1890 gg. (Доклад о путешествии в 1889–1890 гг.), *Izvestiia* IRGO (*Известия* ИРГО), XXVII (2), 97–118.

Grombchevskii, B.L. (Громбчевский Б.Л.). 1891a. Voyage dans l'Asie Centrale (1889–1890), *La Revue russe: Le revue scientifique (de la France et de l'étranger)*, XLIX (4), 101–111.

Grombchevskii, B.L. (Громбчевский Б.Л.). 1891b. Sovremennoe politicheskoe polozhenie Pamirskikh khanstv i pogranichnoi linii s Kashmirom. Voenno-politicheskii ocherk. (Современное политическое положение Памирских ханств и пограничной линии с Кашмиром. Военно-политический очерк), Новый Маргелан: типография Ферганского областного правления.

Grombchevskii, B.L. (Громбчевский Б.Л.). 1891c. Na sluchai voiny s Indiei. Ozero Shiva kak strategicheskii punkt. Voenno-politicheskii ocherk (На случай войны с Индией. Озеро Шива как стратегический пункт. Военно-политический очерк), Новый Маргелан: типография Ферганского областного правления.

Grombchevskii, B.L. (Громбчевский Б.Л.). 1891d. Nashi interesy na Pamire (Наши интересы на Памире), Новый Маргелан: типография Ферганского областного правления.

Grombchevskii, B.L. (Громбчевский Б.Л.). 1897. Predvaritel'nyi otchet o poezdke po Sungari v Girin i Tsitsikar v 1897 g. (Предварительный отчет о поездке по Сунгари в Гирин и Цицикар в 1897 г.), Благовещенск: Private print.

Grombchevskii, B.L. (Громбчевский Б.Л.). 1897a. Kitaiskie zolotye priiski na pravom beregu Amura (Китайские золотые прииски на правом берегу Амура), *Priamurskie Vedomosti* (*Приамурские ведомости*), 245–246.

Grombchevskii, B.L. (Громбчевский Б.Л.). 1900. Karta Mukdenskoi provintsii (Карта Мукденской провинции), n.p.

Grombchevskii, B.L. (Громбчевский Б.Л.). 1923. Podróż do Kaszgarji w Chinach Zachodnich, *Kurjer Polski*, 202.

Grombchevskii, B.L. (Громбчевский Б.Л.). 1923a. Ze wspomnień o Japonji, *Kurjer Polski*, 270, 272, 273.

Grombchevskii, B.L. (Громбчевский Б.Л.). 1923b. Wspomnienie myśliwskie z Azji Środkowej, in *Przegląd Myśliwski*, VI, 2–3; VIII, 1–3; X, 2–3; XI, 2–4; XII, 3–5.

Grombchevskii, B.L. (Громбчевский Б.Л.). 1923c. Zwierzostan Pamiru, in *Przegląd Myśliwski*, XIII, 3–4; XIV, 2; XV, 2–3; XVI, 2–3, XVII, 5–6.

Grombchevskii, B.L. (Громбчевский Б.Л.). 1924. Kaszgarja. Kraj i Ludzie: Podróz do Azji Środkowej, Warszawa: Nakład Gebethnera i Wolffa.

Grombchevskii, B.L. (Громбчевский Б.Л.). [1924]a. Przez Pamiry i Hindukusz do Źródeł Rzeki Indus, Warszawa: Nakład Gebethnera i Wolffa.

Grombchevskii, B.L. (Громбчевский Б.Л.). [1925]. W pustyniach Raskemu i Tybetu, Warszawa: Nakład Gebethnera i Wolffa.

Grombchevskii, B.L. (Громбчевский Б.Л.). 1925a. W głębi chanatu bucharskiego, in *Naokoło Świata*, XII, 59–100.

Grombchevskii, B.L. (Громбчевский Б.Л.). 1925b. Na Pamirach i w pustyni Raskiemu, in *Przegląd Myśliwski i Łowiectwo Polskie*, VI, 82–83; VII, 99–100; VIII, 113–114; IX, 131–132; X, 145–146.

Grombchevskii, B.L. (Громбчевский Б.Л.). 1926. Na Służhbie Rosyjskiej. Fragmenty wspomnień, Warszawa: Nakład Gebethnera i Wolffa.

Grombchevskii, B.L. (Громбчевский Б.Л.). 2010. Podróże po Azji Środkowej: 1885–1890 – Bronisław Grąbczewski. Seria Odkrywanie Świata, Warszawa: Wydawnictwo Naukowe PWN.

Grombchevskii, B.L. (Громбчевский Б.Л.). 2015. Dnevnik ekspeditsii v Kandzhut i Raskem v 1888 g. (Дневник экспедиции в Канджут и Раскем в 1888 г.), in Dervish Gindukusha. Putevye dnevniki tsentral'noaziatskikh ekspeditsii generala B.L. Grombchevskogo (Дервиш Гиндукуша. Путевые дневники центральноазиатских экспедиций генерала Б.Л. Громбчевского), Санкт-Петербург: Нестор-История, 77–185.

Grombchevskii, B.L. (Громбчевский Б.Л.). 2015a. Dnevnik ekspeditsii v Darvaz, na Pamiry, v Raskem i Severo-Zapadnyi Tibet. 1889–1890 gg. (Дневник экспедиции в Дарваз, на Памиры, в Раскем и Северо-Западный Тибет. 1889–1890 гг.), in Dervish Gindukusha. Putevye dnevniki tsentral'noaziatskikh ekspeditsii generala B.L. Grombchevskogo (Дервиш Гиндукуша. Путевые дневники центральноазиатских экспедиций генерала Б.Л. Громбчевского), Санкт-Петербург: Нестор-История, 186–374.

Grombchevskii, B.L. (Громбчевский Б.Л.). 2016. Na sluzhbe rossiiskoi: fragmenty vospominanii (На службе российской: фрагменты воспоминаний), trans. from Polish by M.G, Leonov, ed. by E.G. Korol'kova, Москва: Научный мир.

Grumm-Grzhimailo, A.G. (Грумм-Гржимайло А.Г.). 1947. Dela i dni G.E. Grumm-Grzhimailo (puteshestvennika i geografa), 1860–1935 (Дела и дни Г.Е. Грумм-Гржимайло (путешественника и географа), 1860–1935), Москва: Издательство Московского общества исследователей природы.

Grumm-Grzhimailo, G.E. (Грумм-Гржимайло Г.Е.). 1881. Nesko'ko slov o cheshuekrylykh Kryma (Несколько слов о чешуекрылых Крыма), *Trudy Russkogo entomologicheskogo obshchestva* (*Horae Societatis Entomologicae Rossicae*), XIII, 153–168.

Grumm-Grzhimailo, G.E. (Грумм-Гржимайло Г.Е.), Swiatsky, S. 1882. Ueber einige Lepidoptera von Narva, *Horae Societatis Entomologicae Rossicae*, XVIII, 148–155.

Grumm-Grzhimailo, G.E. (Грумм-Гржимайло Г.Е.). 1884. Kratkii otchet o rezul'tatakh ekspeditsii v Prialaiskie strany (Краткий отчет о результатах экспедиции в Приалайские страны), *Izvestiia* IRGO (*Известия* ИРГО), XX, 660–673.

Grumm-Grzhimailo, G.E. (Грумм-Гржимайло Г.Е.). 1885. Bericht über meine Reise in das Alai-Gebiet (Correspondenz), in N.M. Romanov (ed.) Memoires sur les lepidopteres, II, St. Petersburg: Imprimerie de M.M. Stassulewitch, 212–247.

Grumm-Grzhimailo, G.E. (Грумм-Гржимайло Г.Е.). 1885a. Puteshestvie chlena-sotrudnika G.E. Grum-Grzhimailo v Pripamirskiie strany (Путешествие члена-сотрудника Г.Е. Грум-Гржимайло в Припамирские страны), *Izvestiia IRGO* (*Известия ИРГО*), XXI, 579–582.

Grumm-Grzhimailo, G.E. (Грумм-Гржимайло Г.Е.). 1886. Ocherk Pripamirskikh stran. Doklad chitannyi v Obshchem sobranii IRGO 4 dekabria 1885 (Очерк Припамирских стран. Доклад, читанный в Общем собрании ИРГО 4 декабря 1885 г.), *Izvestiia IRGO* (*Известия ИРГО*), XXII, 82–109.

Grumm-Grzhimailo, G.E. (Грумм-Гржимайло Г.Е.). 1887. Diagnosen einiger neuer Species, in: N.M. Romanov (ed.) Memoires sur les lepidopteres, III, St. Petersburg: Imprimerie de M.M. Stassulewitch, 397–406.

Grumm-Grzhimailo, G.E. (Грумм-Гржимайло Г.Е.). 1888. Novae species et varietates Rhopalocerorum e Pamir, *Trudy Russkogo entomologicheskogo obshchestva* (*Труды Русского энтомологического общества*), XXII, 303–307.

Grumm-Grzhimailo, G.E. (Грумм-Гржимайло Г.Е.). 1890. Le Pamir et sa faune lepidopterologique, in N.M. Romanov (ed.) Memoires sur les lepidopteres, IV, St. Petersburg: Imprimerie de M.M. Stassulewitch.

Grumm-Grzhimailo, G.E. (Грумм-Гржимайло Г.Е.). 1891. Doklad G.E. Grumm-Grzhimailo o puteshestvii v 1889–1890 gg. (Доклад Г.Е. Грум-Гржимайло о путешествии в 1889–1890 гг.), *Izvestiia IRGO* (*Известия ИРГО*), XXVII, 169–180.

Grumm-Grzhimailo, G.E. (Грумм-Гржимайло Г.Е.). 1893. Lepidoptera palaearctica nova, I–II, *Trudy Russkogo entomologicheskogo obshchestva* (*Труды Русского энтомологического общества*), XXVII, 127–129; 379–386.

Grumm-Grzhimailo, G.E. (Грумм-Гржимайло Г.Е.). 1893a. Verreichniss der von D. Glasunov 1892 in Gebiete des Serafschan Thales und in Wüste Kisil-kum gesammelten Lepidopteren, *Trudy Russkogo entomologicheskogo obshchestva* (*Труды Русского энтомологического общества*), XXVIII, 88–95.

Grumm-Grzhimailo, G.E. (Грумм-Гржимайло Г.Е.). 1894. Po stupeniam "Bozhiego trona" (По ступеням «Божьего трона»), *Istoricheskii vestnik* (*Исторический вестник*), 55, 147–163.

Grumm-Grzhimailo, G.E. (Грумм-Гржимайло Г.Е.). 1894a. Opisanie Amurskoi oblasti (Описание Амурской области), Санкт-Петербург: Типо-Литография С.М. Николаева.

Grumm-Grzhimailo, G.E. (Грумм-Гржимайло Г.Е.). 1895. Lepidoptera palaearctica nova, III, *Trudy Russkogo entomologicheskogo obshchestva* (*Труды Русского энтомологического общества*), XXIX, 290–294.

Grumm-Grzhimailo, G.E. (Грумм-Гржимайло Г.Е.). 1896. (With the participation of M.E. Grum-Grzhimailo). 1896. Opisanie puteshestviia v Zapadnyi Kitai, I: Vdol' Vostochnogo Tianshania (Описание путешествия в Западный Китай, I: Вдоль Восточного Тянь-Шаня), Санкт-Петербург: Типо-Литография С.М. Николаева.

Grumm-Grzhimailo, G.E. (Грумм-Гржимайло Г.Е.). 1898. Pochemu kitaitsy risuiut demonov ryzhevolosymi? (Почему китайцы рисуют демонов рыжеволосыми?), Санкт-Петербург: Типография В.С. Балашев & К°.

Grumm-Grzhimailo, G.E. (Грумм-Гржимайло Г.Е.). 1898a. Istoricheskoe proshloe Bei-Shania v sviazi s istoriei Srednei Azii (Историческое прошло Бэй-Шаня в связи с историей Средней Азии), Санкт-Петербург: Типография В. Киршбаума.

Grumm-Grzhimailo, G.E. (Грумм-Гржимайло Г.Е.). 1899. Opisanie puteshestviia v Zapadnyi Kitai, II: Poperek Bei-Shania i Nan-Shania v dolinu Zheltoi reki (Описание путешествия в Западный Китай, II: Поперек Бэй-Шаня и Нань-Шаня в долине Желтой реки), Санкт-Петербург: Типография В. Киршбаума.

Grumm-Grzhimailo, G.E. (Грумм-Гржимайло Г.Е.). 1901. Predislovie i primechaniia (Предисловие и примечания) in: V.V. Rokhill, V stranu Lam. Puteshestvie po Kitaiu i Tibetu (В.В. Рокхилл, В страну лам. Путешествие по Китаю и Тибету) (Preface and commentaries to: W.W. Rochhill, To the Land of the Lamas), transl. by V.K. Agafonov, Санкт-Петербург: Типография И.Н. Скороходова.

Grumm-Grzhimailo, G.E. (Грумм-Гржимайло Г.Е.). 1907. Opisanie puteshestviia v Zapadnyi Kitai, III: Vokrug Kuku-Nora, cherez Nan Shan', Bei-Shan i vdol' Vostochnogo Tian-Shania i obratno na rodinu (Описание путешествия в Западный Китай, III: Вокруг Куку-Нора, через Нань-Шань, Бэй-Шань и вдоль Восточного Тянь-Шаня обратно на родину), Санкт-Петербург: Типография В. Киршбаума.

Grumm-Grzhimailo, G.E. (Грумм-Гржимайло Г.Е.). 1914. Zapadnaiia Mongoliia i Uriankhaiskii Krai, I: Opisanie prirody etikh stran (Западная Монголия и Урянхайский край, I: Описание природы этих стран), Санкт-Петербург: Издание ИРГО.

Grumm-Grzhimailo, G.E. (Грумм-Гржимайло Г.Е.). 1926. Zapadnaiia Mongoliia i Uriankhaiskii Krai, II: Istoricheskii ocherk etikh stran v sviazi s istoriei Srednei Azii (Западная Монголия и Урянхайский край, II: Исторический очерк этих стран в связи с историей Средней Азии), Ленинград: Издание РГО.

Grumm-Grzhimailo, G.E. (Грумм-Гржимайло Г.Е.). 1926a. Zapadnaiia Mongoliia i Uriankhaiskii Krai, III (1): Antropologicheskoe i etnograficheskoe opisanie etikh stran (Западная Монголия и Урянхайский край, III (1): Антропологическое и этнографическое описание этих стран), Ленинград: Издание РГО.

Grumm-Grzhimailo, G.E. (Грумм-Гржимайло Г.Е.). 1930. Zapadnaiia Mongoliia i Uriankhaiskii Krai, III (2): Antropologicheskoe i etnograficheskoe opisanie etikh stran (Западная Монголия и Урянхайский край, III (2): Антропологическое и этнографическое описание этих стран), Ленинград: Издание РГО.

Grumm-Grzhimailo, G.E. (Грумм-Гржимайло Г.Е.). 1931. Review: Vavilov, N.I., Bukinich, D.D. Zemledel'cheskii Afganistan, Leningrad, 1929 (Вавилов Н.И., Букинич Д.Д. Земледельческий Афганистан. Ленинград, 1929), *Izvestiia Vsesoiuznogo Geograficheskogo Obstchestva* (*Известия Всесоюзного географического общества*), VXIII, 97–103.

Grumm-Grzhimailo, G.E. (Грумм-Гржимайло Г.Е.). 1936. Kitai (China). Fiziko-geograficheskii ocherk (Китай. Физико-географический очерк), Bol'shaia Sovetskaia Entsiklopediia, XXXVI, 455–462. Москва: Издательство «Советская энциклопедия».

Gumboldt, A.F. (Гумбольдт, А.Ф.). 1915. Tsentral'naiia Aziia. Issledovaniia o tsepiakh gor i po sravnitel'noi klimatologii (Центральная Азия. Исследования о цепях гор и по сравнительной климатологии), Transl. by P.I. Borozdich; ed. D.N. Anuchin, I, Москва: Типо-литография товарищества И.Н. Кушнерев и К°.

Gusev, B. (Гусев, Б.). 1995. Doktor Badmaev: tibetskaia meditsina, tsarskii dvor, sovetskaia vlast' (Доктор Бадмаев: тибетская медицина, царский двор, советская власть), Москва: Русская книга.

Hedin, Sven. 1886. Das Lop-nor problem. *Zeitschrift,* 31, 305–361.

Hedin, Sven. 1925. General Prschewalskij in Innerasien, Leipzig: Brockhaus.

Holdich, Thomas. 1994. Tibet, the Mysterious; repr., New Delhi (1st ed. 1904).

Hopkirk, Peter. 1990. The Great Game: On secret service in High Asia, London: John Murray.

Hopkirk, Peter. 1992. The Great Game. The Struggle for Empire in Central Asia, London: John Murray.

Horner, Nils G., Chen, Parker C. 1935. Alternating Lakes. Hullningsskrift tillagnad Sven Hedin, Geogr. Ann., Stockholm, 145–166.

Humboldt, Alexander. 1843. Asie centrale: recherches sur les chaines de Montaignes et la climatologie compare, I–III, Paris: Gide.

Isachenko, A.G. (Исаченко, А.Г.). 1995. Russkoe geograficheskoe obshchestvo: 150 let (Русское географическое общество: 150 лет), Ленинград; Москва: АО Прогресс, Пангея.

Its, R.F. (Итс, Р.Ф.). 1989. Kunstkamera (Кунсткамера), Ленинград: Наука.

Kaganovich, B.S. (Каганович, Б.С.). 2006. Sergei Feodorovich Ol'denburg: Opyt biografii (Сергей Фёдорович Ольденбург: Опыт биографии), Санкт-Петербург: Феникс.

Karataev, N.M. (Каратаев, Н.М.). 1948. Nikolai Mikhailovich Przhevalskii – pervyi issledovatel' prirody Tsentral'noi Azii (1839–1888) (Николай Михайлович Пржевальский – первый исследователь природы Центральной Азии (1839–1888)), Москва; Ленинград: Издательство АН СССР.

Katalog. (Каталог). 1887. Katalog zoologicheskoi kollektsii, sobrannoi N.M. Przhevalskim v Tsentralnoi Azii i postupivshei v Zoologicheskii Musei Akademii nauk. (Каталог зоологической коллекции, собранной Н.М. Пржевальским в Центральной Азии и поступившей в Зоологический музей Академии наук), Санкт-Петербург: Типография Императорской Академии наук.

Kazantsev, V.P. (Казанцев В.П.). 2015. "Na linii vse zhivut nervno, v postoiannoi trevoge teper' i s mrachnym predchuvstviem v budushchem": Rossiiskii razvedchik polkovnik B.L. Grombchevskii o problemakh stroitel'stva KVZhD. 1898 g. ("На линии все живут нервно, в постоянной тревоге теперь и с мрачным предчувствием в будущем": Российский разведчик полковник Б.Л. Громбчевский о проблемах строительства КВЖД. 1898 г.), *Istoricheskii arkhiv* (*Исторический архив*), IV, 190–201.

Keay, J. 1979. The Gilgit Game. The Explorers of the Western Himalayas, 1865–95, London: John Murray.

Khmelnitskii, S. (Хмельницкий, С.). 1950. Przhevalskii, 1839–1888 (Пржевальский, 1939–1988), Ленинград: Молодая Гвардия.

Khoroshevskii, A. (Хорошевский А.). 2014. Introduction in Grumm-Grzhimailo G.E. Po stupeniam "Bozhiego trona" (Грумм-Гржимайло Г.Е. По ступеням «Божьего трона»), Москва: ЭКСМО.

Klein, L.S. (Клейн, Л.С). 2002. Drugaia storona svetila: Neobychnaia liubov' vydaiushchikhsia liudei. Rossiiskoe sozvezdie. (Другая сторона светила: Необычная любовь выдающихся людей. Российское созвездие), Санкт-Петербург: Фолио-Пресс.

Kliashtornyi, S.G., Kolesnikov, A.A. (Кляшторный С.Г., Колесников А.А.). 1988. Vostochnyi Turkestan glazami russkikh puteshestvennikov, vtoraia polovina XIX v. (Восточный Туркестан глазами русских путешественников, вторая половина XIX в.), Алма-Ата: Наука.

Kochergin, A.S., Khokhriakov, V.R., Sidenko, M.V. (eds). 2014. Tvorcheskoe nasledie N.M. Przheval'skogo i sovremennost'. Chetvertye mezhdunarodnye nauchnye chteniia pamiati N.M. Przheval'skogo. Materialy konferentsii (Творческое наследие Н.М. Пржевальского и современность. Четвёртые международные научные чтения памяти Н.М. Пржевальского (материалы конференции), Смоленск.

Kolesnik, S.V. (Колесник, С.В.) (ed.) 1970. Geograficheskoe obshchestvo za 125 let (Географическое общество за 125 лет), Ленинград: Наука.

Kolesnikov, A.A. (Колесников, А.А.). 1990. B.L. Grombchevskii i ego puteshestvie v Kashgariiu (Б.Л. Громбчевский и его путешествие в Кашгарию), *Izvestiia Akademii nauk Kazakhskoi SSR,* Seriia obshchestvennykh nauk, 4, 45–50.

Kolesnikov, A.A. (Колесников, А.А.). 1997. Russkie voennye issledovateli Azii (XIX nachala XX vekov). (Русские военные исследователи Азии (XIX – начала XX веков). Душанбе: Издательство «Дониш».

Kolesnikov, A.A. (Колесников, А.А.), Ploskikh, V.M. (Плоских, В.М.) (eds). 2004. N.M. Przheval'skii i russkie issledovateli Kyrgyzstana. Dokumenty, materialy, issledovaniia. (Н.М. Пржевальский и русские исследователи Кыргызстана: Документы, материалы, исследования), Бишкек: КРСУ.

Kolesnikov, A.A. (Колесников, А.А.). 2006. Russkie v Kashgarii (2-ia polovina XIX – nachalo XX veka) (Русские в Кашгарии (2-ая половина XIX – начало XX в.), Бишкек: Раритет.

Kolesnikov, A.A. (Колесников, А.А.). 2008. Genial'nyi puteshestvennik (Гениальный путешественник), *Nashe Nasledie* (*Наше Наследие*), 85, 64–69.

Komarov, V.L. (Комаров, В.Л.). 1908. Vvedenie k floram Kitaia i Mongolii (Введение к флорам Китая и Монголии), *Trudy St.Peterburgskogo Botanicheskogo Sada* (*Труды Санкт-Петербургского Ботанического Сада*), XXIX (1).

Korolev, V.A., Murzin, V.S. (Королев В.А., Мурзин В.С.). 2005. Istoriia lepidopterologicheskih issledovanii v Rossii (История лепидоптерологических исследований в России), Москва.

Kowalski B. 1978. Bronisław Grąbczewski – polski podróżnik i badacz Azji Centralnej, *Wszechświat*, 10.

Kozlov, I.V., Kozlova, A.V. (Козлов, И.В., Козлова, А.В.). 1991. Petr Petrovich Semenov-Tian-Shanskii, 1827–1914 (Пётр Петрович Семёнов-Тян-Шанский, 1827–1914), Москва: Наука.

Kozlov, P.K. (Козлов, П.К.). 1896. Poezdka na reku Konche-darju; Rekognostirovka severnogo berega Bagrash-gulia (Поезда на реку Конче-дарью; Рекогносцировка северного берега Баграш-гуль), in Trudy Tibetskoi ekspeditsii 1889–1890 gg. pod nachal'stvom M.V. Pevtsova (Труды Тибетской экспедиции 1889–1890 гг. под начальством М.В. Певцова), III, 93–116.

Kozlov, P.K. (Козлов, П.К.). 1898. Lob-nor (Лоб-нор), *Izvestiia IRGO* (*Известия ИРГО*), 34 (1), 60–113.

Kozlov, P.K. (Козлов, П.К.). 1899. Otchot pomoshnika nachal'nika ekspeditsii P.K. Kozlova (Отчет помощника начальника экспедиции П.К. Козлова), in Trudy ekspeditsii Russkogo Geograficheskogo Obshchestva po Tsentral'noi Azii v 1893–1895 gg. pod nachal'stvom V.I. Roborovskogo (Труды экспедиции Русского географического общества по Центральной Азии в 1893–1895 гг. под начальством В.И. Роборовского), II, Санкт-Петербург: Издание ИРГО.

Kozlov, P.K. (Козлов, П.К.). 1901. Nachrichten von der Expedition P.K. Koslows, *Petermanns Geographische Mitteilungen*, XXXXVII, 90–94, 160–163, 182–187, 205–208, 237–239.

Kozlov, P.K. (Козлов, П.К.). 1902. The Russian Tibet Expedition, 1899–1901, by Captain Kozlov, *Geographical Journal*, XIX (5), May, 597–598.

Kozlov, P.K. (Козлов, П.К.). 1902a. Pamiati Mikhaila Vasil'evicha Pevtsova (Памяти Михаила Васильевича Певцова), *Izvestiia IRGO* (*Известия РГО*), XXXVIII (6), 615–620.

Kozlov, P.K. (Козлов, П.К.). 1905–1906. Mongolia i Kam. Trudy ekspeditsii Imperatorskogo Russkogo Geograficheskogo Obshchestva, sovershennoi v 1899–1901 godakh pod rukovodstvom P.K. Kozlova (Монголия и Кам. Труды экспедиции Императорского

Русского географического общества, совершенной в 1899–1901 под руководством П.К. Козлова), I (1): Po Mongolii do granits Tibeta (По Монголии до границ Тибета); 1 (2): Kam i obratnyi put' (Кам и обратный путь).

Kozlov, P.K. (Козлов, П.К.). 1909. The Mongolia-Szechuan expedition of the Imperial Russian Geographical Society, by Captain P.K. Kozloff, *Geographical Journal*, XXXIV (4), October, 384–408.

Kozlov, P.K. (Козлов, П.К.). 1910. The Mongolia-Szechuan expedition of the Imperial Russian Geographical Society, *Geographical Journal*, XXXVI (3), September, 288–310.

Kozlov, P.K. (Козлов, П.К.). 1910a. Vsevolod Ivanovich Roborovskii (Всеволод Иванович Роборовский), *Izvestiia IRGO* (*Известия ИРГО*), XLVI (VIII–X), 357–367; separate offprint, 1911.

Kozlov, P.K. (Козлов, П.К.). 1911. Spedizione nella Mongolia e nel Seciuan, *Bulletin della Sozieta Geografico Italiana*, VI, 758–775; VII, 865–885.

Kozlov, P.K. (Козлов, П.К.). 1914. Askania Nova v eë proshlom i nastoiashchem (Аскания-Нова в ее прошлом и настоящем), *Russkaia starina* (*Русская старина*), 7, July, 19–38.

Kozlov, P.K. (Козлов, П.К.). 1919. Sovremennoe polozhenie zooparka Askania-Nova (Современное положение зоопарка Аскания-Нова), *Priroda* (*Природа*) 10–12, 467–482.

Kozlov, P.K. (Козлов, П.К.). 1920. Tibet i Dalai-lama (Тибет и Далай-лама), Петроград: 15-я Государственная типография.

Kozlov, P.K. (Козлов, П.К.). 1923. Mongolia i Amdo i mertvyi gorod Khara-Khoto (Монголия и Амдо и мёртвый город Хара-Хото), Moscow; Petrograd: Gosudarstvennoe izdatel'stvo.

Kozlov, P.K. (Козлов, П.К.). 1924. Mongol'skii zapovednik Bogdo-ula (Монгольский заповедник Богдо-ула), *Izvestiia RGO* (*Известия РГО*), LVI, 169–173.

Kozlov, P.K. (Козлов, П.К.). 1925. Zur Toten Stadt Chara-Choto, Die Expedition der Russischen Geographischen Gesellschaft nach der Mongolei, Amdo und Chara-Choto, Berlin: Verlag Neufeld & Henius.

Kozlov, P.K. (Козлов, П.К.). 1927. Puteshestvennik P.K. Kozlov o sebe, Avtobiograficheskaia zametka (Путешественник П.К. Козлов о себе. Автобиографическая заметка), *Ogoniok* (*Огонёк*), 14, 4–5.

Kozlov, P.K. (Козлов, П.К.). 1928. Kratkii otchet o Mongolo-Tibetskoi ekspeditsii Gosudarstvennogo Russkogo Geograficheskogo Obshchestva, 1923–1926 gg. (Краткий отчет о Монголо-Тибетской экспедиции Государственного Русского Географического общества, 1923–1926 гг.), Ленинград: Издательство АН СССР.

Kozlov, P.K. (Козлов, П.К.). 1928a. Velikii russkii puteshestvennik N.M. Przheval'skii, *Priroda i liudi* (*Природа и люди*), 11, Ленинград.

Kozlov, P.K. (Козлов, П.К.). 1929. Lichnye vospominaniia o Przheval'skom, Sorokaletie so dnia smerti N.M. Przheval'skogo (Личные воспоминания о Пржевальском, Сорокалетие со дня смерти), *Izvestiia Gosudarstvennogo Geograficheskogo obshchestva* (*Известия Государственного Географического общества*), LXI (2), 379–394.

Kozlov, P.K. (Козлов, П.К.). 1935. Kochuiushchie reki Tsentral'noi Azii, *Izvestiia Gosudarstvennogo Geograficheskogo Obshchestva* (*Известия Государственного Географического общества*), 67, 599–601.

Kozlov, P.K. (Козлов, П.К.). 1947. V aziatskikh prostorakh. Kniga o zhizni i puteshestviiakh N.M. Przheval'skogo, pervogo issledovtelia prirody Tsentral'noi Azii (В Азиатских просторах. Книга о жизни и путешествиях Н.М. Пржевальского, первого исследователя природы Центральной Азии), Москва: Молодая гвардия.

Kozlov, P.K. (Козлов, П.К.). 1951. Mŕtve mesto Chara-Choto (Preložil A.M. Ľubomiersky), Matica Slovenska.

Kozlov, P.K. (Козлов, П.К.). 1955. Die Mongolei, Amdo und die tote Stadt Chara-Choto, (Übers. von H. Sträubig), Leipzig: Brockhaus.

Kozlov, P.K. (Козлов, П.К.). 2003. Dnevniki Mongolo-Tibetskoi ekspeditsii, 1923–1926. (Дневники Монголо-Тибетской экспедиции, 1923–1926), eds. T.I. Yusupova, A.I. Andreyev, A.V. Postnikov (ред. Т.И. Юсупова, А.И. Андреев, А.В. Постников), Санкт-Петербург: Наука.

Kozlov, P.K. (Козлов, П.К.). 2004. Tibet i Dalai-Lama (Тибет и Далай-Лама), ed. S.L. Kuz'min (ред. С.Л. Кузьмин), Москва: КМК, 2nd ed.

Kozlov, P.K. (Козлов, П.К.). 2015. Dnevniki Mongolo-Sychuan'skoi ekspeditsii, 1907–1909 (Дневники Монголо-Сычуаньской экспедиции, 1907–1909, eds. T.I. Yusupova, A.I. Andreev, T.Yu. Gnatiuk (ред. Т.И. Юсупова, А.И. Андреев, Т.Ю. Гнатюк), Санкт-Петербург: Нестор-История.

Kozlov, V.P. (Козлов, В.П.) 1947. Zhizn' i deiatelnost' Petra Kuz'micha Kozlova, puteshestvennika, issledovatelia Tsentral'noi Azii (Жизнь и деятельность Петра Кузьмича Козлова, путешественника, исследователя Центральной Азии), in P.K. Kozlov, Mongolia i Kam (Монголия и Кам), 2nd ed., Москва: Географгиз.

Kozlova, E.V. (Козлова Е.В.) 1923. Askaniia-Nova, zoopark v iuzhno-russkich stepiakh (Аскания-Нова, зоопарк в южно-русских степях), Петроград: Начатки знания.

Krasheninnikov, S.P. (Крашенинников, С.П.). 1949. Opisanie Zemli Kamchatki (Описание Земли Камчатки), Москва; Ленинград: Издательство Главсевморпути.

Kravklis, N.N. (Кравклис, Н.Н.). 2006. Zhizn' i puteshestviia Petra Kuz'micha Kozlova (Жизнь и путешествия Петра Кузьмича Козлова), Смоленск: «Полиграмма».

Kravklis, N.N. (Кравклис, Н.Н.). 2007. Spletenie sudeb (Сплетение судеб), Смоленск: «Полиграмма».

Kravklis, N.N. (Кравклис, Н.Н.). 2011. O general'she ne mozhet byt' i rechi! Pochemu Nikolai Przheval'skii tak i ne zhenilsia (О генеральше не может быть и речи!

Почему Николай Пржевальский так и не женился), *Smolenskaia Gazeta* (*Смоленская Газета*), 117 (845), 20 October.

Kravklis, N.N. (Кравклис, Н.Н.). 2013. Nikolai Przhevalskii – velikii syn zemli smolenskoi (Николай Пржевальский – великий сын земли смоленской), Смоленск: «Полиграмма».

Krementsov, Nikolai. 1997. Stalinist Science, Princeton, Jersey: Princeton University Press.

Kulganek, I.V, Zhukov, V.Yu. (eds, comment.). 2006. Zhizn' i nauchnaiia deiatel'nost' S.A. Kondrat'ieva (1896–1970): V Mongolii i Rossii (Жизнь и научная деятельность С.А. Кондратьева (1896–1970): В Монголии и в России), Санкт-Петербург: Петербургское Востоковедение.

Kuropatkin, A.N. (Куропаткин, А.Н.). 1913. Russko-kitaiskii vopros (Русско-китайский вопрос), Санкт-Петербург: Типография товарищества А.С. Суворина "Новое время".

Kuz'min, Yu.V. (Кузьмин, Ю.В). 1997, 1998. Ia.P. Shishmarev – general'nyi konsul Rossii v Mongolii: 1861–1911 gg. (Я.П. Шишмарёв – генеральный консул России в Монголии: 1861–1911 гг.), *Bulletin International Association for Mongol Studies*, Ulaanbaatar, 2 (20), 77–86.

Kuz'min, Yu.V. (Кузьмин, Ю.В). 2007. Vostochnye proekty doktora P.A. Badmaeva (Восточные проекты доктора П.А. Бадмаева), Иркутск: Издательство БГУЭП.

Kychanov, E.I. (Кычанов, Е.И.) 2008. Tangutskii fond Instituta vostochnykh rukopisei Rossiiskoi Akademii nauk i ego izuchenie (Тангутский фонд Института восточных рукописей Российской академии наук и его изучение), in I. Popova (ed.), Rossiiyskie ekspeditsii v Tsentral'noi Azii v kontse XIX – nachale XX veka (Российские экспедиции в Центральной Азии в конце XIX – начале XX века). Санкт-Петербург: Славия.

Latushko, Iu.V., Briantseva, S.O.(Латушко, Ю.В., Брянцева, С.О.). 2010. "Dela i Dni" Grumm-Grzhimailo: K 150-letiiu so dnia rozhdeniia uchenogo («Дела и дни» Грумм-Гржимайло: к 150-летию со дня рождения ученого), *Vestnik Dal'nevostochnogo Otdeleniia RAN* (*Вестник Дальневосточного отделения РАН*), 2, 106–114.

Lavrov, V.I. (Лавров, В.И.). 1962. Petr Petrovich Semenov-Tian-Shanskii (Пётр Петрович Семёнов-Тян-Шанский), in Kuznetsov I.V. (Кузнецов, И.В.) (ed.). Liudi russkoi nauki: Ocherki o vydaiushchikhsia deiateliakh estestvoznaniia i tekhniki: Geologiia, Geografiia (Люди русской науки: Очерки о выдающихся деятелях естествознания и техники. Геология, География), Москва: Госиздат Физико-математической литературы, 460–468.

Lebedev, D.M. (Лебедев, Д.М.). 1949. Geografiia v Rossii XVII veka (dopetrovskoi epokhi). Ocherki po istorii geograficheskikh znanii (География в России XVII века (допетровской эпохи). Очерки по истории географических знаний). Москва; Ленинград: Издательство АН СССР.

Lebedev, D.M. (Лебедев, Д.М.). 1950. Geografiia v Rossii Petrovskogo vremeni (География в России Петровского времени), Москва; Ленинград: Издательство АН СССР.

Lebedev, V.T. (Лебедев, В.Т.). 1898. V Indiiu. Voenno-statisticheskii i strategicheskii ocherk: Proekt budushchego pokhoda (В Индию. Военно-статистический и стратегический очерк. Проект будущего похода), С.-Петербург: Типография А.А. Пороховщикова.

Lencewicz, S. Generał Grąbczewski Bronisław. 1926. *Przegląd Geograficzny*, V, 144–145.

Lialina, M.A. (Лялина М.А.). 1901. Puteshestvie brat'iev G.E. i M.E. Grum-Grzhimailo v Zapadnyi Kitai. Obrabotano po podlinnym G.E. Grum-Grzhimailo sochineniiam M.A. Lialinoi (Путешествие братьев Г.Е. и М.Е. Грум-Гржимайло в Западный Китай. Обработано по подлинным Г.Е. Грум-Гржимайло сочинениям М.А. Лялиной). Санкт-Петербург: Типография А.Ф. Девриен.

Lincoln, W. Bruce. 1982. Petr Petrovich Semenov-Tian-Shanskii: The life of a Russian geographer, DeKalb: Northern Illinois University Press.

Linda, K.P. (Линда К.П.). 1902. M.V. Pevtsov i ego puteshestviia (М.В. Певцов и его путешествия). Iubileinyi sbornik Zapadno-Sibirskogo otdela IRGO (Юбилейный сборние Западно-Сибирского отдела ИРГО), Омск: типография окружного штаба, 1–47.

Lotman, Iu. M. (Лотман, Ю.М.). 1997. Intelligentsia i svoboda (k analizu intelligentskogo diskursa, (Интеллигенция и свобода (к анализу интеллигентского дискурса), Russkaia intelligentsia i zapadnyi intellektualizm: Istoriia i tipologiia. Materialy mezhdunarodnoi konferentsii (Русская интеллигенция и западный интеллектуализм: История и типология. Материалы международной конференции), B.A. Uspenskii (comp.) (Б.А. Успенский, сост.), Неаполь.

Luzianin, S.G. (Лузянин, С.Г.). 2000. Rossia – Mongoliia – Kitai v pervoi polovine XX veka. Politicheskie vzaimootnosheniia v 1911–1946 (Россия – Монголия – Китай в первой половине XX в. Политические взаимоотношения в 1911–1946), Москва: ИДВ РАН.

Luzhetskaia, N.L. (Лужецкая Н.Л.). 1997. Doroga v Hunzu. B.L. Grombchevskii o svoem puteshestvii 1888 g. (Дорога в Хунзу. Б.Л. Громбчевский о своем путешествии 1888 г.), *Kunstkamera: Etnograficheskie tetradi* (*Кунсткамера: Этнографические тетради*), XI, 352–362.

Magidovich, I.P., Magidovich, V.I. (Магидович И.П., Магидович В.И.). 1983–1986. Ocherki po istorii geograficheskikh otkrytii, in 5 vols. (Очерки по истории географических открытий), Москва: Просвещение. (Vol. 3: Novoe vremia: ser. XVII – XVIII v., 1984; Vol. 4: Novoe vremia: XIX – nach. XX v., 1985).

Marciniak, M. 1983. Bronisław Grąbczewski jako badacz kultury ludów Azji Środkowej, in Kultura i życie społeczne Azji Środkowej: z polskich badań dawnych i współczesnych, pod red. Z. Jasiewicza, Poznań: Wydawnictwo Naukowe Uniwersytetu im. Adama Mickiewicza.

Margolin, Ia. (Марголин, Я.). 1951. Ot redaktora (От редактора), in: Pevtsov M.V. Puteshestvie po Kitaiu i Mongolii (Певцов М.В. Путешествие по Китаю и Монголии), Москва: Географгиз.

Marshall, Alex. 2006. The Russian General Stuff and Asia, 1800–1917. London: Routledge.

Matveeva, T.P., Filonovich, T.S., Zharukova, L.I. (comp.). (Матвеева, Т.П., Филонович, Т.С., Жарукова, Л.И.) (сост.). 1971. Russkie geografy i puteshestvenniki. Fondy arkhiva Geograficheskogo Obshchestva (Русские географы и путешественники. Фонды архива Географического общества), Ленинград: Наука.

Mehra, Parshotam. 2005. The Younghusband Expedition (to Lhasa). An Interpretation, New Delhi: Gyan Publishing House, 2nd ed.

Men'shikov, L.N. (Меньшиков, Л.Н.)1984. Opisanie kitaiskoi chasti kollektsii iz Khara-Khoto (Fond P.K. Kozlova) (Описание китайской части коллекции из Хара-Хото (фонд П.К. Козлова)), Москва: Наука.

Miliutin, D.A. (Милютин, Д.А.). 1846. Kriticheskoe issledovanie znachenia voennoi geografii i voennoi statistiki (Критическое исследование значения военной географии и военной статистики), Санкт-Петербург: Военная типография.

Miroshnikov, L.I. 1992. Appendix: A note on the meaning of the term 'Central Asia' as used in A.H. Dani, V.M. Masson (eds), History of civilizations of Central Asia, I, Paris: UNESCO Publishing.

Morgan, E.D. 1893. On the Pevtsov Expedition and M. Bogdanovich's Survey, *Geographical Journal*, 2 (1), 55–56.

Morgan, Gerald. 1981. Anglo-Russian Rivalry in Central Asia: 1810–1895, London: Frank Cass.

Murzaev, E.M. (Мурзаев, Э.М.). 1946. Velikii russkii puteshestvennik Nikolai Mikhailovich Przheval'skii (Великий русский путешественник Николай Михайлович Пржевальский), in N.M. Przheval'skii. Mongolia i strana tangutov (Н.М. Пржевальский. Монголия и страна тангутов), 2nd ed., Москва: Государственное издательство географической литературы.

Murzaev E.M. (Мурзаев, Э.М.). 1947. Lobnorskoe puteshestvie N.M. Przheval'skogo i zagadka Lob-nora (Лобнорское путешествие Н.М. Пржевальского и загадка Лоб-нора), in N.M. Przheval'skii, Ot Kul'dzhi za Tian'-Shan' i na Lob-nor (Н.М. Пржевальский, От Кульджи за Тянь-Шань и на Лоб-нор), 2nd ed., Москва, ОГИЗ.

Murzaev E.M. (Мурзаев, Э.М.). 1947a. Lobnorskoe puteshestvie N.M. Przheval'skogo i zagadka Lob-Nora («Лобнорское путешествие Н.М. Пржевальского и загадка Лоб-Нора), Introduction to: N.M. Przhevalskii. Ot Kuldji za Tian-Shan i na Lob-Nor (Н.М. Пржевальский. От Кульджи за Тянь-Шань и на Лоб-Нор), 2nd ed., Москва: Государственное издательство географической литературы, 3–23.

Murzaev E.M. (Мурзаев, Э.М.). 1948. Geograficheskie issledovaniia Mongol'skoi Narodnoi Respubliki (Географические исследования Монгольской Народной Республики), Москва; Ленинград: Издательство АН СССР.

Murzaev E.M. (Мурзаев, Э.М.). 1995. Russkoe geograficheskoe obshchestvo v istorii Rossiiskoi imperii (Русское географическое общество в истории Российской империи), in Russkoe geograficheskoe obshchestvo: 150 let (Русское географическое общество. 150 лет), Санкт-Петербург; Москва: АО «Издательская группа Прогресс»; «Пангея».

Mushketov, I.V. (Мушкетов, И.В.). 1915. Turkestan. Geologicheskoe i orograficheskoe opisanie (Туркестан. Геологическое и орографическое описание), 1, Санкт-Петербург: Типография М.М. Стасюлевича.

Nauchnye rezul'taty. (Научные результаты). 1888–1912. Nauchnye rezul'taty puteshestvii N.M. Przheval'skogo po Tsentral'noi Azii (Научные результаты путешествий Н.М. Пржевальского по Центральной Азии). Санкт-Петербург: Типография Академи наук. Published in several volumes and separate installments, edited by: E. Büchner (Mammals, I (1), 1888); V. Zalenskii (Ungulates, I (2), 1902); F.D. Pleske (Birds, II (1–3), 1889–1894); V. Bianki (Birds, II (4), 1905); A.V. Bedriaga (Amphibians and raptiles, III (1–4), 1898–1912); S.M. Gerzenshtein (Fishes, III (2), 1888–1891); K. Maksimovich (Plants: Flora Tangutica; Plants of Mongolia and Chinese Turkestan, in 2 vols., 1889); A.I. Voeikov (Meteorological data and routes, 1895).

Nepomniashchii, N.N. (Непомнящий, Н.Н.). 1913. M.V. Pevtsov: raznostaronnii praktik i geograficheskii issledovatel' (М.В. Певцов: разносторонний практик и географический исследователь), in: 100 velikikh russkikh puteshestvennikov (100 великих русских путешественников), Москва: Вече, 266–272.

Niebrzydowski, W. 1926. Bronisław Grąbczewski, *Wiadomości Meteorologiczne*, II, 60–61.

Nikolaev, N.M. (Николаев Н. М.). 1893. M.V. Pevtsov i 5-ia ekspeditsia v Tsentral'nuiu Aziiu (М.В. Певцов и 5-я экспедиция в Центральную Азию), *Razvedchik*, 137 (468–470); 138 (484–486).

Oberman, P. (Оберман, П.). 1981. Aleksandr Gumbol'dt i ego dnevnik o puteshestvii v Rossiiu (Александр Гумбольдт и его дневник о путешествии в Россию), *Voprosy istorii estestvoznaniia i tekhniki* (*Вопросы истории естествознания и техники*), 4, 87–91.

Obruchev, V.A. (Обручев, В.А.). 1901. Tsentral'naia Aziia, Severnyi Kitai i Nan-Shan (Центральная Азия, Северный Китай и Нань-Шань), II, Санкт-Петербург: Типография М.М. Стасюлевича.

Obruchev, V.A. (Обручев, В.А.). 1924. Zadachi issledovaniia Mongolii (Задачи исследования Монголии), *Novyi Vostok* (*Новый Восток*), 6, 287–290.

Obruchev, V.A. (Обручев, В.А.). 1926. Estestvennye bogatstva Tannu-Tuvinskoi respubliki i stepen' izuchennosti poslednei (Естественные богатства Танну-Тувинской республики и степень изученности последней), *Novyi Vostok* (*Новый Восток*), 13–14, 261–270.

Obruchev, V.A. (Обручев, В.А.). 1947. Po Vnutrennei Azii: M.V. Pevtsov (По Внутренней Азии: М.В. Певцов), Москва: Географгиз.

Ogurtsov, S.A. (Огурцов С.А.). 1960. Mikhail Vasilievich Pevtsov – geograf i puteshestvennik (Михаил Васильевич Певцов – географ и путешественник), Омск: Омское книжное издательство.

Oldenburg, S.F. (Ольденбург, С.Ф.). 1922. Peshchery Tysiachi Budd (Пещеры Тысячи Будд), *Novyi Vostok* (*Новый Восток*), 1, 57–66.

Oldenburg, S.F. (Ольденбург, С.Ф.). 1914. Materialy po buddiiskoi ikonografii Khara-Khoto (Материалы по буддийской иконографии Хара-Хото), Санкт-Петербург: Товарищество Р. Голике и А. Вильборг.

Olszewicz, B. 1926. General Bronisław Grąbczewski, Bronisław Grąbczewski, podróżnik polski po ązji Środkowej, in *Naokoło Świata*, 25, 89–103.

Ovchinnikova, T.N. (Овчинникова, Т.Н.). 1964. P.K. Kozlov, issledovatel' Tsentral'noi Azii (П.К. Козлов, исследователь Центральной Азии), Москва: Наука.

Pevtsov, M.V. (Певцов, М.В.). 1879. Putevye ocherki Dzhungarii (Путевые очерки Джунгарии), *Zapiski Zapadno-Sibirskogo Otdela IRGO* (*Записки Западно-Сибирского отдела ИРГО*), I, 1–140.

Pevtsov, M.V. (Певцов, М.В.). 1880. Kratkii ocherk puteshestviia po Mongolii i Vnutrennemu Kitaiu v 1878 i 1879 gg. (Краткий очерк путешествия по Монголии и Внутреннему Китаю в 1878 и 1879 гг.), *Izvestiia IRGO* (*Известия ИРГО*), XVI (5), 435–457.

Pevtsov, M.V. (Певцов, М.В.). 1881. Nachal'nye osnovaniia matematicheskoi i fizicheskoi geografii (Начальные основания математической и физической географии), С.-Петербург: Типография В. Безобразова.

Pevtsov, M.V. (Певцов, М.В.). 1883. Ocherki puteshestviia po Mongolii i severnym provintsiiam Vnutrennego Kitaia (Очерк путешествия по Монголии и северным провинциям Внутреннего Китая), *Zapiski Zapadno-Sibirskogo Otdela IRGO* (*Записки Западно-Сибирского отдела ИРГО*), V, 1–354.

Pevtsov, M.V. (Певцов, М.В.). 1884. Astronomicheskie nabliudeniia g. Lazareva dlia opredeleniia geograficheskogo polozheniia mest v Altae, na Verkhnei Bukhtarme i ego zhe barometricheskie opredeleniia vysot v iugo-vostochnoi chasti Semipalatinskoi oblasti (Астрономические определения г. Лазарева для определения географического положения мест в Алтае, на Верхней Бухтарме и его же барометрические определения высот в юго-восточной части Семипалатинской области), *Zapiski Zapadno-Sibirskogo Otdela IRGO* (*Записки Западно-Сибирского отдела ИРГО*), VI, 1–6.

Pevtsov, M.V. (Певцов, М.В.). 1884a. Nasha torgovlia v Zapadnom Kitae. Iz otcheta General'nogo shtaba podpolkovnika Pevtsova o puteshestvii v Mongoliiu i Vnutrennii

Kitai v 1878–1879 gg. (Наша торговля в Западном Китае. Из отчета Генерального штаба подполковника Певцова о путешествии в Монголию и Внутренний Китай в 1878–1879 гг.), *Vostochnoe Obozrenie* (*Восточное обозрение*), 3, 2–4, 7–8.

Pevtsov, M.V. (Певцов, M.B.). 1885. Pamiatnaia knizhka Stepnogo general-gubernatorstva za 1883 god (Памятная книжка Степного генерал-губернаторства за 1883 год). Омск: Типография окружного штаба.

Pevtsov, M.V. (Певцов, M.B.). 1887. Ob opredelenii geograficheskoi shiroty po sootvetstvennym vysotam dvukh zvezd (Об определении географической широты по соответственным высотам двух звезд), *Zapiski IRGO po obshchei geografii* (Записки ИРГО по общей географии), XVII (5), 1–115.

Pevtsov, M.V. (Певцов, M.B.). 1887a. Zametka o zemletriaseniiakh v 1885 i 1887 gg. v Semirechenskoi oblasti (Заметка о землетрясениях в 1885 и 1887 гг. в Семиреченской области), *Russkii Invalid*, 120 (7 June).

Pevtsov, M.V. (Певцов, M.B.). 1889. Rezul'taty astronomicheskikh nabliudenii dlia opredeleniia geograficheskogo polozheniia mest i izmerenii barometrom vysot v Severo-Zapadnoi Dzhungarii (Результаты астрономических наблюдений для определения географического положения мест и измерений барометром высот в Северо-Западной Джунгарии), *Izvestiia IRGO* (*Известия ИРГО*), XXV (1), 97–104.

Pevtsov, M.V. (Певцов, M.B.). 1891. Predvaritel'nyi otchet o puteshestvii cherez Kashgariiu v Severo-Zapadnyi Tibet (Предварительный отчет о путешествии через Кашгарию в Северо-Западный Тибет), *Izvestiia IRGO* (*Известия ИРГО*), XXVII (6), 461–479.

Pevtsov, M.V. (Певцов, M.B.). 1892. Otzyv byvshego nachal'nika Tibetskoi ekspeditsii o zaslugakh sotrudnikov i sputnikov ego po ekspeditsii (Отзыв бывшего начальника Тибетской экспедиции о заслугах сотрудников и спутников его по экспедиции), Otchet IRGO za 1891 g. (Отчёт ИРГО за 1891 г.), С.-Петербург, 31–32.

Pevtsov, M.V. (Певцов, M.B.). 1894. Klimat Kashgarii (Vostochnogo Turkestana) (Климат Кашгарии (Восточного Туркестана), *Meteorologicheskii Vestnik* (*Метеорологический Вестник*), 9, 345–348.

Pevtsov, M.V. (Певцов, M.B.). 1895. Puteshestvie po Vostochnomu Turkestanu, Kun'luniu, severnoi okraine Tibetskogo nagor'ia i Chzhungarii v 1889 i 1890 godakh (Путешествие по Восточному Туркестану, Куньлуню, северной окраине Тибетского нагорья и Чжунгарии в 1889 и 1890 годах). С.-Петербург: Типография М.М. Стасюлевича.

Pevtsov, M.V. (Певцов, M.B.). 1895a. Sifonno-reservuarnyi barometr dlia meteorologicheskikh stantsii i puteshestvii (Сифонно-резервуарный барометр для метеорологических станций и путешествий), *Meteorologicheskii Vestnik* (*Метеорологический Вестник*),12, 421–428.

Pevtsov, M.V. (Певцов, M.B.). 1895b. Pribor dlia napolneniia priamykh barometricheskikh trubok rtut'iu kholodnym sposobom (Прибор для наполнения прямых барометрических трубок ртутью холодным способом), *Meteorologicheskii Vestnik* (*Метеорологический Вестник*),12, 429–431.

Pevtsov, M.V. (Певцов, М.В.). 1895c. Spisok punktov Vnutrennei Azii, vysoty kotorykh opredeleny neodnokratno posredstvom barometrov razlichnymi puteshestvennikami (Список пунктов Внутренней Азии, высоты которых определены посредством барометров различными путешественниками), *Izvestiia IRGO* (*Известия ИРГО*), XXXI (1), 44–56.

Pevtsov, M.V. (Певцов, М.В.). 1895d. Vostochnyi Turkestan (Восточный Туркестан), *Voennyi Sbornik* (*Военный сборник*), 222 (4), 389–436; 223 (5), 216–236; 224 (6), 437–467.

Pevtsov, M.V. (Певцов, М.В.). 1895e. O barometricheskom nivelirovanii (О барометрическом нивелировании), *Zapiski IRGO po obshchei geografii* (*Записки ИРГО по общей географии*), XXIX (2), 1–77.

Pevtsov, M.V. (Певцов, М.В.). 1896. Zametka po voprosu o periodicheskom izmenenii vysoty poliusa (Заметка по вопросу о периодическом изменении полюса), *Izvestiia Russkogo Astronomicheskogo Obshchestva* (*Известия Русского астрономического общества*), 5, 222–227.

Pevtsov, M.V. (Певцов, М.В.). 1896a. Instruktsiia dlia opredeleniia vysot posredstvom posledovatel'nogo barometricheskogo nivelirovaniia (Инструкция для определения высот посредством последовательного барометрического нивелирования). С.-Петербург: Типография Академии наук.

Pevtsov, M.V. (Певцов, М.В.). 1896b. Ob ustranenii nekotorykh neudobstv barometra Parrota dlia puteshestvii (Об устранении некоторых неудобств барометра Паррота для путешествий), *Meteorologicheskii Vestnik* (*Метеорологический Вестник*), 10, 281–284.

Pevtsov, M.V. (Певцов, М.В.). 1899. Ob opredelenii geograficheskoi shiroty po sootvetstvennym vysotam dvukh zvezd (Об определении географической широты по соответственным высотам двух звезд), *Zapiski IRGO po obshchei geografii* (*Записки ИРГО по общей географии*), XXXII (2), 1–126.

Pevtsov, M.V. (Певцов, М.В.). 1901. Sokrashchennyi sposob predvychislenii pokrytii nepodvizhnykh zvezd lunoi i solnechnykh zatmenii dlia dannykh mest (Сокращенный способ предвычислений покрытий неподвижных звезд луной и солнечных затмений для данных мест), *Izvestiia Russkogo Astronomicheskogo Obshchestva* (*Известия Русского астрономического общества*), VIII (7–9), 106–126.

Pevtsov, M.V. (Певцов, М.В.). 1905. Ob opredelenii parallaksa luny iz nabliudenii ee na ravnykh vysotakh so zvezdami bliz pervogo vertikala (Об определении параллакса луны из наблюдений ее на равных высотах со звездами близ первого вертикала), *Izvestiia Russkogo Astronomicheskogo Obshchestva* (*Известия Русского астрономического общества*), XI (1), 16–34.

Pevtsov, M.V. (Певцов, М.В.). 1905a. Ob opredelenii geograficheskoi dolgoty i nabliudenii luny na ravnykh vysotakh so zvezdami bliz pervogo vertikala (Об определении географической долготы и наблюдений луны на равных высотах со звездами близ первого вертикала), *Izvestiia Russkogo Astronomicheskogo Obshchestva* (*Известия Русского астрономического общества*), 2–3.

Pevtsov, M.V. (Певцов, М.В.). 1949. Puteshestvie v Kashgariiu i Kun-Lun' (Путешествие в Кашгарию и Кун-Лунь), 2nd ed., Москва: Государственное издательство географической литературы.

Pevtsov, M.V. (Певцов, М.В.). 1951. Cesta de Kašgarska a Kuen-Lunu. Praha: Přirodovědecké vyd.

Pevtsov, M.V. (Певцов, М.В.). 1953. Wo man mit Zegeltee bezahlt: Bericht einer Reise durch die Mongolei und die nördichen Provinzen des inneren China, Leipzig: Brockhaus.

Pevtsov, M.V. (Певцов, М.В.). 2010. Puteshestvie po Kitaiu i Mongolii; Puteshestvie v Kashgariiu i Kun'lun': dokumental'nye ocherki (Путешествие по Китаю и Монголии; Путешествие в Кашгарию и Куньлунь: документальные очерки). Москва: Издательство «Дрофа».

Pevtsov, M.V. (Певцов, М.В.). 2015. Altai. Mongoliia. Kitai. Tibet: puteshestviia v Tsental'noi Azii (Алтай. Монголия. Китай. Тибет: путешествия в Центральной Азии). Москва: Издательство «Эксмо».

Pelliot, Paul. (1914). Les documents chinois trouves par la mission Kozlov a Khara-Khoto, *Journal Asiatique*, Mai – Juin, Paris, 1–20.

Poliakov, I.S. (Поляков И.С.). 1881. Loshad' Przheval'skogo (Equus Przewalskii п. sp.) (Лошадь Пржевальского (Equus Przewalskii п. sp.), *Izvestiia IRGO* (*Известия ИРГО*), XVII, 1–21.

Popova, I.F. (ed.). (Попова, И.Ф., ред.). 2008. Rossiiskie ekspeditsii v Tsentral'nuiu Aziiu v kontse XIX – nachale XX veka (Russian Expeditions to Central Asia at the Turn of the 20TH Century) (Российские экспедиции в Центральную Азию в конце XIX – начале XX века), Санкт-Петербург: Славия.

Poppe, N.N. (Поппе, Н.Н.). 1936. Znachenie puteshestvii P.K. Kozlova dlia archeologicheskogo izuchenia Tsentral'noi Azii (Значение путешествий Р.К. Козлова для археологического изучения Центральной Азии). Nekrolog (*Obituary*), *Izvestiia Gosudarstvennogo Geograficheskogo Obstchestva* (*Известия Государственного Географического общества*), LXVIII (5), 748–753.

Postnikov, A.V. (Постников, А.В.). 2005. Skhvatka na "Kryshe Mira": Politiki, razvedchiki, geografy v bor'be za Pamir v XIX veke (Схватка на "Крыше Мира": Политики, разведчики, географы в борьбе за Памир в XIX веке), Москва: ИД РИПОЛ классик, (2nd ed).

Postnikov, A.V. (Постников, А.В.). 2007. Stanovlenie rubezhei Rossii v Tsentral'noi i Srednei Azii (XVIII – XIX vv.). Rol' istoriko-geograficheskikh issledovanii i kartografirovaniia. Monografia v dokumentakh (Становление рубежей России в Центральной и Средней Азии (XVIII – XIX вв.). Роль историко-географических исследований и картографирования. Монография в документах), Москва: Памятники исторической мысли.

Potanin, G.N. (Потанин, Г.Н.). 1893. Put' iz Urgi v Hlassu cherez kochev'iia Golukov (Путь из Урги в Хлассу через кочевья Голуков), *Izvestia Vostochno-Sibirskogo otdela IRGO* (*Известия Восточно-Сибирского отдела IRGO*), XXIV (2), 38–43.

Potanin, G.N. (Потанин, Г.Н.). 1893a. Tangutsko-Tibetskaia okraina Kitaia i Tsentral'naia Mongoliia (Тангутско-тибетская окраина Китая и центральная Монголия), Санкт-Петербург: Типография А.С. Суворина.

Potapov, Eugene, Potapov, Roald. 2006. Nikolay Przhevalskiy (1839–1888), *Birding Asia*, 6, 86–91.

Praag, M.C. van Walt. 1987. The status of Tibet: History, Rights and Prospects in International Law, Boulder, Colo: Westview Press.

Prejevalsky, N. 1876. Mongolia, the Tangut Country, and the solitudes of northern Tibet, being a narrative of three years travel in eastern High Asia. In 2 vols, vol. 2, London: Sampson, Low, Marston, Searle & Rivington.

Prokhorov, A. (Прохоров, А.). 1975. K voprosu o sovetsko-kitaiskoi granitse (К вопросу о советско-китайской границе), Москва: Международные отношения.

Przhevalskaia, L.K. (Пржевальская, Л.К.). 2008. Vydaiushchiisia predstavitel' roda – puteshestvennik N.M. Przhevalskii (Выдающийся представитель рода – путешественник Н.М. Пржевалский), *Smolenskoe dvorianstvo* (*Смоленское дворянство*), 9, Москва: Издательство Смоленского Дворянского Землячества, 9–15.

Przhevalskaia, L.K. (Пржевальская, Л.К.). 2014. Osnovatel' roda Przhevalskikh, vitebskii konnyi meshchanin (Основатель рода Пржевальских, витебский конный мещанин), in Tsentral'naia Aziia v issledovaniiakh XIX – XXI vekov. Sbornik statei (Центральная Азия в исследованиях XIX – XXI веков. Сборник статей), A collection of articles, Bishkek.

Przhevalskaia, L.K. (Пржевальская, Л.К.), Przhevalskii, N.M. (Пржевальский, Н.М.). 2008. Pervyi dvorianin roda (Первый дворянин рода), *Smolenskoe dvorianstvo* (*Смоленское дворянство*), 9, Москва: Издательство Смоленского Дворянского Землячества, 5–8.

Przhevalskaia, L.K. (Пржевальская, Л.К.). 2016. Vladimir Mikhailovich Przhevalskii (Владимир Михайлович Пржевальский), Москва: Издательство РГАУ – МСХА им. К.А. Тимирязева.

Przhevalskii, N.M. (Пржевальский, Н.М.). 1862. Vospominaniia okhotnika (Воспоминания охотника), *Konnozavodstvo i Okhota* (*Коннозаводство и охота*), 6–8.

Przhevalskii, N.M. (Пржевальский, Н.М.). 1867. Zapiski vseobshchei geografii po programme iunkerskikh uchilishch (Записки всеобщей географии по программе юнкерских училищ), Varshava; 2nd ed., with additions from I.L. Fateev, 1870.

Przhevalskii, N.M. (Пржевальский, Н.М.). 1869. Inorodcheskoe naselenie v iuzhnoi chasti Primorskoi oblasti (Инородческое население в южной части Приморской области), *Izvestiia IRGO* (*Известия ИРГО*), V (5).

Przhevalskii, N.M. (Пржевальский, Н.М.). 1870. Puteshestvie v Ussuriiskom krae, 1867–1869 g. (Путешествие в Уссурийском крае, 1867–1869 г.), Санкт-Петербург: Типография Н. Неклюдова.

Przhevalskii, N.M. (Пржевальский, Н.М.). 1875–1876. Mongoliia i strana tangutov. Trekhletnee puteshestvie v Vostochnoi Nagornoi Azii. (Монголия и страна тангутов. Трёхлетнее путешествие в Восточной Нагорной Азии), 1–2, Санкт-Петербург: Типография В.С. Балашева.

Przhevalskii, N.M. (Пржевальский, Н.М.). 1879. Neskol'ko slov po povodu zamechanii barona Rikhtgofena na stat'iu "Ot Kul'dzhi za Tian'-shan' i na Lob-nor" (Несколько слов по поводу замечаний барона Рихтгофена на статью «От Кульджи за Тянь-шань и на Лоб-нор»), *Izvestiia IRGO* (*Известия ИРГО*), 15 (1), 1–6.

Przhevalskii, N.M. (Пржевальский, Н.М.). 1883. Iz Zaisana cherez Khami v Tibet i na verkhov'ia Zheltoi reki (Из Зайсана через Хами и на верховья Жёлтой реки), Санкт-Петербург: Типография В.С. Балашева.

Przhevalskii, N.M. (Пржевальский, Н.М.). 1883a. O vozmozhnoi voine s Kitaem; Razbor pogranichnykh raionov Pritian-Shanskogo, Urginskogo i Amurskogo: Dopolneniia k obeim zapiskam polkovnika Przheval'skogo (О возможной войне с Китаем; Разбор пограничных районов Притян-Шанского, Ургинского и Амурского; Дополнение к обеим запискам полковника Пржевальского), in Sbornik geograficheskikh, topograficheskikh i statisticheskikh materialov po Azii (Сборник географических, топографических и статистических материалов по Азии), 1, Санкт-Петербург: Издание ВУК Главного Штаба.

Przhevalskii, N.M. (Пржевальский, Н.М.). 1883b. Iz Zaisana cherez Khami i Tibet na verkhoviia Zheltoi reki (Из Зайсана через Хами и Тибет на верховья Жёлтой реки), Санкт-Петербург: Типография В.С. Балашева.

Przhevalskii, N.M. (Пржевальский, Н.М.). 1886. Sovremennoe polozhenie Tsentral'noi Azii (Современное положение Центральной Азии), *Russkii Vestnik* (*Русский Вестник*), Т. 182, 12, 472–524.

Przhevalskii, N.M. (Пржевальский, Н.М.). 1887. Novye vidy ptits Tsentral'noi Azii (Новые виды птиц Центральной Азии), *Zapiski Imperatorskoi Akademii nauk* (*Записки Императорской Академии наук*), LV (1), 76–95.

Przhevalskii, N.M. (Пржевальский, Н.М.). 1888. Ot Kiakhty na istoki Zheltoi reki. Issledovanie severnoi okrainy Tibeta i put' cherez Lob-nor po basseinu Tarima (От Кяхты на истоки Жёлтой реки. Исследование северной окраины Тибета и путь через Лоб-нор по бассейну Тарима), Санкт-Петербург: Типография В.С. Балашева.

Przhevalskii, N.M. (Пржевальский, Н.М.). 1888a. Avtobiografiia (Автобиография), *Russkaia starina*, 11, November, 528–540.

Przhevalskii, N.M. (Пржевальский, Н.М.). 1937. Puteshestvie v Ussuriiskom krae, 1867–1869 (Путешествие в Уссурийском крае, 1867–1869), 2nd ed., Москва: Соцэкгиз.

Przhevalskii, N.M. (Пржевальский, Н.М.). 1940. Avtobiograficheskii rasskaz (Автобиографический рассказ); Vospominaniia okhotnika (Воспоминания охотника); Dnevnik vtorogo puteshestviia v Tsentral'nuiu Aziiu (Дневник второго путешествия в Центральную Азию); Pis'ma k kazaku Teleshovu (Письма к казаку Телешову); Dnevnik poslednego puteshestviia v 1888 g. (Дневник последнего путешествия в 1888 г.), *Izvestiia Vsesoyuznogo Geograficheskogo Obshchestva* (*Известия Всесоюзного географического общества*), 72, 4–5.

Przhevalskii, N.M. (Пржевальский, Н.М.). 1946. Mongolia i strana tangutov (Монголия и страна тангутов), 2nd. ed., Москва: Государственное издательство географической литературы.

Przhevalskii, N.M. (Пржевальский, Н.М.). 1948. Iz Zaisana cherez Khami v Tibet i na verkhov'ia Zheltoi reki (Из Зайсана через Хами и на верховья Жёлтой реки), 2nd ed., Москва: Государственное издательство географической литературы.

Przhevalskii, N.M. (Пржевальский, Н.М.). 1948a. Iz Zaisana cherez Khami i Tibet na verkhoviia Zheltoi reki (Из Зайсана через Хами и Тибет на верховья Жёлтой реки), 2nd ed., Москва: Государственное издательство географической литературы.

Rayfield, Donald. 1976. The Dream of Lhasa: the Life of Nikolay Przhevalsky (1839–88), Explorer of Central Asia. Athens, Ohio: Ohio University Press.

Ritter, K. 1856. Zemlevedenie Azii (Землеведение Азии), I, Санкт-Петербург: Издание ИРГО.

Roborovskii, V.I. 1888. Poslednie chasy zhizni N.M. Przheval'skogo (Последние часы жизни Н.М. Пржевальского), *Izvestiia IRGO* (*Известия ИРГО*), XXIV, 277–280.

Roborovskii, V.I. 1892. N.M. Przhevalskii v 1878–1888 g. Vospominaniia (Н.М. Пржевальский в 1878–1888 гг. Воспоминания), *Russkaia Starina* (*Русская старина*), 73 (1), 217–238; (3), 653–674.

Roborovskii, V.I. 1896. Ekskursii v storonu ot putei Tibetskoi ekspeditsii V.I. Roborovskogo i P.K. Kozlova (Экскурсии в сторону от путей Тибетской экспедиции В.И. Роборовского и П.К. Козлова), Trudy Tibetskoi Ekspeditsii 1889–1890 pod nachal'stvom M.V. Pevtsova (Труды Тибетской экспедиции 1889–1890 под началом М.В. Певцова), III, Санкт-Петербург: Издание ИРГО, 1–92, 119–122.

Roborovskii, V.I. 1897. O drevnostiakh Lukchunskoi kotloviny (О древностях Люкчунской колтловины), *Izvestiia IRGO* (*Известия ИРГО*), XXXIII, 435–437.

Roborovskii, V.I. 1897a. Predvaritel'nyi otchet ob ekspeditsii v Tsentral'nuiu Aziiu kapitana V.I. Roborovskogo v 1893–1895 gg. (Предварительный отчёт об экспедиции в Центральную Азию капитана В.И. Роборовского в 1893–1895 гг.), *Izvestiia IRGO* (*Известия ИРГО*), XXXIV, 1–59.

Roborovskii, V.I. 1899–1900. Trudy Ekspeditsii Imperatorskogo Russkogo Geograficheskogo Obshchestva po Tsentral'noi Azii, sovershennoi v 1893–1895 gg. pod nachal'stvom V.I. Roborovskogo (Труды Экспедиции Императорского Русского Географического Общества по Центральной Азии, совершённой в 1893–1895 гг. под начальством В.И. Роборовского), I: Otchet nachal'nika ekspeditsii V.I. Roborovskogo (Отчёт начальника экспедиции В.И. Роборовского), Санкт-Петербург: Издание ИРГО, 1900; II: Otchet P.K. Kozlova (Отчёт П.К. Козлова), Санкт-Петербург: Издание ИРГО, 1899; III: Nauchnye rezul'taty ekspeditsii V.I. Roborovskogo (Научные результаты экспедиции В.И. Роборовского), Санкт-Петербург: Издание ИРГО, 1899.

Roborovskii, V.I. 1949. Puteshestvie v Vostochnyi Tian' Shan' i Nan'-Shan' (Путешествие в Восточный Тянь Шань и Нань Шань), Москва: Географгиз, 2nd ed., prefaced by B.V. Yusov.

Rousseau, Jean-Jacques. 1995. Dialogues, 1, Paris: Gallimard.

Rudnitskii, A.Iu. (Рудницкий А.Ю.). 2013. Etot groznyi Grombchevskii… Bol'shaia igra na granitsakh imperii (Этот грозный Громбчевский… Большая игра на границах империи), Санкт-Петербург: Алетейя.

Rustamov, U.A. (Рустамов У.А.). 1956. Prigindukushskie kniazhestva Severnoi Indii v kontse XIX – nachale XX v. (Пригиндукушские княжества Северной Индии в конце XIX – начале XX в.), Ташкент: Издание АН Узбекской ССР.

Rymiński, L. 1930. Wielcy polscy podróżnicy. Generał Bronisław Grąbczewski, *Wschód*, 1, 11–16.

Samarukha, I.V. et al. (eds). (Самаруха, И.В. и др.). 2004. Tret'i vostokovednye chteniia Baikal'skogo gosudarstvennogo universiteta ekonomiki i prava, posviashchennye zhizni i deiatel'nosti P.A. Badmaeva. Materialy mezhdunarodnoi nauchnoi konferentsii (Третьи востоковедные чтения БГУЭП (Байкальский государственный университет экономики и права), посвященные жизни и деятельности П.А. Бадмаева. Материалы международной научной конференции), Иркутск.

Samosiuk, K.F. (Самосюк, К.Ф.) 2006. Buddiiskaia zhivopis' iz Khara-Khoto, XII–XIV vv. (Буддийская живопись из Хара-Хото, XII–XIV вв.), Санкт-Петербург: Издательство Государственного Эрмитажа.

Schimmelpenninck van der Oye, D. (Схиммельпеннинк ван дер Ойе, Д.). 1997. Neizvestnyi Przeval'skii (Неизвестный Пржевальский), *Арьяварта. Запад и Восток*: История исследований Центральной Азии (*Aryavarta*: *East and West*. History of Science, Philosophy and Literature of Central Asia), Санкт-Петербург, 1, 207–226.

Schimmelpenninck van der Oye, D. (Схиммельпеннинк ван дер Ойе, Д.). 2001. Toward the Rising Sun: Russian Ideologies of Empire and the Path to war with Japan. Northern Illinois University Press.

Schimmelpenninck van der Oye, D. (Схиммельпеннинк ван дер Ойе, Д.). 2014. Paul's Great Game: Russia's plan to invade British India, *Central Asian Survey*, XXXIII (2), 143–152.

Selihanovich, V.G. (Селиханович, В.Г.). 1956. Pevtsov. Puteshestvennik, geograf i astronom (Певцов. Путешественник, географ и астроном), Москва: Геодезиздат, 1956.

Semennikov, V.N. (ed.) (Семенников, В.Н.). 1925. Za kulisami tsarizma. Arkhiv tibetskogo vracha Badmaeva (За кулисами царизма. Архив тибетского врача Бадмаева), Leningrad.

Semenow, P.v[on] 1858. P.v[on] Semenow's Erforschungsreisen in Inner Asien im Jahre 1857, *Petermanns Geographische Mittheilungen*, Heft IX, 351–369.

Semenov, P.P. (Семёнов, П.П.). 1856. O vulkanicheskikh iavleniiakh vo vnutrenei Azii (О вулканических явлениях во внутренней Азии), *Vestnik IRGO* (*Вестник ИРГО*), XVII (2), 243–258.

Semenov, P.P. (Семёнов, П.П.). 1856a. Dopolneniia (Дополнения) (Supplementary Notes) in: Karl Ritter. Zemlevedenie Azii (Землеведение Азии), I, Санкт-Петербург.

Semenov, P.P. (Семёнов, П.П.). 1858. Pervaia poezdka na Tian' Shan', ili Nebesnyi Khrebet, do verkhov'ev reki Iaksarta, ili Syr Dar'i, v 1857 godu (Первая поездка на Тянь Шань, или Небесный Хребет, до верхо-вьев реки Яксарта, или Сыр Дарьи, в 1857 году), *Vestnik IRGO* (*Вестник ИРГО*), XXIII (5), 1–25.

Semenov, P.P. (Семёнов, П.П.). 1867. Poezdka iz ukrepleniia Vernogo cherez gornyi pereval u Suok-Tiube i ushchel'e Buam k zapadnoi okonechnosti oz. Issyk-Kul' v 1856 g. Otryvok iz putevykh zapisok (Поездка из укрепления Верного через горный перевал у Суок-Тюбе и ущелье Буам к западной оконечности оз. Иссык-Куль в 1856 г. Отрывок из путевых записок), *Zapiski IRGO po obshchei geografii* (*Записки ИРГО по общей географии*), I, 181–254.

Semenov, P.P. (Семёнов, П.П.). 1892. Otzyv o trudakh G.E. i M.E. Grum-Grzhimailo (Отзыв о трудах Г.Е.и М.Е. Грум-Гржимайло), Otchet IRGO za 1891 g. (Отчет ИРГО за 1891 г.), Санкт-Петербург, 14–17.

Semenov, P.P. (Семёнов, П.П.). 1894. Introduction in Grum-Grzhimailo G.E. Opisanie Amurskoi oblasti (Описание Амурской области), Санкт-Петербург: Типо-Литография С.М. Николаева, I-V.

Semenov, P.P. (Семёнов, П.П.). 1896. Istoriia poluvekovoi deiatel'nosti IRGO 1845–1895 (История полувековой деятельности ИРГО 1845–1895), in 3 vols., Санкт-Петербург: Типография В. Безобразова и К°.

Semenov, P.P. (Семёнов, П.П.). 1896–1898. Preface to: A.M. Pozdneev, Mongolia i mongoly (Предисловие к: А.М. Позднеев. Монголия и монголы), 1–2, Санкт-Петербург.

Semenow, P.v[on] 1869. P. v[on] Semenow's Forschungsreisen in den Trans-Ilischen Alatau und zum Issyk-kul ausgeführt in der Jahren 1856 und 1857, *Zeitschrift der Gesellschaft fur Erdkunde zu Berlin*, IV, 116–136.

Semenov-Tian-Shanskii, A.P. (Семенов-Тян-Шанский, А.П.). 1937. Zoologicheskie rezul'taty ekspeditsii P.K. Kozlova (Зоологические результаты экспедиции П.К.

Козлова), *Izvestiia Gosudarstvennogo geograficheskogo obstchestva* (*Извеcтия Государственного Географического общества*), 69 (1), 131–137.

Semenov-Tian-Shanskii, P.P. (Семенов-Тян-Шанский, П.П.). 1958. Puteshestvie na Tian-Shan' v 1856–1857 gg. (Путешествие в Тянь-Шань в 1856–1857 гг.), repr., 2nd ed., Москва: ОГИЗ.

Sergeev, E.Yu. (Сергеев, Е.Ю.). 2012. Bol'shaiia Igra, 1856–1907: Mify i realii rossiisko-britanskikh otnoshenii v Tsentral'noi Azii (Большая Игра, 1856–1907: Мифы и реалии российско-британских отношений в Центральной Азии), Москва: Товарищество научных изданий КМК.

Sharngorst, K.V. 1886. Otzyv K.V. Sharngorsta o zaslugakh polkovnika M.V. Pevtsova (Отзыв

Л.В. Шарнгорста о заслугах М.В. Певцова), Otchet IRGO za 1885 god (Отчет ИРГО за 1885 год), Санкт-Петербург, 14.

Sharngorst, K.V. (Шарнгорст, К.В.). 1889. Astronomicheskie opredeleniia kapitana B.L. Grombchevskogo vo vremia puteshestviia v Kandzhut i Raskem (Астрономические определения капитана Б.Л. Громбчевского во время путешествия в Канджут и Раскем), *Izvestiia IRGO* (*Известия РГО*), XXV (6), 437–439.

Shastina N.P. (Шастина, Н.П.). 1958. Russko-Mongol'skie posol'skie otnosheniia XVIII veka (Русско-Монгольские посольские отношения XVIII века), Москва: Издательство восточной литературы, 22–26.

Shchukina, N.M. (Щукина Н.М.). 1954. Vklad russkikh geografov v kartografiiu Tsentral'noi Azii (vtoraia polovina XIX veka) Вклад русских географов в картографию Центральной Азии (вторая половина XIX века), *Voprosy geografii* (*Вопросы географии*), XXXV, 260–267.

Shchukina, N.M. (Щукина Н.М.). 1955. Kak sozdavalas' karta Tsentral'noi Azii (Как создавалась карта Центральной Азии), Москва: Государственное издательство географической литературы.

Sheremetev, S.D. (Шереметев, С.Д.). 2005. Memuary grafa Sheremeteva (comp. R.A. Vakh and L.I. Shokhin; trans. from French by K.N. Sokolova) (Мемуары графа С.Д. Шереметева), Москва: Индрик.

Shishmarev Y.P. (Шишмарёв, Я.П.). 1873. Marshrut iz Urgi v Hlassu (Маршрут из Урги в Хлассу), *Izvestiia IRGO* (*Известия ИРГО*), IX (6), 185–191.

Shcherbakov, D.I. (ed.). (Щербаков, Д.И., ред.). 1962. Perepiska Aleksandra Gumbol'dta s uchenymi i gosudarstvennymi deiateliami Rossii (Переписка Александра Гумбольдта с учёными и государственными деятелями России), Москва: Издателсьтво АН СССР.

Shokal'skii, Iu.M. (Шокальский, Ю.М.) 1936. Pamiati Grigoriia Efimovicha Grum-Grzhimailo (Памяти Григория Ефимовича Грумм-Гржимайло). *Izvestiia Geograficheskogo Obshchestva* (*Известия Географического общества*), 5, 754–755.

Shtubendorf, O.E. (Штубендорф О.Э.). [1895]. Astronomicheskie opredeleniia podpolkovnika B.L. Grombchevskogo vo vremia puteshestviia na Pamiry, v Darvaz, Raskem, Kashgariiu i Severo-Zapadnyi Tibet v 1889–90 godakh (Астрономические определения подполковника Б.Л. Громбчевского во время путешествия на Памиры, в Дарваз, Раскем, Кашгарию и Северо-Рападный Тибет, в 1889–90 годах), Санкт-Петербург: Типорафия А.С. Суворина.

Simukov, A.D. (Симуков, А.Д.). 2007–2008. Trudy o Mongolii i dlia Mongolii (Труды о Монголии и для Монголии), comp. Yu. Konagaia, S. Bayaraa, I. Lhagvasuren (сост. Ю. Конагая, С. Баяраа, И. Лхагвасурэн), Т. 1–3. Осака: Национальный музей этнологии.

Słabczyński W. 1973. Polscy podróżnicy i odkrywcy, Warszawa: Państwowe Wydawnictwo Naukowe.

Slepkova, N.V. (Слепкова, Н.В). 2014. "Blagodarit' imenem Akademii": Tsentral'noaziatskie kollektsii Zoologicheskogo Muzeiia v khronike zasedanii fiziko-matematicheskogo otdeleniia IAN, 1874–1894 («Благодарить именем Академии»: Центральноазиатские коллекции Зоологического музея в хронике заседаний Физико-математического отделения ИАН, 1874–1894), *Istoriko-biologicheskie issledovaniia* (*Историко-биологичесские исследования*), 6 (4), 80–104.

Sokolov, B.S. (Соколов, Б.С.). 2002. Vsevolod Ivanovich Roborovskii – vydaiushchiisia issledovatel' Vnutrennei Azii (Всеволод Иванович Роборовский – выдающийся исследователь Внутренней Азии), *Voprosy istorii estestvoznaniia i tekhniki* (*Вопросы истории естествознания и техники*), 3, 507–517.

Sokolov, N.N. (Соколов, Н.Н.). 1897. V debriakh Azii. Ocherki puteshestviia general-maiora Pevtsova (В дебрях Азии. Очерки путешествия генерал-майора Певцова), Санкт-Петербург: типография П.П. Сойкина.

Spafarii, N.G. (Спафарий, Н.Г.). 1882. Puteshestvie cherez Sibir' ot Tobol'ska do Nerchinska i granits Kitaia russkogo poslannika Nikolaia Spafariia v 1675 g. Dorozhnyi dnevnik Spafariia (Путешествие через Сибирь от Тобольска до Нерчинска и границ Китая русского посланника Николая Спафария в 1675 г. Дорожный дневник Спафария), *Zapiski IRGO po otdeleniiu etnografii* (*Записки ИРГО по отделению этнографии*), 10 (1).

Stebnitskii, I.I. (Стебницкий, И.И.). 1892. Otzyv deistvitel'nogo chlena IRGO I.I. Stebnitskogo o puteshestviiakh i trudakh M.V. Pevtsova, in Otchet IRGO za 1891 god (Отчет ИРГО за 1891 год), Санкт-Петербург: Типография А.С. Суворина, 4–8.

Suckow, Christian. 2001. Alexander von Humboldt und Russland, in: Alexander von Humboldt – Aufbruch in die Moderne, Berlin: Akademie Verlag.

Suess, E. 1901. Das Anblltz der Erde, III, Prag; Wien: F. Tempsky; Leipzig: G. Freytag.

Tachalov, N.A. (Тачалов, Н.А.) 1907. Astronomicheskie nabliudeniia P.K. Kozlova (Астрономические наблюдения П.К. Козлова), in: Mongolia i Kam. Trudy ekspeditsii Imperatorskogo Russkogo Geograficheskogo Obshchestva, sovershennoi v

1899–1901 godakh pod rukovodstvom P.K. Kozlova (Монголия и Кам. Труды экспедиции Императорского Русского Географического общества, совершенной в 1899–1901 годах под руководством П.К. Козлова), III (1), Санкт-Петербург, 1–3.

Tikhvinskii, S.L., Miasnikov, V.S., Ignatov, A.S., Dionisii Sv. (eds) (С.Л. Тихвинский, В.С. Мясников, А.С. Игнатов, св. Дионисий, ред.). 1997. Istoriia Rossiiskoi dukhovnoi missii v Kitae (История Российской духовной миссии в Китае), Москва: Издательство Свято-Владимирского братства.

Tillo, A.A. (Тилло, А.А.). 1892. Absoliutnye vysoty, opredelennye B.L. Grombchevskim vo vremia puteshestviia na Pamir, v Raskem, Kashgariiu i Severo-Zapadnyi Tibet v 1889–1890 gg. (Абсолютные высоты, определенные Б.Л. Громбчевским во время путешествия на Памир, в Раскем, Кашгарию и Северо-Западный Тибет в 1889–1890 гг.), in *Izvestiia IRGO* (*Известия ИРГО*), XXVIII (1), 89–100.

Titov, V.B. (Титов, В.Б.). 2012. Voenno-sluzhilye liudi Przheval'skiie (Военно-служилые люди Пржевальские), Москва: Граница.

Tucholska, K. 2014. Dwa obrazy przemian Japonii w epoce Meiji we wspomnieniach Rudyarda Kiplinga oraz Bronisława Grąbczewskiego, *Gdańskie Studia Azji Wschodniej*, VI, 98–109.

Tumarkin, D.D. (Тумаркин, Д.Д.) (comp., ed.). 2002. Repressirovannye etnografy (Репрессированные этнографы), Москва: Восточная литература.

Valskaia, B.A. (Вальская, Б.А.). 1967. Neopublikovannaia rukopis' N.M. Przheval'skogo 'O sushchnosti zhizni na zemle'(Неопубликованная рукопись Н.М. Пржевальского «О сущности жизни на земле»), in Doklady Vostochnoi komissii Geograficheskogo obshchestva SSSR (Доклады Восточной комиссии Географического общества СССР), 4, Ленинград, 94–104.

Vinogradov, B.K. (Виноградов, Б.К.). 2003. Vsevolod Ivanovich Roborovskii, Vyshnevolotskii kraevedcheskii al'manakh (Вышневолотский краеведческий альманах), Вышний Волочок, 37, 127–162.

Vinogradov, B.K. (Виноградов, Б.К.). 2012. Issledovatel' Tsentral'noi Azii V.I. Roborovskii, Тверь, Альфа-Пресс.

Volkov, S.V. (Волков, С.В.). 2010. Mikhail Vasilievich Pevtsov (Михаил Васильевич Певцов), Generalitet Rossiiskoi Imperii. Entsiklopedicheskii slovar' (Генералитет Российской империи. Энциклопедический словарь), II, Москва: Центрполиграф, II, 288.

Waller, Derek J. 2004. The Pundits: British exploration in Tibet and Central Asia, The University Press of Kentucky.

Wąsowicz, J. 1926. Bronisław Grąbczewski, *Przyroda i Technika*, V (5), 226.

Wąsowicz, J. 1927. Generał Bronisław Grąbczewski, polski badacz Azji Środkowej (1855–1926), Poznań.

Wąsowicz, J. 1927a. O pracach Grąbczewskiego w kaszmirskiem dorzeczu Indusu, in *Czasapismo Geograficzne*, 1, 26.

Wąsowicz, J. 1960. Grąbczewski Bronisław in: *Polski Słownik Biograficzny*, VIII, 557–559.

Wood, Frances. 2002. The Silk Road: Two Thousand Years in the Heart of Asia. Berkeley; Los Angeles: University of California Press.

Yetts, W.P. 1926. Discoveries of the Kozlov Expedition, *Burlington Magazine*, April, 168–185.

Younghusband, Colonel Sir F. 1926. The Problem of the Shaksgam Valley, *Geographical Journal*, LXVIII (3), 225–230.

Yusov, B.V. (Юсов, Б.В.). 1949. V.I. Roborovskii – issledovatel' Tsentral'noi Azii (В.И. Роборовский – исследователь Центральной Азии), Preface in V.I. Roborovskii. Puteshestvie v Vostochnyi Tian' Shan' i Nan'-Shan' (В.И. Роборовский. Путешествие в Восточный Тянь-Шань и Нань-Шань), 2nd ed., Москва: Географгиз, 3–32.

Yusov, B.V. (Юсов, Б.В.). 1985. N.M. Przhevalskii (Н.М. Пржевальский), Москва: Просвещение.

Yusupova, T.I. (Юсупова, Т.И.). 2003. Neosushchestvlennaia mechta P.K. Kozlova (Неосуществленная мечта П.К. Козлова), *Priroda* (*Природа*), 9, 52–56.

Yusupova, T.I. (Юсупова, Т.И.). 2006. Mongol'skaia komissiia Akademii nauk: Istoriia sozdaniia i deiatel'nosti (1925–1953 gg.) (Монгольская комиссия Академии наук: История создания и деятельности (1925–1953 гг.)), Санкт-Петербург: Нестор-История.

Yusupova, T.I. (Юсупова, Т.И.). 2008. Mongolo-Sychuan'skaia ekspeditsiia (1907–1909) P.K. Kozlova i otkrytie Khara-Khoto (Монголо-Сычуаньская экспедиция (1907–1909) П.К. Козлова и открытие Хара-Хото), in: I. Popova (ed). Rossiyskie ekspeditsii v Tsentral'noi Azii v kontse XIX – nachale XX veka (Российские экспедиции в Центральной Азии в конце XIX – начале XX века). Санкт-Петербург: Славия, 112–129.

Yusupova, T.I. (Юсупова, Т.И.). 2010. Sluchainosti i zakonomernosti v arkheologicheskikh otkrytiiakh: Mongolo-Tibetskaia ekspeditsiia P.K. Kozlova i raskopki Noin-Uly (Случайности и закономерности в археологических открытиях: Монголо-Тибетская экспедиция П.К. Козлова и раскопки Ноин-Улы), *Voprosy istorii estestvoznaniia i tekhniki* (*Вопросы истории естествознания и техники*), 4, 26–67.

Yusupova, T.I. (Юсупова, Т.И.). 2010a. Premiia Frantsuzskoi Akademii nauk russkomu puteshestvenniku P.K. Kozlovu (Премия Французской академии наук русскому путешественнику П.К. Козлову), *Istoricheskii arkhiv* (*Исторический архив*), 6, 200–204.

Yusupova, T.I. (Юсупова, Т.И.). 2011. Istoriia ne sovsem obychnogo arkheologicheskogo otkrytiia (История не совсем обычного археологического открытия), in N.V. Polos'miak, E.S. Bogdanov, D. Tsevendorzh (Н.В. Полосьмак, Е.С. Богданов, Д.

Цевендорж), Dvadtsatyi noin-ulinskii kurgan (Двадцатый ноин-улинский курган), Новосибирск: Инфолио, 11–53.

Yusupova, T.I. (Юсупова, Т.И.). 2012. Russian-American Researches of Central Asia in 1920s: R.Ch. Andrews and P.K. Kozlov (Российско-американские исследования в Центральной Азии: Р.Ч. Эндрюс и П.К. Козлов), in Iu.P. Tretiakov (ed.) Russian-American Links: Leaps Forward and Backward in Academic Cooperation, Санкт-Петербург: Нестор-История, 226–231.

Yusupova, T.I. (Юсупова, Т.И.). 2016. Puteshestvie kak obraz zhizni: issledovatel' Tsentral'noi Azii P.K. Kozlov (Путешествие как образ жизни: исследователь Центральной Азии П.К. Козлов), Санкт-Петербург, Нестор-История.

Yusupova, T.I. (Юсупова, Т.И.). 2016a. O nerealizovannom proekte po podgotovke uchebnika mongol'skoi istorii G.E. Grumm-Grzhimailo (О нереализованном проекте по подготовке учебника монгольской истории Г.Е. Грумм-Гржимайло), *Mongolica–XVI*, Санкт-Петербург, 79–90.

Zagorodnikova, T.N. (Загородникова, Т.Н.) (comp., preface, comment.). 2014. 'Bol'shaia igra' v Tsentral'noi Azii: 'Indiiskii pokhod' russkoi armii ("Большая игра" в Центральной Азии: 'Индийский поход' русской армии), A collection of archival documents (Сборник архивных документов), Москва: Новый хронограф.

Zelenin, A.V. (Зеленин, А.В.). 1900. Puteshestviia N.M. Przheval'skogo (Путешествия Н.М. Пржевальского), 2 vols., Санкт-Петербург: Издание П.П. Сойкина.

Zhitomirskii, S.V. (Житомирский С.В.). 1989. Issledovatel' Mongolii i Tibeta P.K. Kozlov (Исследователь Монголии и Тибета П.К. Козлов), Москва: Знание.

Zieliński, S. 1932. Grąbczewski Bronisław, in Mały słownik pionierów polskich kolonialnych i morskich. Podróżnicy, odkrywcy, zdobywcy, badacze, eksploratorzy, emigranci, pamiętnikarze, działacze i pisarzeemigracyjni. Warszawa: Nakładem Instytutu Wydawniczego Ligi Morskiej i Kolonjalnej, 135–138.

Index

www.ingramcontent.com/pod-product-compliance
Lightning Source LLC
LaVergne TN
LVHW010541300926
849162LV00002B/203